Programming for TV, Radio, and Cable

AF712488

Programming for TV, Radio, and Cable

Edwin T. Vane
Lynne S. Gross

Focal Press

Boston London

Some products are claimed as trademarks by their manufacturers. Where a trademark appears in this work, and Butterworth-Heinemann was aware of the trademark claim, the product has been printed with an initial capital letter.

Focal Press is an imprint of Butterworth-Heinemann.

Copyright © 1994 by Butterworth-Heinemann
A member of the Reed Elsevier group
All rights reserved.

No part of this publication may be reproduced, stored in a retrieval system, or transmitted, in any form or by any means, electronic, mechanical, photocopying, recording, or otherwise, without the prior written permission of the publisher.

Recognizing the importance of preserving what has been written, it is the policy of Butterworth-Heinemann to have the books it publishes printed on acid-free paper, and we exert our best efforts to that end.

Library of Congress Cataloging-in-Publication Data

Vane, Edwin T.
Programming for TV, radio, and cable / by Edwin T. Vane and Lynne S. Gross.
p. cm.
Includes bibliographical references and index.
ISBN 0-240-80128-8 (acid-free paper) :
1. Television programs—Planning. 2. Radio Programs—Planning. 3. Cable television—Planning. I. Gross, Lynne S. II. Title.
PN1992.55.V36 1994
791.45′0236′0973—dc20

93-21554
CIP

British Library Cataloguing-in-Publication Data
A catalogue record for this book is available from the British Library.

Butterworth-Heinemann
313 Washington Street
Newton, MA 02158

10 9 8 7 6 5 4 3 2 1

Printed in the United States of America

To Claire and Paul

CONTENTS

Foreword

By Grant A. Tinker

Less than fifty years ago, television was an infant and radio was king. No one quite believed that this new thing, radio-with-pictures, would ever catch on. At NBC many employees had both radio and TV assignments, the former occupying most of their day, the latter being fit in as time permitted. In fact, a few years later, when volunteers were sought to work exclusively in television, there were very few takers. The company had to pay employees a premium to persuade them to make the move.

Today, television reigns and radio is a largely local and secondary medium. But the television landscape is quite different from what it was a few short years ago, and it continues to change. The market share of the traditional networks has declined by one-third. Cable, initially a technique to extend television coverage, is now a powerful and expanding competitor in its own right. Syndication, once a quiet cottage industry, has become a major force. VCRs enable every viewer to become a program scheduler, and satellites bring us wars and invasions, live and in color.

If all that can happen in a half century, what does the future hold? Already scientists are able to put in our hands a receiver that can summon 500 channels with the tap of a button. Every TV hardware convention offers yet another mind-numbing advance in the surreal world of "virtual reality," a process through which viewers can create their own effects to excite all their senses. (Yes, *all* their senses.) When Aldous Huxley predicted that the movies would become the "feelies" in his 1932 book *Brave New World*, he was denounced as a hallucinator. It turns out Aldous's only failing was a shortage of imagination. We may well ask, if all of these developments are either here or in the pipeline, what will the unknown bring us?

Amidst all this technical razzle-dazzle, one fundamental question is frequently overlooked: What will people be looking at? What are the shows going to be? A bad program is still going to be a bad program no matter what the quality of the picture or the means of delivery. The thought of 500 channels blazing away each day from dawn until midnight leaves me dazed. I know how hard it is to program successfully one night on a single network. How can the creators of the future satisfy such a massive and relentless appetite for software? Who will collect enough dollar resources, in one service, to produce "Hill Street Blues"?

I have spent my professional life buying, selling, and making television programs, and I'm ever fascinated by their mysteries. Why do shows that seem to have the ingredients of a smash hit become instant disasters? Why does a throwaway program in a murderous time period become a top-ten winner overnight? How can a performer be an embarrassment in one role and a national icon in another? Why is the audience entranced by a certain genre of show in one decade and bored with it in the next? Why do well-paid and highly motivated program executives have such a poor success rate? (I once heard a programmer whose schedule was in trouble respond to that question by saying, "Even the best hitter in the history of baseball failed 63 percent of the time, chief." His boss's stony silence was deafening.)

Even though hits are elusive and largely unpredictable, there is one rule that I believe gives programmers their best chance for success: strive for quality. That may sound like a preachment, but it is really a piece of practical advice. If you examine the consistently high-rated programs down through the decades, you will find the vast majority of them have been crafted by the medium's most gifted people. From the inspired genius of Lucille Ball through "Gunsmoke," "The Dick Van Dyke Show," "All in the Family," "M*A*S*H," "Mary Tyler Moore," "60 Minutes," "The Cosby Show," "Cheers," and myriad dramatic shows, these ratings champions have all cared scrupulously about the writing, directing, and casting of every episode. It is simply not true that the way to the Nielsen promised land is via the lowest common denominator approach to programming.

For aspiring programmers who wish to plunge into the electronic maelstrom, I suggest the following: learn the history of the business; find out how each of the media functions; discover the elements that tilt shows toward success or failure. The industry is only going to get more complex, and the earlier the basics are mastered, the better.

This book is an excellent way to start. The authors bring to their task both academic and practical credentials. I have known Ed Vane and observed his work for more than forty years; he is a consummate, proven professional. Lynne Gross has had an exemplary career teaching at various universities and working in cable and broadcasting.

Their book should leave you with one strong signpost of advice: when you are finally in a position to influence the content of programs, take the high road. Even if you fail, you'll feel better.

Preface

In the electronic media nothing is forever. New technologies force change. Radio executives discovered the truth of this maxim in the late 1940s when their cozy world was abruptly exploded by an invention called television. Thirty years later TV was turned upside down by a new use of an old distribution form named cable. As the 1990s unfold, a host of new scientific advances (compression, DBS, addressability, etc., etc., etc.) are changing the face of electronic media almost daily.

But through it all, one thing remains constant: the need for programming. No matter how state-of-the-art the delivery system, it doesn't mean a thing if the consumers aren't interested in what's being conveyed. The product must be good, it must be presented in an attractive manner, and it must be equal to the challenges placed against it by all of the forces that vie for the attention of the public.

The goal of this book is to help current and would-be programmers to succeed in a restless, competitive environment. The authors are not clairvoyant and are no more able to tell readers what TV and radio will look like in ten years than anyone else in the business. But the programmer who knows the past and who is aware that certain guidelines reduce the odds against failure is far more likely to prosper in virgin territory than one who has no touchstones.

The intent of the book is to expose readers to the fundamentals of programming in the various electronic media. To that end we have used four media forms—commercial television, cable television, commercial radio, and public broadcasting—as overall organizational entities. Within these forms we discuss various programming elements such as syndication, networking, and local production. In many respects, the rudiments are the same for all media. For example, all radio and television programs face competition and must plan strategies to preserve their share of the audience. But in many other respects, the media differ in their requirements; e.g., commercial broadcast networks almost automatically have their programs aired by their affiliated stations but cable networks must compete with each other for channel space on local cable systems.

To keep these similarities and differences sorted out, we have structured the book by programming function, e.g., sources of shows, elements of successful programs, competitive strategies. Because network commercial television is the most familiar form to most readers, we begin each chapter by showing how the function operates in that medium. Much of what is said about network TV is also applicable to the other media, and, as we go through each one, we point out the modifications and variations. We believe this panoramic look at how programming's elements are put to use throughout the industry will provide the reader with the basic knowledge required by the business.

We feel programmers can be more effective if they have at least a broad, general knowledge of the industry's history. In Chapter 1 we outline the colorful twists and turns that the business has taken over the last seventy years to indicate that this year's strange and seemingly unprecedented events are not so unusual after all.

The creation of programming does not take place in a vacuum. All shows are developed to appear, at least initially, on a specific delivery system. It is vital that programmers know the strengths, weaknesses, eccentricities, and curiosities of each of these media forms lest they design a beachball for a tennis match. In Chapters 2 and 3 we describe the various marketplaces for which the programmer can construct shows and point out how they differ in terms of creative needs, business arrangements, and distribution.

Where do program ideas come from? How do you search them out? And what do you do with them once they come to you? The whole development process is discussed in

Chapters 4 through 6. As neoprogrammers will discover, the job doesn't end with the construction of the show. It also has to be sold. This is a process that has proved difficult for many, but there are techniques that can make the task simpler and more effective.

There are no surefire methods of producing a hit show. But there are a number of creative elements which, if properly incorporated into a program, can give it a better chance to succeed. Their inclusion does not guarantee a winner, but the exclusion of too many of them will virtually assure a failure. These ingredients are identified and reviewed in Chapters 7 and 8.

Programmers are not alone when they create their shows. With them in spirit and influence, if not in body, are a host of forces that have much to say about the content and placement of programs. In Chapters 9 and 10 we talk about the many voices that must be heard before shows are seen. These include, but are not limited to, advertisers, pressure groups, government agencies, in-house departments, and professional critics. They are important influences on the programmer and are ignored at great peril.

One of the most crucial processes in programming involves the scheduling of shows. Over the years a number of programming strategies have proved successful (and unsuccessful). Programmers must adjust their strategies to changing times and must look ever vigilantly at the plans and procedures being used by the competition. The vast array of strategies available to a programmer is the topic of Chapters 11 and 12.

Once a program is on the air, there is a need to evaluate its performance. It's not as simple as merely looking at a Nielsen rating. Many other factors determine the success or failure of a show, and these considerations are reviewed in Chapter 13.

Unfortunately, not every concept works. Cancellations are a painful but unavoidable part of the business. When the deed must be done, there is a right way and a wrong way to go about it, as is explained in Chapter 14. Sometimes, however, a show can be saved by creative changes, or by a new location on the schedule. These possibilities are examined in the same chapter.

No doubt the electronic media business has its fair share of expedient corner-cutters. But for the greater part, we believe industryites behave honorably. In Chapter 15 we present a variety of incidents that involve ethical considerations. It is our hope that these examples will encourage programmers to look beyond the purely commercial concerns of a matter before reaching a decision.

In almost every chapter we have included at least one story, a real-life incident, that illuminates the issues or principles under discussion. The reader may be surprised to see that at times the deportment of some important individuals is decidedly unbusinesslike. All we can say is that the radio and TV business is entertainment. It tends to attract colorful and unorthodox personalities, and they help to make the business fun, unpredictable, and, at times, a bit bewildering.

The authors are greatly indebted to the many radio and TV professionals who gave willingly of their time and knowledge to supply facts and anecdotes. We also appreciate the suggestions given by the reviewers. And we are grateful for the support offered by the staff at Focal Press.

Edwin T. Vane and Lynne S. Gross

1 A Concise History of Programming

The year was 1921. Thomas ("Tommy") H. Cowan, an employee at Westinghouse's Newark, New Jersey, plant, was told to report to a shack on the roof of the plant building and "make this thing talk." The "thing" turned out to be radio station WJZ, and Tommy's qualifications were that at age 14 he had been an extra in the Metropolitan Opera.

As the early "program director," Tommy convinced actors who were performing in Newark (or sometimes even as far away as New York) to come to his WJZ shack to sing, recite, play a musical instrument, or in some other way entertain—all without pay, of course.

One who volunteered to come was Olga Petrova, who was appearing at the Broad Street Theater in Newark. Tommy and other Westinghouse executives were a bit nervous about her because Olga was a major advocate of birth control and was constantly making speeches on the subject. But everyone was relieved when she promised all she was going to do was read some of her own versions of nursery rhymes. She then went on the air and broadcast:

> There was an old woman who lived in a shoe;
> She had so many children because she didn't know what to do.

Everyone was petrified. A short "program policy" meeting was held, and the decision was made to remove Olga from the airwaves and not invite her back.[1]

EARLY RADIO PROGRAMMING

No one knows for sure, but perhaps the removal of Olga Petrova was the first programming decision—and the beginnings of censorship. At any rate, early radio did not have programming departments as such. The novelty of the medium was exciting enough that people would stay glued to their earphones and huge battery-operated sets just to "hear Pittsburgh" (i.e., hear the call letters and phonograph music coming from station KDKA in Pittsburgh, generally regarded to be the first radio station).

The early stations were supported primarily by companies that manufactured and sold radio sets, and they programmed whatever free talent wandered into the studio. For most stations, this included a preponderance of would-be operatic sopranos. The goal was to have something on the air to encourage people to buy radios. Eventually the novelty of radio wore off and performers wanted to be

FIGURE 1.1 KDKA in Pittsburgh, Pennsylvania, launched its radio programming on November 2, 1920, with this broadcast of the Warren Harding–James Cox election results. (Photo courtesy of Westinghouse Broadcasting Company.)

FIGURE 1.2
KFI, Los Angeles's first radio station, which began broadcasting in 1922, had this rather elaborate studio. Most early studios, although unseen by the listening audience, had elaborate decors that included potted palms, a common foliage of the 1920s. (Photo courtesy of KFI.)

paid, so some economic means of supporting radio had to be found.

American Telephone and Telegraph (AT&T) hit upon an economic idea that was based on its telephone experience. It established station WEAF in New York as a "toll station." People would be required to pay a toll to broadcast some message to all radio listeners in the same way that they paid a toll to send a private message from a phone booth. The company built a studio about the size and shape of a phone booth and waited for people to come and pay to send their messages. No one did. After a long struggle, WEAF, in August of 1922, finally sold its first message, a ten-minute announcement from a Long Island real estate company that paid $50.

The first commercial message indicated the sort of excess and hyperbole that was to characterize electronic advertising for the next 70 years. It was a ten-minute message from the Queensboro Corporation to promote the sale of apartments in Jackson Heights, Queens. The apartments were called "Hawthorne Court" and were named after the American novelist Nathaniel Hawthorne, whom the sponsor anointed as the "greatest of American fictionists." Warming to his task, the developer, Mr. Blackwell, stated that Mr. Hawthorne had "analyzed with charming keenness the social spirit of those who had thus happily selected their homes, and he painted the people who had inhabited those homes with good-natured relish." Now aloft, Mr. Blackwell began to pick up speed: "There should be more Hawthorne sermons preached about the utter inadequacy and the general hopelessness of the congested city home. The cry of the heart is for more living room, more chance to unfold, more opportunity to get near the Mother Earth, to play, to romp, to plant and to dig." And finally, Mr. Blackwell left earth and soared with the eagles: "Let me enjoin upon you as you value your health and your hopes and your home happiness, get away from the solid masses of brick where the meager opening admitting a slant of sunlight is mockingly called a light shaft and where children grow up starved for a run over a patch of grass and the sight of a tree."[2]

But even after this initial commercial, advertising was not viewed as a primary source of income for broadcasting. Herbert Hoover, who was secretary of commerce during the 1920s, said that "ether advertising" was possible, but he quickly dismissed the idea. "It is inconceivable that we should allow so great a possibility for service to be drowned in advertising chatter," he stated.[3] As late as 1930, by which time the airwaves were drenched with advertising, the industry was still professing its commercial virginity. Before a Senate committee, Merlin Aylesworth, president of NBC, testified, "I am opposed to direct advertising on the air." When a senator asked him what he meant by direct advertising, he replied, "I mean stating prices."[4] The following year even that distinction fell by the wayside.

THE BEGINNING OF NETWORKING

Another method for making radio economically sound involved producing programs that could be used by a number of stations, thus reducing the cost to each station. Wires could carry the signal of a program produced in New York, networking it to stations in Boston, Philadelphia, Washington, and beyond. The first network to be established was the National Broadcasting Company (NBC), which, in 1926, broadcast its debut program (an orchestra from New York, a singer from Chicago, Will Rogers from Kansas City, and dance bands from other cities) to twenty-two local stations. In 1927, flushed with success, NBC started a second network. To distinguish the two, the older was known as the "Red" network and the new one was dubbed the "Blue" network. NBC was owned by Radio Corporation of America (RCA), whose dynamic president, David Sarnoff, became involved in most aspects of network programming.

In 1928 William S. Paley's family bought out the floundering beginnings of another network, and Paley built it into the Columbia Broadcasting System (CBS). The American Broadcasting Company (ABC) grew out of an early 1940s ruling that said no one company could own more than one network. As a result NBC had to sell gits Blue network, which eventually became ABC. A fourth network, Mutual, was formed in the mid-1930s but never achieved the power and prestige of the other three.

THE BEGINNING OF REGULATION

With advertising and networking established, radio grew phenomenally. By 1927, there were over 700 stations on the air. The problem was that most of them transmitted on the same frequency, and technical interference abounded. So many signals were bouncing around at the same time that nothing could be heard distinctly. As a result, Congress passed the Radio Act of 1927, which set up the Federal Radio Commission (FRC) to study the radio interference problem and serve as a guiding hand and watchdog over the radio industry.

The FRC enlarged the frequency band for radio to what is now the AM band and issued specific licenses to specific stations. In 1934 Congress passed a new act that solidified and somewhat expanded the provisions of the earlier 1927 act and replaced the FRC with the Federal Communications Commission (FCC). This Communications Act of 1934 still governs broadcasting today, and the FCC is still the main government body that oversees the technical and programming aspects of modern electronic media.

THE ADVERTISING AGENCY YEARS

When NBC was formed, it purchased WEAF from AT&T and continued a variation of the "toll station" concept, wherein advertising agencies bought blocks of time for their clients and filled this time with programming and short, unobtrusive mentions of the sponsor and its product line. The main value to the advertisers was that their name was associated with the program—"Maxwell House Hour," "Mercury Theater," "Eveready Hour," "Lux Radio Theater." Sometimes the product became part of the story line—the announcer would visit Fibber McGee and Molly and talk about waxing the floor with Johnson's Wax.

The advertisers and their agencies made most of the programming decisions, in conjunction with top-level radio network executives. The advertisers came up with the concepts, hired the talent, and oversaw the production. The networks provided the studio where the program was performed live, and they also provided the facilities for distributing the programming around the country. Of course, the advertisers paid the networks for these services.

Under this arrangement, the networks did very little in the way of program decision-making. As long as the advertiser was happy with the program and its time period and conformed to the network's standards and policies, it was left alone. The networks distributed the programs and collected their money. As a result, many programs aired on radio for years and years at the same time. Jack Benny was on radio from 1932 until the mid-1950s at 7:30 Sunday evening. For much of his reign he was sponsored by Jell-O. George Burns and Gracie Allen were on from 1933 to 1951, primarily sponsored by Robert Burns cigars. Dramas, comedies, children's programs, soap operas—all were handled by advertising agencies.

Networks did produce and control their own news, and they occasionally made concerted attempts to steal each other's shows. In 1936, William Paley conducted a well-planned raid on some of NBC's top programs and succeeded in obtaining Al Jolson, Nelson Eddy, and Major Edward Bowes' popular "Amateur Hour." He managed to attract the latter by actually going to the NBC studios when the Bowes program was broadcasting and talking to the star. Paley also approached Chrysler, the program's sponsor, and convinced its executives that CBS was the right place for Bowes. Right after World War II, NBC took "Your Hit Parade," "Take It or Leave It," and "Ozzie and Harriet" from CBS. In the biggest coup of all, Paley, in 1948, managed to lure Jack Benny from NBC to CBS.

Since most stations of the day were affiliated with one of the four networks (NBC, CBS, ABC, or Mutual), they mostly just transmitted the network programming. What little they did produce on their own was mainly of a public service or phonograph music nature.

THE INTRODUCTION OF TV AND ITS EFFECT ON RADIO

When the networks started TV programming in the late 1940s and early 1950s, they

FIGURE 1.3 The Tommy Dorsey Band was obviously sponsored by cigarettes. Again, although the home audience couldn't see them, the enlarged cigarette boxes were predominantly placed on stage. Since this program was performed before a live audience, the sponsor's product was constantly seen by the studio audience. (Photo courtesy of KFI.)

adopted the same programming methods that had been used in radio. Once again ad agencies provided programs and paid for them in their entirety—"Philco Playhouse," "Kraft Television Theater," "Texaco Star Theater." TV adopted not only the programming process of radio but also its stars and its advertisers. Jack Benny, Burns and Allen, Bob Hope, and many others left radio for the medium that featured both sight and sound. Radio had to reinvent its programming.

According to radio lore, radio group station owner Todd Storz was in a bar one night in the early 1950s trying to drown his sorrows about the decreasing income of his radio stations. He noticed that the same musical selections from the jukebox were played over and over. After almost everyone had left, one of the bar waitresses went over to the jukebox and, instead of playing something that hadn't been heard all evening, inserted her nickel (yes, *nickel*) and played one of the songs that had been heard over and over. This gave Storz the idea for Top 40 radio, which he and several other station owners (most notably Gordon McLendon) used to bring radio back to health.

Obviously Storz' "vision," all by itself, did not revive radio. Another important factor was the rise of rock and roll music, which gave radio a new sound and a new specific audience—teenagers. Gradually radio switched to a new programming form that was dependent on the disc jockey and music. Disc jockey programs were not new to radio. In fact, an argument could be made that the early radio stations that programmed call letters and phonograph music were disc jockey shows. But in the heyday of the networks, disc jockeys and music were not a dominant form of radio programming.

In the 1950s, however, recorded music became the primary radio fare and led to a whole new structure for radio programming. For one thing, the networks became mere shells of their former selves, providing primarily news and very little in the way of entertainment. Radio became a local, rather than national, medium. Characteristics of the local community and the selection of a rigid, daylong format became major factors in programming decisions. Advertisers no longer supplied entire programs; they merely bought commercials (advertising spots) within news or music programs. Programming decision-making rested with the local program managers and station managers, not with networks or advertising agencies.

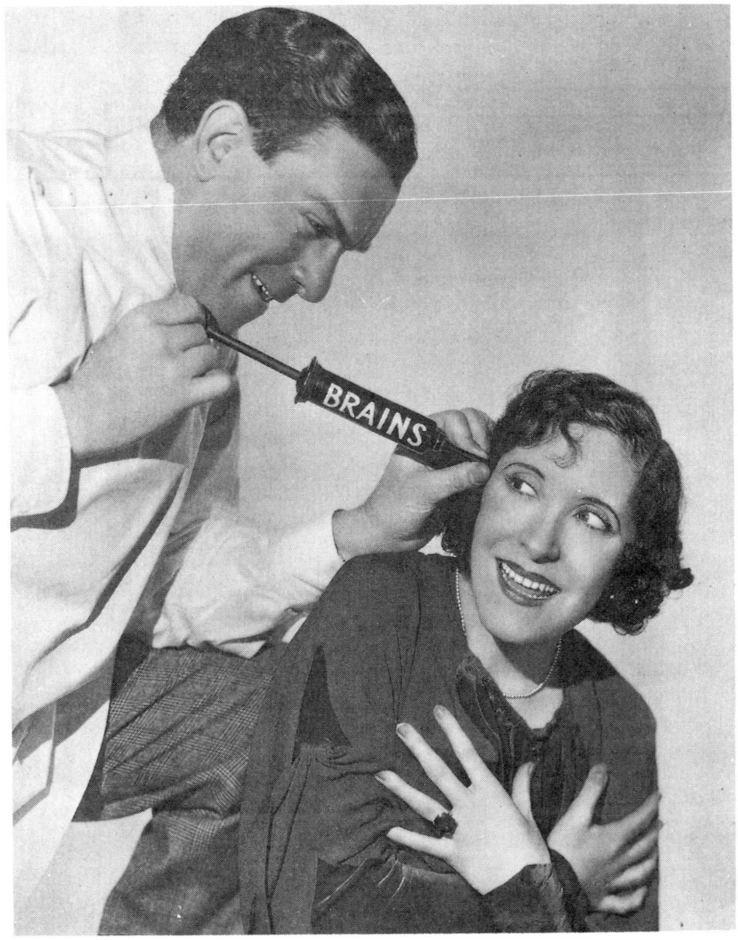

FIGURE 1.4
George Burns and his supposedly dippy wife, Gracie Allen. The two were sponsored for most of their radio life by Burns cigars, which capitalized on George's last name and on his cigar smoking habit. (Photo courtesy of KFI.)

TV PROGRAMMING CHANGES

Meanwhile, television was changing, too. The idea of advertisers supplying programming wasn't working as well as it had in radio. There were several reasons. For one, television programming was much more expensive than radio programming. Advertisers who had easily been able to underwrite the costs of several actors capable of changing voices to play several parts, a sound effects person, a small core of audio technicians, a few writers, and a director found the visual demands for scenery, props, and additional actors and behind-the-scenes personnel more than they could handle financially. As a result, co-sponsorship sprang up. Two or more companies would join to share the costs of producing and distributing a television series. Programs would be "brought to you by Colgate toothpaste, Oldsmobile, and Marlboro cigarettes." This form of advertising made it harder for the viewer to identify the program with a specific product, as had been the case with radio, but it temporarily solved the budget problem.

In addition, strong-minded TV network executives were becoming more interested in controlling their own programs. Sylvester L. "Pat" Weaver, while he was president of NBC from 1953 to 1955, devised what he called the "magazine concept." Advertisers bought commercial insertions in programs such as "Today" and "Tonight" but had no say about program content. Those decisions were in the hands of the networks.

The trend toward such network control was accelerated in 1959 when the quiz scandals broke. Quiz programs on which contestants won large amounts of money had become very popular. Contestants on programs such as "The $64,000 Question" and "Twenty-One" were locked in soundproof booths where they agonized and perspired as they tried to answer very difficult questions. The programs were so popular that Revlon, the company that sponsored and produced "The $64,000 Question," found its products had sold out nationwide. (So much for the theory that broadcast advertising would never sell products.) The problem was that some of the contestants who were popular with the public were "coached" ahead of time about the questions they were to answer and about how to act to build suspense and sympathy. This was done in the name of entertainment—so that lively, personable contestants could remain on the air and the duller, less likable ones would be defeated. When the rigging of the programs was disclosed, the network executives claimed they knew nothing of the irregularities and placed the blame on the ad agency executives who brought the programs to them.

Networks, already on the road to taking control of programming, now hastened the process and began selling commercial spots within most programs rather than having them totally sponsored by advertisers.

THE RISE OF PROGRAMMING STRATEGIES

This change led to a whole new philosophy of program decision-making. Networks now began to consider their programming schedule as an overall entity. By the 1960s, the amount of money a network could charge for a commercial spot was dependent upon its ratings. In the past a network could generally continue to air a program as long as the advertiser was satisfied with it. But now networks (and stations) were dependent upon the fickle public. Even if an advertiser wanted to pay the total cost of a program, the network might not want the program because it proved to be a poor lead-in to another program that was being sold on a spot basis. Such did happen in 1963 when the "Voice of Firestone" was canceled. Firestone Tire Company had totally sponsored this classical music TV program since 1949. The audience was small but appreciative and Firestone wanted to continue the program. But the network canceled it, amid much controversy, because having it in the lineup cut down on the size of the audience for programs before and after it.

When advertising agencies were phased out as suppliers of programs, the networks had to find other means of obtaining program material. They wound up producing a great deal of it themselves and buying it from production companies, many of which (MCA Universal, Paramount, Fox) had been engaged in producing feature films. They also started buying from new independent production companies (MTM, Tandem-TAT, Aaron Spelling Productions) that had formed specifically to produce programs for TV. The networks underwrote all or most of the cost of production in exchange for the right to air the program and sell commercials within it.

In addition, the networks received part of the profit from the sale of the show in the syndication market (details on this process are offered in Chapter 2). The income from the commercials, all of which the network kept for itself, and the revenue from syndication made networking an extremely profitable business.

In 1970, the FCC took a hard look at this situation and decided the networks now had too much control over their programming and were accumulating too powerful a base. As a result, the FCC instituted financial interest–domestic syndication (fin-syn) rules that barred networks from having a financial interest in programs produced by outside production companies. They could no longer receive part of the profit when the programs were sold to stations as syndicated reruns. Rules were also instituted that limited the amount of programming that networks could produce themselves.

During the 1960s and 1970s, a number of programming strategies were devised as networks competed with each other to attract audiences and to capture the best programming from the production companies. These strategies are discussed in detail in this book because they are still prevalent today, despite the constantly changing electronic media world.

THE GROWTH OF INDEPENDENT STATIONS

In the early 1960s, the programming dominance of NBC, CBS, and ABC in TV was akin to their power in radio of the 1940s. Most stations were affiliated with one of the networks and presented primarily network programming. Their local programming choices consisted largely of public service shows and some old movies.

However, during the 1960s, 1970s, and 1980s, many independent TV stations were authorized and began broadcasting. These stations did not affiliate with the networks, mainly because other stations in town had already done so. In 1961 there were only 28 stations that were not affiliated with one of the three networks, but by 1979 that number had swelled to 103, and by 1989 it had reached 339.[5]

These stations had to devise entirely different forms and strategies of programming. Part of what they scheduled was network reruns—the programs supplied by production companies to the networks and then sold to the independent stations, first with network profit-participation and, after the fin-syn rules, with the profits going solely to the production companies. But this was not enough to fill all the hours of the broadcast day. A large business sprung up to supply these stations with movie packages and with various forms of programs produced specially for syndication. Independent stations also produced some of their own material and, in some cases, acquired material from foreign countries.

Then in 1987, a new source of programming, Fox Broadcasting Company, arose for some of the independent stations. Not officially a network because it did not broadcast more than fifteen hours of programming a week (the legal amount needed to be consid-

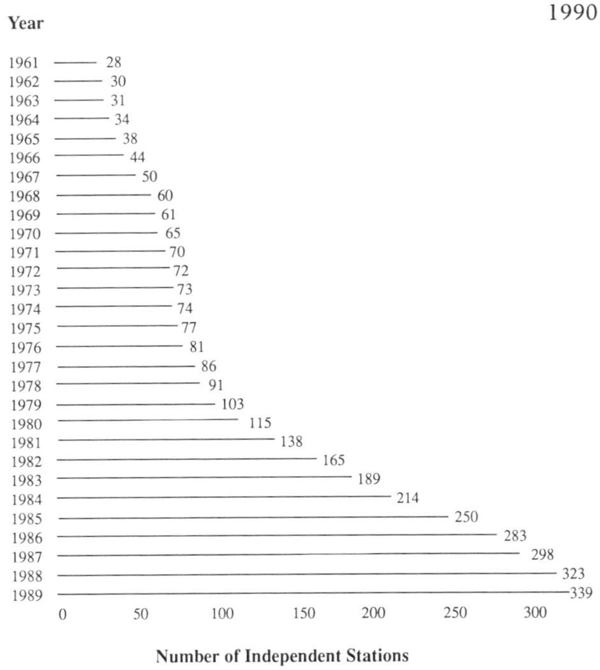

FIGURE 1.5 The growth of independent stations over the years. Source: *1990 INTN Census* (Washington, DC: Association of Independent Television Stations, 1990): 18.

ered a network), it nevertheless provided a large block of programming for many of the independent stations and cut down on their need for syndicated material. Through innovative programming, Fox succeeded in capturing a fairly large portion of the viewing audience for its stations and cutting into the audience size of the three traditional networks.

THE DEVELOPMENT OF PUBLIC BROADCASTING

Another programming change of the 1960s and beyond was the rise of public radio and TV. Both of these had been around earlier in a weak form usually called "educational broadcasting." The government had set aside part of the FM band, 88.1 to 91.9, for noncommercial radio services and had reserved specific TV channels in each major market for educational TV. Programming decisions for all these educational stations were made locally. Most of the radio stations played classical music and produced local public affairs shows. The TV stations aired locally produced "talking head" programs that featured one or two people directly addressing the camera plus some programs from other local educational stations supplied through an exchange that mailed programs from station to station. No actual networks existed. The programs were generally, in a word, dull.

Then in 1967, Congress passed the Public Broadcasting Act, which implemented most of the recommendations that had been made by a blue-ribbon Carnegie Commission set up to develop improvements for educational (now called "public") broadcasting. The Corporation for Public Broadcasting was devised to receive funding from the government and then apportion it among public radio and TV stations and to a TV network, Public Broadcasting Service (PBS), and a radio network, National Public Radio (NPR). With this infusion of money from the government, the quality of public broadcasting programming increased significantly.

The noncommercial programming structure is very different from that of the commercial networks. PBS does not produce any programming itself but rather acquires it from public TV stations, foreign countries, and independent producers. NPR produces much of its own programming but also acquires material from other sources. NPR's reluctance to air material from public radio stations led several of these stations to form, in 1982, a competing network, American Public Radio (APR), that primarily distributes programs from its member stations. Of course, the networks do not produce all of the programming needed by the stations; they still fill many hours with material they produce or acquire on their own.

The audience for public broadcasting is fairly upscale. That and the noncommercial nature of the system lead the networks and stations to approach program decision-making quite differently from the commercial networks and stations.

EARLY CABLE TV

Probably the most significant of all the developments in television in the past few decades has been the development of cable TV. Actually, cable had been around since the early days of television broadcasting. No one knows exactly how it began, but one folklore story says that it was started by the owner of a little appliance store in central Pennsylvania in about 1947. He noticed that he was selling sets only to people who lived on one side of town. When he investigated, he found that people on the other side couldn't receive the signal very well. So he placed an antenna on top of a hill and ran the signals through a cable down the hill to the homes with poor reception. When someone on the "weak" side bought a TV set from him, he hooked them to the cable.

Cable grew during the early days because the only way communities without TV stations were able to obtain TV programs was to put up an antenna, catch the signals as they traveled through the air, and run a cable, often strung from tree to tree, to the individual homes. From the 1940s to the 1970s, cable TV was mainly used to retransmit signals from existing TV stations. Usually these were local stations, but as time passed cable companies imported station signals from distant areas to provide their customers with a wider variety of programming.

Because the cable signals traveled through wire and not through the airwaves, cable systems were able to have more useable channels than broadcasting could accommodate. For example, no local area could air broadcast stations on both channels 5 and 6 because the two signals would interfere. But on a cable, where the signals were shielded, all the channels (then 2 through 13) could be used. As a result, some of the cable systems provided viewers with inexpensive "local programming" that was placed on one of the extra unused channels. The most common form of this "programming" was a thermometer and barometer with a camera focused on them to allow local residents to see the temperature and barometric pressure. Not much was needed in the way of program decision-making.

Not much was needed in the way of regulation either. Broadcast stations complained to the FCC when some cable systems showed a distant channel that was playing the same program (e.g., the same "I Love Lucy" episode) as the local station because this practice obviously lessened the audience size for the local station. The FCC responded by establishing a rule called "syndicated exclusivity," which said that cable systems had to black out the distant station when the programming was the same. But generally the FCC left cable alone to grow on the fringes of the rapidly expanding television universe.

What little regulation there was came from local governments. Some cable systems installed their wires underground, which meant they had to dig up city streets. Since the city streets were overseen by the local city council, this body would grant a "franchise" to a cable company and then keep an eye on it to make sure it didn't create potholes or damage the shrubs. From the late 1940s to the mid-1970s, the cable TV business was a sleepy little adjunct to broadcasting.

FIGURE 1.6
Public television's first big hit was "Sesame Street," the children's program produced by Children's Television Workshop, which began airing in 1969. (Photo courtesy of Children's Television Workshop/Richard Termine.)

THE BURGEONING OF CABLE TV

But all this changed in the late 1970s when Home Box Office (HBO) began selling a satellite-delivered movie service to local cable systems. Although this was a difficult sell at first, mainly because cable systems did not want to invest in the satellite dish needed to receive the signal, once the idea caught on, the floodgates opened. A large variety of cable networks sprung up offering programming to be placed on the spare channels of the cable systems. So much programming became available that engineers devised ways to show more than twelve channels through the TV set.

Between 1980 and 1989, the number of households subscribing to cable TV grew from 22 percent to 59 percent, the advertising dollars spent on cable went from $53 million to $1.5 billion, the number of programming networks rose from 8 to 45, and the number of channels available to the average TV home rose from 8 to 28.[6] Cities that had not needed cable TV because the reception was excellent suddenly found numerous companies pounding on the doors of city hall—all begging for franchises to wire the area and collect their part of the pot of gold at the end of what seemed to be a very promising rainbow. These companies promised not only to pre-

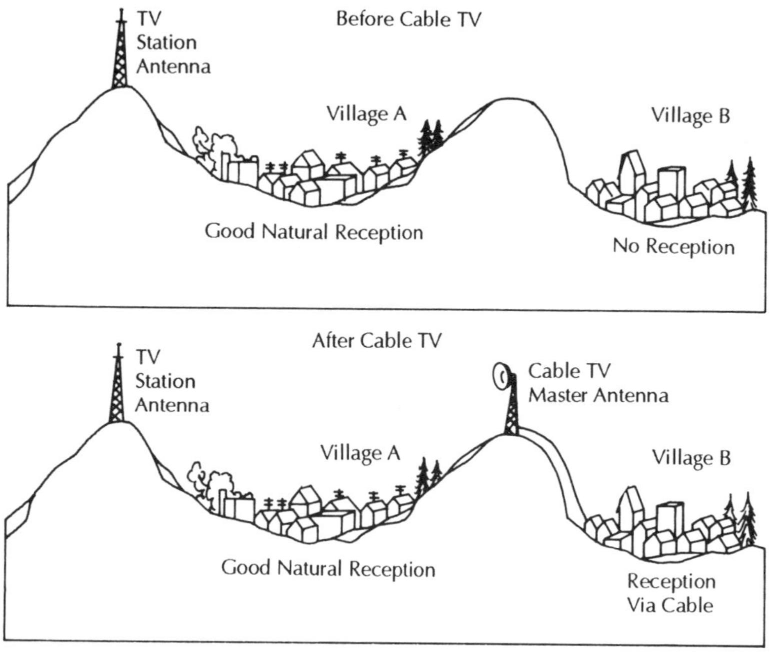

FIGURE 1.7
Early cable TV systems placed an antenna on top of a hill to catch television station signals. Then a wire, with the station signals in it, was run down the hill and attached to various homes.

vent potholes and damaged shrubs but also to provide the city with an almost infinite array of services—television equipment for local universities and community groups, public access channels that would highlight local events (and maybe even make a star out of a city council person or two), interactive security systems for every home in town. The cable frenzy of the early 1980s was intense.

All the different program services that sprung up needed programming. Some networks, such as ESPN and CNN, produced all or most of their own material. But many went to the same sources networks and stations had been using—film and TV production companies and syndicators. Most cable networks claimed they were not "broadcasting" but rather were "narrowcasting," aiming their programming at a certain subsection of the TV audience—sports fans (ESPN), young people who like rock music (MTV), children (Nickelodeon), movie fans (HBO). New programming strategies were needed by the cable companies to compete with broadcasting and with each other. Many of the cable services actually adopted some of the strategies started in broadcasting of the 1960s because they still worked, but they also had to give thought to their own unique conditions.

THE INFLUENCE OF THE VCR

Another form of electronic distribution of programming has been very significant, especially as it has impacted on both traditional broadcast and cable TV programming practices: the videocassette. Videocassette recorder ownership has grown from 1 percent of homes in 1980 to 75 percent by 1991.[7] With the VCR, members of the public can become their own programmers. They do not need to watch programs when the network executives want them to; they can tape them and watch at their own convenience. They also have programming options far beyond the myriad of channels provided by broadcasters and cablecasters; they can go to the local video store and rent from a huge catalogue of programs.

Although the videocassette recorder is a significant electronic media form, its importance to the process of programming relates mostly to its effects on other media. Viewers can take the schedules that have been painstakingly designed by network and station executives and totally rearrange them. They can also skip through the commercials. Programmers now need to take the impact of the VCR into account in their program decision-making.

MODERN-DAY RADIO

Meanwhile, back in the radio corral, changes have been taking place, too. For the most part, local stations still dominate, and decision-making revolves around format and features of local stations. But networks are reemerging in a different form. Some provide satellite-delivered music that stations are free to use as entire programs or as portions of locally produced programs. Others provide news and/or features. Radio syndicators also exist to provide various programming and marketing services to radio stations.

Radio stations select material provided by networks, syndicators, and their own local programmers and mesh it together into a unified whole. Radio, at present, is trying to maintain a local feel while obtaining at least some of its programming from national sources.

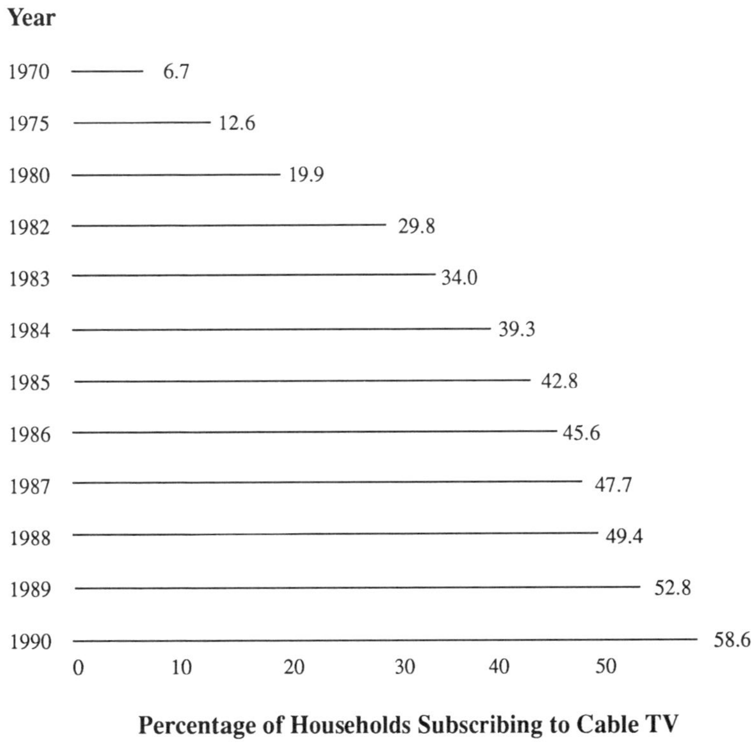

FIGURE 1.8 The growth of cable TV in terms of the percentage of homes subscribing. Sources: Bureau of the Census, *Statistical Abstracts of the United States, 1990* (Washington, DC: U.S. Department of Commerce, 1990): 550; and *The Broadcasting Yearbook* (Washington, DC: Broadcasting Publications, Inc. 1991): D-3.

PROSPECTS FOR THE FUTURE

The near future is likely to see the art of programming undergoing a continued state of flux—nothing new for an industry that has experienced a variety of programming alterations during its relatively brief history.

All of the changes of recent times—the growth of independent TV stations and Fox, the development of public broadcasting, the popularity of cable TV, the flexibility of the VCR, and the local-national interchange of radio—have altered the basic underpinnings of the old network-station structure. Radio networks have lost their identity and TV networks, which in 1980 could claim that 85 percent of American households were tuned to their programs during primetime, now have to settle for about 60 percent.[8] All the forms of electronic media need to take a new look at how they program. Many of the old strategies will still work in their original or a revised form, but new ideas will need to surface to meet changing situations.

New Television Delivery Methods

Although commercial television, cable television, commercial radio, and public broadcasting are the major forms of electronic media available today, there may be a variety of new forms in the future. For example, direct broadcast satellite (DBS) is currently in operation in several other countries and has been proposed for the United States. DBS is a system whereby programming is delivered directly from satellite to very small dishes mounted on customers' homes. If this becomes popular, it could greatly alter the programming structure of the United States, making obsolete the network-station relationship wherein stations act as intermediaries between the network programming and the consumer.

The telephone companies are also chomping at the bit, trying to enter the video production and delivery business. At present they are allowed to transmit and provide data services—written words that give the latest news, weather, stock market quotes, recipes, encyclopedia information, and the like—often with the aid of a computer. But the descendants of the almighty AT&T want to provide the "one wire" into the home. The phone line, which is attached to virtually every home in the country, would provide not only the usual audio-related phone services and the computer-based data services, but a vast array of television channels. This, of course, has alarmed the cable TV folks who are attempting to persuade Congress to preserve the existing rules and prevent the phone companies from entering the TV business.

One of the delivery advances that faces radio involves digital audio broadcasting (DAB), which will bring compact disc (CD) quality audio through the radio. Present plans are to

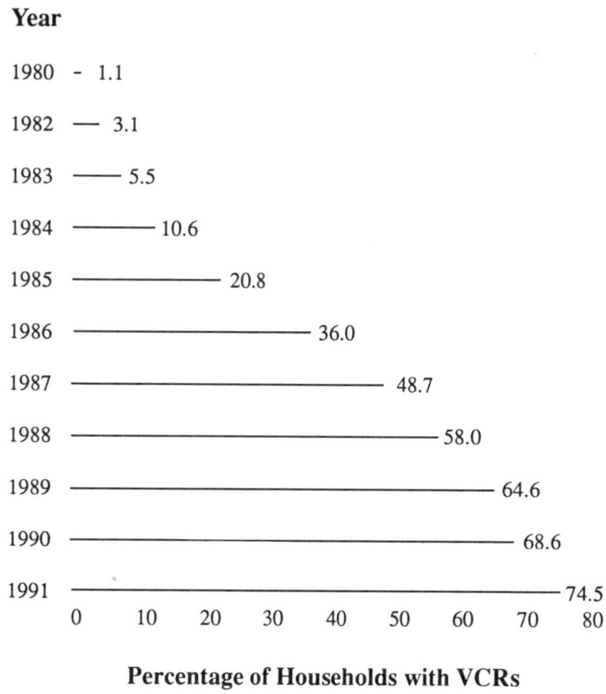

FIGURE 1.9
The growth of videocassette recorder ownership in terms of percentage of homes with VCRs. Sources Bureau of the Census, *Statistical Abstracts of the United States, 1991* (Washington, DC: U.S. Department of Commerce, 1991): 555; and "VCR Penetration Up to 74.5% in U.S.," *Variety* (January 29, 1991): 26.

develop a whole new service that will not be part of the AM and FM bands. DAB would resemble DBS in that it would come directly from satellite to receiver. This, of course, would have a great impact on the present radio station and network structure.

Growth (or Decline) of Channels

No matter who delivers the radio and television signals, there are likely to be more of them. A new technology called "compression" allows for a great deal more video information to be transmitted, either over the airwaves or through wires. Time Warner is already experimenting with an upgrade of its 75-channel cable TV system in Queens, New York, to 150 channels, mainly by using fiber optics. Their goal: to provide their subscribers with limitless program choices ranging from the hottest movies to neighborhood news shows. When Time Warner begins using compression, it figures the system will be capable of 600 to 1200 channels.[9] Where the talent and money will come from to produce all these programs, nobody is quite sure. However, history suggests that if something is scientifically possible, someone will figure out a way to use it profitably.

Of course, what really may be needed is fewer channels. There are so many audio and video choices now that many outlets are experiencing financial difficulties. The viewer, too, is faced with overload. There is, after all, a limit to how much radio and TV a poor individual can absorb during the course of the day. The future may see a shakedown in the number of services available. One or more of the major TV networks may go out of business; more cable networks may fail or merge; AM radio stations may disappear for lack of audience and lack of technical quality.

The International Impact

The "global village" predicted in the 1960s is quickly arriving. The whole international scene for buyers and sellers is fluid and uncertain due mainly to changes in the radio and TV systems of other countries. The United States, which has been by far the largest supplier of programming to the rest of the world, is receiving competition from other countries as they develop and refine their electronic media structures and set their sights on a piece of the international programming pie.

Historically, the United States has imported the least programming of any country—basically just a few British Broadcasting Corporation (BBC) programs for public television. But a small crack has appeared in that dike, and programming from other countries is beginning to appear with greater regularity on cable TV and even commercial station TV. Also, American production companies are engaging in an increasing number of co-productions with media entities from other countries.

Further VCR Impact

The VCR may further erode the ability of programmers to set schedules. The main thing that keeps people from recording programs and watching them at a later time is that most people have never figured out how to set the timers on their VCRs. New inventions to make the VCRs more user-friendly may lead to a greater degree of time shifting. At one time ABC experimented with broadcasting scrambled signals of movies and other programming in the middle of the night. The idea called for subscribers to set their timers on their VCRs and record the programs. There weren't many takers, but maybe it is now time to resurrect that idea.

● ● ● ● ● ● ● ● ● ● ● ● ● ● ● ●

The rich and dynamic history of broadcasting has been nothing more than a prelude to the changes about to come. The multiplicity of channels, the new high-tech delivery systems, and other scientific advances that arrive almost daily will cause major adjustments in the production and content of programming. As veteran writer Barry Kemp ("Coach," "Taxi") sees it, "This expansion of the television landscape has led to a greater competition for viewers than ever before. The result is that we may be finding ourselves in a decade that has the potential to be as creatively exciting as any in our history."[10]

But these changes will almost surely be modifications of our current TV universe, whether it be the form and styling of programs or the manner in which they are distributed. Therefore, it is essential that readers have a solid grasp of the media that now represent the numerous marketplaces for shows. In Chapter 2 we describe how three of these commercial outlets operate: network television, syndication, and stations (both independent and affiliated).

● ● ● ● ● ● ● ● ● ● ● ● ● ● ● ●

NOTES

1. Erik Barnouw, *A Tower In Babel* (New York: Oxford University Press, 1966): 83–87. This well-researched work contains many delightful anecdotes about early radio programming.
2. Ibid., 110–111.
3. Ibid., 96. Most direct quotes in this book were obtained by the authors interviewing the people quoted. These will not be cited. The quotes that will be footnoted will be those obtained from books, magazines, and newspapers. Obviously, the authors did not interview Herbert Hoover.
4. Ibid., 238.
5. *1990 INTV Census* (Washington, DC: Association of Independent Television Stations, 1990): 18.
6. "Decade of Change, Decade of Choice," *TV Guide* (December 9, 1989): 29; and Bureau of the Census, *Statistical Abstracts of the United States, 1991* (Washington, DC: U.S. Department of Commerce, 1991): 563.
7. Bureau of the Census, *Statistical Abstracts of the United States, 1991*: 556; and "VCR Penetration Up to 74.5% in U.S.," *Variety* (January 29, 1991): 26.
8. "Tangled TV Webs Begin to Unravel," *Los Angeles Times* (August 8, 1991): A-16.
9. "Time Warner Unveils 150 Channels," *Broadcasting* (December 23, 1991): 18.
10. "TV, an Eroding Medium? Creatively, It's Exploding," *Los Angeles Times* (September 9, 1992): F-1

2 The Marketplace: Part I

One of the sadder sights in life is to watch someone design a solution for a nonexistent need: the chimney sweep who makes a brilliant pitch to a homeowner only to learn the house has no fireplace; the deluxe, multicolor brochure about the glories of Las Vegas that is dropped into the mailbox of a convent.

To be effective in television, programmers must know who the buyers are, how they operate, where they're located, what turns them on and off, what they can and cannot do. In short, they must know the marketplace.

COMMERCIAL TELEVISION

Commercial television is that area of broadcasting that is privately owned, operated for profit, and offered to the public without charge. Its entities include networks, individual stations, and syndicators.

Networks

A TV network is a delivery system that transmits programs and advertisements from a central source via satellite to an interconnection of stations that retransmit the pictures to their coverage areas at a specified day or time.

TV networks are the most glamorous and far-reaching of all the electronic media. Although their audience has diminished considerably in the past few years, they are still watched by about 60 percent of the homes in the country on any given evening, and they are the medium most favored by national advertisers.

The three traditional networks—ABC, CBS, and NBC—have been in existence since television became a source of public entertainment in the 1940s. A fourth network, the Fox Television Network, joined the original trio in 1987.

Relationships with Affiliates. A network is simply the combining of a central corporate entity with a number of individual stations for the purpose of distributing programs. In its basic form the relationship is very straightahead. The company agrees to supply programming to each of these affiliated stations for fixed periods of time each day. It further agrees to pay the station an agreed-upon amount for every program that the station sends out to its coverage area. The station, in turn, agrees to broadcast these shows. According to FCC rules, a network may not prevent a station from broadcasting programs from "any other network organization." Gen-

> Ken Auletta in his book *Three Blind Mice* describes the network-affiliate relationship this way: a network is really a tenant pretending to be a landlord. Since government regulations did not permit the networks to own more than a handful of the stations that distributed their programs, the networks were at the mercy of the six hundred or so TV stations that agreed to call themselves affiliates. Having stitched together a national network of local stations that wanted to show the programs produced for the networks by Hollywood or the independent producers, the networks then went to advertisers and sold t hem mass audiences that were, in effect, rented from local stations. A network was, in fact, nothing more than a broker. It was an odd business.[1]

erally stations broadcast programs from only one network, but in small markets with only two or three stations, one station may affiliate with two networks.

The network and the stations have worked out a simple method of remuneration. The network sells commercials within the programs and keeps the revenue. They also deliver their half hour and hour programs 60 to 90 seconds short of full time. The "missing" time is available to the stations for sale to local advertisers, with all the income remaining with the stations. This "station break" revenue plus the compensation from the network provides the incentive for a station to affiliate. In more recent years, networks, on occasion, have turned over some of their commercial positions to affiliates for local sale to make up for eliminated or reduced compensation.

Within this broad framework variations are negotiated. But the fundamental spirit of the contract is to develop a family relationship for the strength and prosperity of all. By the early 1990s, each of the three original networks was affiliated with slightly more than 200 stations, and Fox counted about 130 stations as affiliates.[2] These interconnections allow television programs to penetrate virtually every home in the United States.

Each of the four networks owns and operates a group of stations. In addition to supplying programs to these owned and operated (O&O) stations, the network exercises financial control over them and determines many of their general policies. CBS has five of these O&Os, Capital Cities/ABC has eight, NBC six, and Fox seven. These stations are located in major markets and are very profitable because much of their programming is supplied by their parent company. However, no network organization is permitted to own (as opposed to affiliate with) stations whose coverage exceeds 25 percent of U.S. households.

As TV rolled into the 1990s, the network-affiliate system began to fray. The dramatic

THE MYSTERY OF STATION COMPENSATION

Almost all newcomers to the network business are puzzled by the concept that a network pays a station to present a program, especially hits such as "Roseanne" or "Murphy Brown." One would think the station would be so delighted to obtain these audience attractions they would be willing to pay for the privilege. Not so.

The station compensation system harkens back to the radio days of the 1920s when Radio Corporation of America (RCA) made a deal with American Telephone and Telegraph. AT&T stepped out of active broadcasting and became a carrier only. RCA, under the corporate title National Broadcasting Company, set up a "chain" of stations and used the lines owned by AT&T. At that time there were 26 million homes in the country, five million of which were already equipped with radios. This meant there were 21 million to go, a splendid marketplace for RCA, already the premier manufacturer of radios in the world.

Although there was money to be made in the newly developed airing of "tolls" (commercials), the real profit was in the sale of the radios themselves. Thus, RCA had a major stake in the dissemination of quality programs. The better the shows the greater the clamor for radios. To assure that these superior programs were heard by the populace, RCA needed to develop relationships with stations that would guarantee the airing of NBC shows. The only way to achieve this was to pay for it. In the original contracts, NBC gave the affiliates $50 per hour for all sponsored programs, but charged the stations $45 for every sustaining (unsponsored) show they carried.

The ball game changed dramatically in 1929. Under the guidance of William Paley, CBS made a bold move to wrest leadership from NBC. The upstart network delivered the sustaining shows free to their affiliates and increased the compensation for sponsored shows. Now it was no longer necessary for CBS to consult the stations before feeding a program. The clearance could be assumed. NBC eventually remodeled its station agreements along the same lines.

Some 60 years later, CBS was once again in the vanguard on station relations, this time less benevolently. Its new owner, Lawrence Tisch, alarmed at the precipitous drop in network profits, raged at the cost of station compensation. If the relationship was a true partnership, he argued, why wouldn't the affiliates be willing to share the economic burden and agree to a reduction of compensation. Because, he was told by colleagues, they do not want to suffer the loss of revenue and, if forced to, will retaliate by dropping clearances of network shows. Although this is a divisive development in network-affiliate relationships, it appears to be a trend in the industry's never ending effort to grapple with the new economic realities.

reduction in network audiences caused a severe downturn in network revenue. By 1992 all three broadcasting companies were either losing money or edging barely into the black. Draconian cost-cutting measures were introduced, including new approaches to station compensation. The networks refused to pay stations to carry certain sports events and gave them additional commercial time instead. This new economic relationship was seen as a forerunner to an eventual elimination of all station compensation. In the next few years this aspect of the network-station association will in all likelihood be totally renegotiated and a new system developed.

Of course, the stations have not stood idly by as their income from networks has diminished. They responded by preempting (refusing to carry) network programs that were performing weakly. They simply substituted a show that they purchased elsewhere and kept the revenue from all the commercials sold in the program. Many times this was substantially greater than the compensation that was sacrificed. These preemptions have been particularly pronounced in daytime, Saturday morning, and late night when attractive program alternatives are available.

Although the prospects for networks have their ominous tone, this chapter will assume that the traditional terms of a network-station association will continue to apply. Readers should be familiar with the basic structure in order to understand how the deviations modify old agreements.

Selection of Network Shows. When stations agree to become affiliates, they turn over the choice of programs to the network for fixed periods of the daily schedule. As long as the network is filling a time period that has been agreed to as "network time," the selection of the show is a unilateral decision. It is obvious that a network with over 200 stations cannot make program choices a democratic process. However, the stations, not the networks, are licensed to broadcast by the FCC. Therefore, it is the stations who run the risk of penalties or even loss of license if some show runs afoul of federal regulations. Since each network owns and operates a number of stations they, too, will suffer penalties—not directly but through the punishment meted out to the O&Os.

Because the affiliates have to be mindful of their vulnerability, they have set up systems with the networks that assure they will be informed in advance of any controversial programming. This gives them an opportunity to prescreen a questionable show and decide not to broadcast it if they feel it is inappropriate for their market.

Around midsummer of every year, each of the networks convenes with its affiliates to view the new programs on the fall lineup. On some occasions the stations have made known their displeasure with certain productions. The objections rarely result in the elimination of the show from the schedule, but the networks usually agree to modifications and reassure the affiliates they'll "keep a close eye" on the series.

Network Time Periods. Although the specific hours vary among the three traditional networks, the typical affiliation agreement calls for the network to supply programs on the following basis:

Monday through Friday
1:00– 6:00 a.m. Local shows
6:00– 9:00 a.m. Network
9:00–10:00 a.m. Local
10:00– 4:00 p.m. Network
4:00– 6:30 p.m. Local
6:30– 7:00 p.m. Network news
7:00– 8:00 p.m. Local
8:00–11:00 p.m. Network
11:00–11:30 p.m. Local
11:30 p.m.–12 midnight Network
12 midnight–1:00 a.m. Either network or local

Saturday
8:00 a.m.–12 noon Network
12 noon–8:00 p.m. Occasional network programs, mostly sports events; varies by season
8:00–11:00 p.m. Network
11:00 p.m.–12 midnight Local

Sunday
12 midnight–9:00 a.m. Local
9:00a.m.–7:00 p.m. Occasional network programs, mostly news and sports; varies by season
7:00–11:00 p.m. Network
11:00 p.m.–12 midnight Local

In recent years a fluidity in the schedule has developed that was virtually unknown for four decades. Although the primetime, early morning, and Monday–Friday 12 noon to 3:00 p.m. hours continue to be near sacrosanct, the more fringelike periods have been gradually drifting away from the networks. The weekday 10:00 a.m. to 12 noon periods have been deeply eroded by station defections, and some of the networks are no longer even feeding some of the half hours. Scores of CBS and ABC stations became disenchanted with their networks' late-night programming in the early 1990s and simply devoted the time to other offerings. As the advertising market dried up for sports events, the networks stopped buying them, leaving the affiliates with huge hunks of time to fill on weekends.

The motivation on the part of both networks and stations is, not surprisingly, money. Networks don't like to be saddled with production costs and rights fees for shows and events that aren't going to be profitable. It is easier just to turn the time back to the stations. Similarly, stations don't want to carry programs that are so unpopular the local commercial positions surrounding them are worthless. They would much prefer to drop the program, commit to a show in which they keep the revenue from all the commercials, and hope for higher ratings. On the downside, the stations lose their station compensation, and they have to pay for the show they substitute. On the upside, they keep every dime's worth of revenue that the substitute show generates. If the program is even halfway decent, the result is a better profit performance for the time period.

The Fox network is a separate matter. Over the years, it has gradually been increasing its program service. By September of 1993 it was supplying its stations with programs all seven nights a week. In addition, the network offered its affiliates a daily block of children's programming.

The purpose of the foregoing is to demonstrate the size and shape of the network program marketplace. It is designed to help programmers decide what types of shows to develop and where to take them. For example, if a programmer has an idea for a situation comedy or hour-long dramatic series, the networks are excellent candidates. They need those kinds of programs to fill the primetime periods for which they are responsible to their affiliates. But a programmer who designs a show that can only work weekdays at 4:00 p.m. or 5:00 p.m. cannot even think of a network as a potential buyer because the networks don't service that time period.

Ad Hoc and Regional Networks

From time to time suppliers and stations will combine to form a network for a special purpose (ad hoc). In 1976 a collection of 95 affiliated and independent stations coalesced into an organization entitled "Operation Prime Time" (OPT). The purpose was to provide all participants with a movie-of-the-week or miniseries once a month. A committee out of the membership was formed to screen submissions and select the best for production commitments. The plan was only moderately successful and it faded away in the mid-1980s.

More recently Warner Brothers Domestic Television Distribution sparked a consortium with independent stations to produce and distribute 22 original hours of two new series: "Kung Fu: The Legend Continues" and "Time Trax." In addition, the arrangement involved a two-hour telefilm and a ten-hour docudrama, "The Wild West." All production fell under the supervision of Warner Brothers. Each station was permitted to sell five minutes in every hour, with Warner Brothers keeping the remaining seven minutes. The appeal to all participants was that Warner could produce each show for approximately two-thirds the cost of a comparable hour for a network.

In the past few years movie ad hoc networks have been set up because of changes in the marketplace. The common thinking among network executives is that popular theatrical titles, which are available through pay cable and home video outlets months before a network premiere, have become "overexposed" and therefore do not attract the audiences they once did. This has left movie companies with a huge backlog of pictures that have gone unseen on the networks. Such a situation led MCA Universal to form the Debut Network, a service that supplies its stations with 24 advertiser-supported titles during the year. MGM/UA formed a similar consortium and called it the Premiere Network.

Regional networks are those that are put together to provide programs of special interest to a particular area. For example, the California Farm Network feeds shows to fifteen stations ranging from Sacramento to San Diego. Many sports fans get to see their favorite teams through the existence of regional hookups.

Syndication

Syndication is a program delivery system through which shows are provided to stations by the producing company or its distributor. There is no preset interconnection of outlets. Distributors must contact each market to see if they can find one station willing to schedule the show. This is in stark contrast to the network system, in which a program is assigned to a time period and virtually all affiliated stations automatically transmit it.

There are slightly more than 1100 commercial television stations in the United States. A little over 600 of them are affiliated with the traditional three networks. The other 500 are independent stations, about 130 of which are allied with Fox for the limited programs it supplies. Stations affiliated with NBC, CBS, ABC, or Fox must obtain programming for those time periods not filled by the networks. The other 370 stations must obtain program material for the entire day, every day of the year.

Aside from local newscasts and a few local public affairs programs, most stations are not equipped to develop their own programs to complete their schedules. To do so would be too expensive. They must look to outside sources that can furnish material to a number of stations, thus reducing the cost per station. Syndicators fill that need. They supply programming to occupy those time periods that networks don't cover and stations themselves can't fill because of insufficient resources.

Types of Syndicated Programming. Stations can buy three types of programming from syndicators:

1. First-run—Production companies such as Paramount, MCA Universal, Time Warner, Group W, and King World are very active in the development of programs that are designed wholly and exclusively for syndicated distribution. This is called "first-run" product; that is, the shows have not appeared on any other medium prior to their availability in syndication.

Before the explosive proliferation of independent stations in the 1970s, first-run programming was a rather sleepy business. Aside from three principal talk/variety programs (hosted by Merv Griffin, Mike Douglas, and Dinah Shore) and a few game shows, there wasn't much production. But when the number of indies zoomed to over 100, the clamor for more programming became thunderous.

Every major production company joined the existing little band of practitioners to make syndication one of the most vigorously contested areas in programming. Now each year, hundreds of new shows are offered for stations' consideration. Such series as "Wheel of Fortune," "Jeopardy," "Oprah Winfrey," "Entertainment Tonight," and "Family Feud," have been multimillion dollar profitmakers for their production companies. (It should be noted that although "Wheel," "Jeopardy," "Feud" and many other syndicated series have appeared on the networks, the episodes in syndication are brand new and qualify as first run.)

2. Off net—When a network television series has accumulated 80 to 100 episodes (approximately four years' worth of shows),

FIGURE 2.1
"Family Feud," with host Ray Combs, has been one of syndication's most successful series, pleasing audiences over three different decades. (Photo courtesy of Mark Goodson Productions.)

it is ready for distribution in syndication. In the shorthand of the business, these episodes are called "off net" programs.

There are two reasons why a package of at least 80 to 100 episodes is required. First, a four-year run on the network indicates that the series has enduring appeal and the reruns on a local station will probably attract satisfactory ratings. Secondly, stations schedule these shows as five-a-week strips; they show the same series at the same time, Monday through Friday. If there is a reservoir of 100 episodes, a show won't be repeated until the 21st week. This wide space between showings keeps the programming relatively fresh. If there are only 50 episodes in a package, each show would be seen every ten weeks, and the repetition would soon burn out interest.

3. Movie packages—One of the staples of station programming, particularly for independents, is feature films. All the major production companies (Paramount, MCA Universal, Warner Brothers, etc.) as well as some lesser distributors have put together movie "packages" that are available for sale in syndication. These consist of anywhere from 8 or 10 titles (including some TV movies-of-the-week) to 25 or more. Usually the buyer is given a fixed amount of time in which to show the films.

Sometimes syndicators try to salvage episodes of weak network series by offering them under some hyped-up generic banner. A half dozen hour adventure shows that ran only a season or less will be packaged under a title such as "Murder Classics" or "Great Moments in Crime" and be made available at attractive prices. Usually it doesn't take viewers long to discover that these are nothing more than relabeled shows they found wanting once before.

At any one time there are about 100 American companies syndicating television programming. Following are a few of the major syndicators and a sample list of the programs each distributes.[3]

Buena Vista Television
- "The Golden Girls"
- "Siskel and Ebert"
- "Chip 'n' Dale"

Fremantle/Talbot TV
- "Curious George"
- "The Price Is Right"
- "TV 101"

Hearst Entertainment
- "Flash Gordon"
- "Inner Sanctum"
- "Hearst Reports"

King World Productions
- "The Oprah Winfrey Show"
- "Wheel of Fortune"
- "Inside Edition"

LBS Communications
- "Baywatch"
- "Inspector Gadget"
- "The Elvis Conspiracy"

New World Television
- "Santa Barbara"
- "The Wonder Years"
- "Dungeons and Dragons"

Paramount Television
- "Cheers"
- "The Arsenio Hall Show"
- "The Complete Star Trek"

Republic Pictures
- "Bonanza"
- "Victory at Sea"
- "Get Smart"

Twentieth Television
- "A Current Affair"
- "Small Wonder"
- "M*A*S*H"

Viacom Int.
"The Cosby Show"
"I Love Lucy"
"Hawaii Five-O"
Warner Brothers
"Dallas"
"Chico and the Man"
"The Waltons"
Worldvision Enterprises
"Tarzan"
"Love Boat"
"Amahl and the Night Visitors"

Forms of Syndicated Revenue. The marketplace has developed three methods of compensation for the syndicator: cash, barter, and cash plus barter.

1. Cash—This is the simplest transaction of all and, up until the 1970s, was the only way that business was done in syndication. The syndicator agrees to supply a certain number of weeks of programming, and the buyer agrees to pay a fixed amount of cash for those shows. On occasion the parties may develop a sophisticated arrangement based on the show's rating performance, but the basic deal still involves the transfer of cash from the buyer to the seller.

2. Barter—An increasingly popular form of compensation is barter, whereby, in lieu of cash, the station turns over a certain amount of its commercial time to the syndicator. In every program, provision is made for the insertion of commercial messages, usually 6 or 6 1/2 minutes in half hour shows and 12 to 13 minutes in hour-long productions. In a straight cash arrangement, the station retains all this commercial time for local sale. However, in a barter deal, the station does not pay any cash to the syndicator but gives up an agreed-upon number of commercial minutes to the syndicator, who in turn sells them to national advertisers. The station keeps the remaining minutes for local sale.

3. Cash plus barter—This is an amalgam of the two compensation methods. The station pays less money to the syndicator than in a straight cash arrangement, but it also forfeits some of the commercial time for national sale. The amount that the station gives up is less than it would be if the sale were made on a straight barter basis.

Programming for Syndication. When developing first-run programs for syndication, production companies place their greatest emphasis on five-a-week strips. The distribution of syndicated shows is a very expensive process, and it behooves a supplier not only to keep production costs low but also to offer a product that brings with it many opportunities to generate revenue. Clearly a five-a-week series with its numerous commercial messages presents a greater revenue source than a weekly show with just a handful of commercial minutes available for sale.

Also, the time periods that stations look to syndicators to fill tend to be weekly strips. Once-a-week offerings are generally limited to weekends or those occasional time periods in which an affiliated station is disenchanted with the network program. To be sure, there have been many weekly successes in syndication ("Solid Gold," "Star Trek," "Hee Haw"), but by and large the foundation of first-run syndication is strip programming.

For a syndicated program to get on the air and become profitable, three steps are necessary:

1. A program must be developed and produced.

2. Representatives of the program must obtain commitments (clearances) from stations to place the show on their schedules.

3. Sales personnel must obtain national advertisers for the barter time that has been made available to the syndicator (assuming that the program is being sold on a barter or cash-plus-barter basis).

Many companies that are active in the syndication business are structured to handle all three of these activities. All the major production firms, e.g., Paramount, Disney, Time Warner, as well as many independents, e.g., Group W, King World, Tribune, maintain program and sales staffs to keep every phase of the enterprise under one roof.

The overhead for such an organization is substantial and the company must maintain a steady flow of productions in order to stay profitable.

Consequently, not all companies that produce programs choose to maintain a distribution arm. For example, Mark Goodson Productions turns out game shows (such as "Family Feud") for syndication but does not distribute them. In these cases the producer will request bids from syndication firms. The winning bidder will generally have to offer a cash guarantee to the production company as well as a percentage of the profit once the guarantee has been recouped. The risks to the syndicator can be substantial. A program such as "Family Feud" with its long and very successful run on the network will command a hefty guarantee. The distributor must pay that amount to the producer even if the income from stations falls short of the mark.

Typically, a syndicator's fee will be 35 percent of the gross revenue in order to distribute the program, and a barter representative will look for 10–15 percent of the gross to sell the time to national advertisers. Life in syndication is one big negotiation, and these numbers are subject to modification.

Physical Distribution of Syndicated Shows. Programs are distributed to stations via two methods: tape and satellite. The tape system requires a syndicator to run off quantities of the show on tape and mail a copy to each station. If the program is not sensitive to day and date, the syndicator will set up a "bicycle" distribution in which a third of the stations will play the show on week #1, then pass them along to the second third of stations who will run them on week #2. When the last third of the lineup receives and plays the episode, they return the tapes to the distributor, who erases them for further use.

The satellite system is now the preferred route because virtually every station in the country is equipped with receiver dishes. The syndicator will work out a feed schedule with the show's station lineups. In some cases, the feed time will coincide with the station's airing of the program, and the station will simply send the program straight to the viewers with the commercials rolled in live. In most cases the station will tape the feed, edit in the local commercials, and play the program at the appropriate time. In the jargon of the business, this process is listed on a cost sheet as "physical distribution" to distinguish it from expenses related to sales distribution (obtaining clearances).

When Should a Syndicator Announce a "Go"? At what point is the market response to a show sufficiently favorable to embolden the syndicator to commit to production? The standard answer is: when the station clearances account for at least 70 percent of the homes in the country. Why 70 percent? Because that's the coverage advertisers and their agencies use to determine that their commercial messages will be seen nationally. Anything less than that and the commercials are termed "local" or "regional."

The distinction is more than just academic. Advertisers divide their budgets into national buys and spot (primarily local) buys. If a program falls short of the magic 70 percent figure, it no longer qualifies for national budget consideration. The advertiser normally won't buy a show that is 40 percent cleared because those same markets can be reached more efficiently with the purchase of local spot time in the same markets. Like any other rule in television, this one is far from set in cement. Sometimes advertisers will waive the definition of national if they like a show and think it has potential, or if the demographics are attractive despite the limited reach.

In the case of cash-plus-barter programs, the production company sometimes may accumulate sufficient revenue through license fees to justify going forward even though the barter money may be skimpy or nonexistent. The hope with all syndicators is that the program will rate very well in the markets in which it is seen and two positive results will quickly occur: new markets will be added, and old time periods will be upgraded, thus substantially improving the profitability of the venture.

This is a very risky judgment to make. Programs that are in marginal time periods or on weak stations have great difficulty registering good ratings. No matter what the merit of the show, the weaknesses of the time and place of its exposure virtually guarantee an unimpressive rating performance. The syndicator is more than likely pouring good money after bad and is doomed to disappointment.

Almost all programs that have enjoyed success in syndication have been launched under the most favorable of circumstances (clearances of over 80 percent on strong stations in excellent time periods) and have scored reasonably good to exceptional ratings right from the start. The most notable exception to this rule was "Wheel of Fortune," which premiered with a relatively short lineup of stations. However, times were better then (the early 1980s), and competitors were far less numerous. In the financially depressed era of the 1990s, station operators have been notoriously impatient with inferior performers, and any show that fails to get off the mark promisingly will be quickly relegated to the scrap heap or a 2:00 a.m. time period.

SCHEMATIC OF THE SYNDICATION PROCESS

Since the syndication process differs from the more familiar network system, it might be useful to see in outline form how a typical syndicated show proceeds from conception to airing.

1. A program idea strikes the fancy of the syndicator. The idea might have been generated internally or come from an outside source.
2. The concept is researched, developed, and eventually goes to pilot.
3. The sales department develops a strategy for obtaining station clearances and national advertisers (if the program is offered on a barter basis).
4. The sales department fans out across the country to seek clearances. The barter sales unit makes presentations to advertising prospects.
5. The program is presented to buyers. Effort is made to persuade stations that the program is the hottest item ever produced.
6. The syndicator assesses clearances and advertiser response.
7. The syndicator announces a go/no go decision.

Affiliated Stations

As noted above, an affiliated station is one that allies itself with a network. Aside from early evening and late night news, the affiliate program director's principal concerns are a weekday morning hour, a late afternoon block, a primetime access half hour (see next page), and possibly a late-late night series. On weekends the director has daytime and early evening hours to fill, plus late night.

Typically the director will attempt to fill the morning strip with either a talk show or games, the afternoon periods with talk or off-net productions, and the late night with off-net dramas or comedies. The primetime access half hour has its own peculiarities because it is potentially a station's most lucrative period and because it has particular requirements mandated by government.

Primetime Access. Primetime access started in 1971 when the Federal Communications Commission determined that viewers of the three networks should receive programs from a larger list of production companies. The commissioners argued that if more companies were producing programs, not only would the air be filled with a greater richness and diversity of information, entertainment, and culture, but the industry would present many more job opportunities to the young and creative.

The FCC passed a ruling entitled the primetime access rule (PTAR) that prohibited network-affiliated stations in the top fifty markets from airing more than three hours of network-supplied programming (including reruns of old network shows) between 7:00 and 11:00 p.m. Stations were permitted to schedule the network's half hour news strip at 7:00 p.m. provided it was juxtaposed to a local news program. An exception was made on Sunday, where the period 7:00 to 8:00 p.m. could be network-supplied if the programs were public service or if they were primarily designed for children ages 2 to 11. In practice the primetime access rule meant that network-affiliated stations in the top fifty markets could not air network programs or reruns between 7:30 and 8:00 p.m. Monday through Saturday.

It was clear the commissioners envisioned an explosion of "quality" programming: documentaries, recitals, mind-stretching cultural journeys, and the like. What they got instead was a cascade of game shows. General managers were not about to turn over their most profitable strip to Shakespeare and Plato.

In the early days of PTAR, most programmers scheduled a different show each night. Very quickly this changed to a "checkerboard" strategy in which two, sometimes three, programs alternated throughout the week. Inevitably one show outscored the others and station executives simply extended the strongest to a five-a-week strip and canceled the others. Not only did this maximize the audience size, but it also saved money: five episodes of one series are cheaper to produce (and therefore offered at a lower price) than the sum of three of one and two of another.

FIGURE 2.2
Each local edition of "PM Magazine" featured attractive male and female cohosts such as Bob Seagren and Pam Thompson (top) and a variety of national "tipsters," of which Chef Tell (bottom) was the most popular. Photos courtesy of Group W Productions.

The only series ever to realize the FCC's dream was "PM Magazine," an ingenious concept developed by the five Group W owned and operated stations. Each program consisted of three or four human interest stories, plus a few 90-second "Tips" about cooking, medicine, exercise, or life-style. Although each station required about twenty stories and fifteen tips a week, no one station had to produce that many. Every day the stations exchanged the material they had produced and presented locally the previous day. Each station was free to take any of the stories that its sister stations had developed and use them in its own market. This was an inventive way to obtain a wide variety of material at minimal cost.

Eventually the idea was made available throughout the country, and by the mid-1980s almost 100 stations were participating. One of the rules was that no station could become a member of the cooperative unless it contributed two local stories a week, which were then sent to "PM" headquarters for possible recycling to the full cooperative. That meant every one of the 100 stations had to hire a production staff, one of the goals that the FCC had in the first place.

Eventually "PM" ran its course. The personal stories and "Tips" became repetitive, the ratings fell off, and the consequent drop in commercial revenue made it impossible for stations to maintain extensive production staffs. Unfortunately, no other company or producer could come up with a similarly effective concept, and the access time became a quagmire of bells, whistles, flashing scoreboards, and "Tell 'em what they've won, Johnny!"

Locally Produced Shows. The main locally produced shows that stations air are their local newscasts. Affiliated stations usually program this material right before the evening network news is shown, and many also have local newscasts later at night after the network primetime programs have aired. News, in addition to serving the community, is of interest to advertisers and usually brings the station a profit. A special news department may be responsible for this offering, or news may be produced under the auspices of the programming department.

There was a time when many stations around the country produced not just local news but a large variety of local programs—talk and public service programs, game shows, documentaries, and ambitious primetime specials. Although oftentimes these programs were unprofitable, they were considered worthwhile in terms of providing a service to the local market. Now in the mid-1990s only a handful of stations are still actively engaged in the production of nonnews local programs. The high cost of development and production makes local programming a risky venture in a period of tight budgets. Furthermore, the superabundant availability of syndicated product allows stations a wide range of choice without the expense of development.

Independent Stations

An independent is a station that has no relationship with a network. The definition has become somewhat blurred since the formation of the Fox Television Network. All stations that have affiliated with Fox were originally independent. Although they now receive programs from the network for a limited number of time periods, they still must fill the bulk of their daily schedules on their own. Therefore, they are still referred to in the trade as independents, although the phrase "modified independents" might be more accurate.

The program decisions confronting the program director of an independent station are considerably more numerous than those facing his or her counterpart at an affiliate. Aside from the Fox-associated stations, every time period on the schedule must be filled either by a locally produced show or by product from an outside source.

Up until the mid-1980s, independents were generally the least-watched stations in a market. They were usually the last to go into business, they did not have the major stars and shows on their primetime schedule, and, because they were frequently UHF stations, they had dial numbers that required a lot of twisting to locate. Consequently their ratings were low-to-modest and syndicators brought them product only after the network affiliates had passed.

In more recent times many independents have become fully competitive. In fact, in some respects an independent is in a better bargaining position for syndicated programming than its affiliate competitor. Because an independent has so much need for programming, it can, in one sitting, make a deal for several of the syndicator's product line. The affiliates, more limited in need, are often unable to match the total value of the independent's offer.

Independents make their living by scheduling shows that attract the audience that their affiliate opponents are not reaching. A typical weekday primetime schedule for an independent (except for the Fox-supplied nights) will call for off-network sitcoms from 7:00 to 8:00 p.m. (because affiliates are forbidden by PTAR to use off-net shows in this hour), a feature film from 8:00 to 10:00 p.m. (because networks usually don't begin movies until 9:00 p.m.), and a local news hour from 10:00 to 11:00 p.m. (beating the network affiliates' late night news by one hour).

On weekday afternoons many independents schedule a one-hour or two-hour children's block. Sometimes these shows are straight animated cartoons and occasionally they are locally produced shows that might include a lively host interacting with an audience of children and introducing cartoon segments. The network affiliates usually schedule serials or adult talk shows at this time, and the cartoons attract the children who are just home from school and available for viewing. The whole idea for the indy is to zig when the affiliate zags.

As with affiliated stations, local production has eased off during recent times. The independents find news production to be in their best interest from a financial and service point of view, but, like their affiliated neighbors, they cannot afford expensive local programming that does not attract advertisers.

Some independent stations have specialized programming strategies that do not require them to obtain programming from either Fox or the traditional syndication companies. For example, a station that programs all-Spanish material might purchase most of its programming from Mexican and South American stations. All-religious stations obtain programs from one religious sect or from a variety of religious organizations that share the TV station time.

Station Groups

Many stations, whether affiliate or independent, are owned and operated by parent companies. All four networks own a cluster of stations. So do the Tribune Company, Westinghouse, Cox Enterprises, Gannett Broadcasting, Great American Broadcasting, Hearst Broadcasting, McGraw-Hill, Maltrite Communications, and many more. In fact, the 1991 *Broadcasting Yearbook* listed 116 companies that own two or more television stations. Some are highly specialized (religion, foreign language) and others are extremely limited in their reach. But scores of them are powerful groups that attract large audiences and are significant forces in their regions. They represent an important marketplace for the suppliers of programming.

The number and organization of affiliated, independent, and O&O stations varies greatly from one city to another. For example:[4]

In Chicago there are eleven stations:
- One affiliated with ABC and owned by it
- One affiliated with CBS and owned by it
- One affiliated with NBC and owned by it
- One affiliated with Fox and owned by it
- Two affiliated with PBS
- Five independents, one of which is religious and one of which is Spanish.

In Philadelphia there are seven stations:
- One affiliated with ABC and owned by it
- One affiliated with CBS and owned by it
- One affiliated with NBC but not owned by it
- One affiliated with Fox but not owned by it
- One affiliated with PBS
- Two independents.

In Fargo, North Dakota, there are five stations:
- One affiliated with NBC but not owned by it
- One affiliated with CBS but not owned by it
- One affiliated with ABC but not owned by it
- One affiliated with Fox but not owned by it
- One affiliated with PBS.

In Marquette, Michigan, there are two stations:
- One affiliated with PBS
- One affiliated with both CBS and NBC.

Following is a representative list of station group owners, the TV stations they own, and the cities in which these stations are located.[5]

Cox Enterprises
- WSB—Atlanta, Georgia
- WSOC—Charlotte, North Carolina
- WHIO—Dayton, Ohio
- WKBD—Detroit, Michigan
- KTVU—Oakland, California
- WFTV—Orlando, Florida
- WPIX—Pittsburgh, Pennsylvania

Dix Communications
- KAAL—Austin, Minnesota
- KURL—Billings, Montana
- KFBB—Great Falls, Montana
- KTWO—Casper, Wyoming
- KKTU—Cheyenne, Wyoming

Gannett Broadcasting
- KPNX—Mesa, Arizona
- KUSA—Denver, Colorado
- WUSA—Washington, DC
- WTLV—Jacksonville, Florida
- WXIA—Atlanta, Georgia
- WLVI—Cambridge, Massachusetts
- KARE—Minneapolis, Minnesota
- WFMY—Greensboro, North Carolina
- KOCO—Oklahoma City, Oklahoma
- KVUE—Austin, Texas

Spartan Broadcasting Company
- KIMT—Mason City, Iowa
- WBTW—Florence, South Carolina
- WSP—Spartanburg, South Carolina

Group ownership presents special problems in the area of program decision-making. General managers are fiercely protective of their right to select the shows for their schedules. Their standard argument is, "If you are going to demand that I achieve certain profit objectives, then you must permit me to choose the shows that I think will get me there." A very defensible position.

However, many large groups have program specialists on staff who do nothing but study programs day in and day out and work closely with the producers of syndicated product. Their feel for programs is sensitive and their knowledge encyclopedic. They frequently exercise a major influence on the president of the group (who usually features himself or herself as no mean program authority either). Also, a company that is willing to make a group deal for all of its stations is in a position to negotiate some economies that three, four, or five individual contracts wouldn't deliver. To the corporate executive who is responsible for the overall profit performance of the group, this has significant appeal.

But, in addition to the bruised feelings of the individual general managers and their program directors, there is another drawback to group deals. Very rarely do all the stations in the group have the same program problem. Two stations might be struggling in a particular time period against talk shows while another has a need to counter a competitor's cartoon block. Making wholesale changes in a successful schedule just to accommodate a group buy is not advisable. In most instances, the group program specialist urges, cajoles, pleads, and roars in behalf of some favorite, but the final selection is left to the general manager and the program director, the two people who have to do all the explaining if their schedule underdelivers.

Syndicators generally call on the group executives first, not only as a courtesy to headquarters personnel but also to take a shot at making a group sale. A program receives a tremendous cachet when it is bought by a major group, and this approval can frequently start a sales stampede for the show.

• • • • • • • • • • • • • • • •

Networks, syndication, and stations represent the traditional marketplaces for program suppliers, and that is why we have discussed these areas first.

But recently cable has opened its doors to fresh programming and become a genuine outlet for new entertainment ideas. In addition, public broadcasting and commercial radio are possible marketplaces for programmers. These are explored in the following chapter.

• • • • • • • • • • • • • • • •

NOTES

1. Ken Auletta, *Three Blind Mice* (New York: Random House, 1991): 37–38.
2. *The Broadcasting Yearbook* (Washington, DC: Broadcasting Publications, Inc., 1991): F-32, F-35, F-36, F-38, and F-39.
3. This material was obtained from the list of exhibitors at the 1992 National Association of Television Program Executives as reported in "The Floor Show in New Orleans," *Broadcasting* (January 21, 1992): 82–90.
4. *The Broadcasting Yearbook*, C-20, C-32, C-45, and C-51.
5. Ibid., A-42–A-44 and A-55.

3 The Marketplace: Part II

CABLE TELEVISION

The basic overall structure of cable television is quite different from that of commercial broadcast television. However, much of the programming that appears on cable TV is similar to that on commercial television. As a result, the differences between cable TV and commercial TV are often unclear to the viewer. Someone flipping through channels and deciding to stop and watch *Rambo* might not know whether the movie is on CBS, the local independent TV station, or one of the myriad of cable networks.

But someone wishing to work in the television industry or someone trying to sell a program idea within the cable marketplace will quickly become aware that there are many differences between commercial broadcast TV and cable TV.

Cable Systems

For starters, locally based cable TV systems are among the main players in the cable marketplace. They supply programming to residences in a small geographic area such as a neighborhood or a city. In that way, they are somewhat akin to commercial TV stations. However, a station sends its programming through the airwaves on one channel (e.g., KVTV is on channel 13 in Laredo, Texas), and this programming is free of charge to anyone in the area who wants to tune a TV set to that channel. Cable systems send numerous channels of programming through wires. Only households (subscribers) who pay to have a cable system wire connected to their TV set (and then continue to pay a regular monthly fee) can receive the programming.

Cable systems receive the bulk of their programming from cable networks. However, the systems affiliate with cable networks in a way that is quite different from the commercial world. While commercial stations generally affiliate with one network and program on only one channel, cable systems affiliate with many networks and distribute a large number of channels. The number of channels a cable system has depends on how and when it was built. Older systems have twenty or so channels whereas newer ones have over a hundred. This difference has arisen because engineers are constantly figuring out ways to squeeze more video information into wires—or fiber optics.

The various cable networks distribute their programming by satellite, and cable systems all have satellite dishes to receive the signals. They then transfer these signals to the wires that they have placed (underground or on telephone poles) throughout the community. When the residents of a particular household decide they wish to receive the cable programming, someone from the cable company connects an offshoot of the wire to the home TV set(s). The household is then charged a monthly fee for the programming received. The main source of income for the cable system comes from subscriber fees. Some cable systems sell a limited amount of advertising, but, unlike commercial TV stations, commercials are not a major revenue-generating source.

The area in which a particular cable system can place its wires is determined by its franchise. City governments grant franchises; some towns have decided to allow one company to wire the entire area while others have divided the city into several franchise areas.

Although theoretically two or more companies could wire the same area and offer competing cable services to the residents, this has not been undertaken to any degree. The cost of running wires throughout an area is considerable, and if two companies wired the same neighborhood and then competed for subscribers, neither would make money. In essence, each cable system has a monopoly in its own franchise area.

So if you live in Poorville where the cable system consists of only twenty channels, some of which malfunction a great deal of the time, and you "want your MTV" but the system doesn't carry it, you can't buy from the competition. You can, however, complain to the city council, which, in turn, can threaten to take away the cable company's franchise.

In television's early days, the companies that ran cable TV systems were of the ma-and-pa variety. They were owned by several local citizens and were only marginally profitable. When cable TV became a big, glamorous business in the early 1980s, most of these ma-and-pa organizations were bought out by big companies such as Time, Inc. (now Time Warner), that put all of its cable systems in various parts of the country under an umbrella organization called American Television and Communications (ATC). ATC and other large companies such as Tele-Communications Inc. (TCI), Comcast Cable, and Jones Cable became known as multiple system owners (MSOs) because they owned many cable systems in many parts of the country.

Individual cable systems are free to fill their channels however they want. However, MSOs sometimes try to dictate how all their systems will program. Hence arises the same friction that occurs between commercial stations and group owners. Most cable systems fill their channels with some combination of local cable channels, local commercial and public television stations, basic cable networks, pay cable networks, and pay-per-view networks.

The top twenty MSOs and the approximate number of subscribers each has are as follows:[1]

MSO	Subscribers
TCI	8,400,000
ATC	4,600,000
Continental	2,710,000
United Artists	2,181,000
Time Warner	1,868,000
Comcast	1,657,000
Cox	1,616,000
Storer	1,600,000
Liberty Media	1,582,000
Jones	1,573,000
Cablevision Systems	1,559,000
Newhouse	1,283,000
Times Mirror	1,100,000
Cablevision Industries	1,085,000
Viacom	1,058,000
Adelphia	1,024,000
Sammons	906,000
Century	875,000
Paragon	830,000
Falcon	769,000

Local Cable Channels

Local cable channels are basically of two types: local origination and public access. Local origination channels are the closest thing the cable world has to a commercial TV station in that the programming is devised and produced by people who work for the cable company. Usually the programs are produced in-house and the content is very local in nature—city council meetings, high school sports, franchise area news. If the cable company can secure advertising, commercials will be run within and around the programming.

Public access channels are programmed by individuals and groups located in the franchise area. The cable company has no say about (or control over) what is shown over access. The nature, number, and operation of these channels varies greatly. A system with a small number of channels may allocate one channel to be used for both local origination and public access programming. A large system may use four or five channels for access programming, allowing the city government to program one, the local college another, nonprofit charity groups another, and members of the general public another. Sometimes the cable company provides the equipment for these groups to use and sometimes they must find their own equipment. The tapes that are cablecast might be played from the cable company headquarters or they might be cablecast from the college, city hall, or other remote location.

Access channels are mainly a product of the bitter franchise wars that took place when cable TV was burgeoning in the early 1980s. Companies eager to obtain the franchises would promise truckloads of TV equipment and access channels to every influential group they could find. One company went so far as to promise to have a channel devoted entirely to lost dogs because the daughter of a councilperson was upset about losing her dog. Many of the promises never became reality, but most cable systems have some vestiges of access programming.

Rarely does a cable system make money on either local origination or public access. Local origination may bring in some money from advertising, but the money devoted to public access must be chalked up to public service—and the cost of obtaining and maintaining a franchise.

Local Stations

One of the original purposes for cable TV was to bring local TV channels to parts of town with poor reception. In fact, in 1965 the FCC established a rule, called "must-carry," that said all cable systems were required to carry all TV stations within sixty miles of their system. Although this rule no longer exists, most systems still devote channels to local TV stations because they are popular with subscribers.

Some systems do not carry stations that would be of little interest to the people who live in their franchise area. For example, a cable system in an area where there are no people of Oriental descent would not carry a station that broadcasts all-Japanese programming. Other systems will only carry one public TV station even though two or three may be located in the local area. Many cable systems have audio channels as well as video channels and, therefore, carry local radio stations.

At no time in the history of cable did the cable systems pay the local stations for their programming, nor did the local stations pay the cable systems to be carried. Both sides always felt that the other party should pay for the service and the battle raged for decades in the press, the courtrooms, and in front of the FCC. In 1993 the FCC had finally had enough of the squabbling and made a ruling: if a station insisted upon being carried on cable, the system had to go along. However, if the station wanted compensation the system could either pay the fee or drop the station from its spectrum. As this book went to publication, stations around the country were agonizing over their choices.

Basic Networks

The majority of channels on most cable systems are devoted to what are referred to as basic cable networks. These are termed "basic" because the subscriber can obtain a large number of them for a low price. They are in contrast to the pay networks (see below) that are much more expensive. Also, most of the basic channels include commercials whereas the pay channels are commercial-free.

Cable systems "affiliate" with many of these networks, the exact number being a function of how many channels the system has. Cable networks employ staff people, usually called affiliate relations representatives, whose job it is to get as many local cable systems as possible to program the network. The cable system affiliates, unlike commercial stations, do not produce any of their own programs to air around the network feed. Most basic cable networks program 24 hours a day. Systems receive the programming via satellite and then send it by wire into homes. They have no control over what is programmed on the networks; they can simply choose to accept or reject the network offering.

Also, unlike the commercial structure, cable networks charge the cable systems for af-

Below is a list of major basic cable channels and the number of households subscribing to them. There are approximately 93.1 million households in the United States.[2]

	In Millions
Arts and Entertainment	51.0
Black Entertainment Television	31.3
Bravo	7.0
Cable News Network	54.4
CNBC	44.3
Comedy Central	21.0
Country Music TV	14.8
Courtroom TV	5.0
C-SPAN	54.6
Disability Channel	14.2
The Discovery Channel	56.0
E! Entertainment TV	19.0
The Education Net	15.5
ESPN	56.8
The Family Channel	53.7
Headline News	47.6
Home Shopping Network	68.0*
The Jukebox Network	13.0
The Learning Channel	15.6
Lifetime Television	53.0
Movie Classics	35.0
MTV	59.2
Nickelodeon	56.1
Nostalgia Television	12.0
Prime Network	23.0
QVC Fashion Channel	7.5
QVC Network	41.0
The Silent Network	14.2
Sportschannel America	15.0
TNN: Nashville Net	53.6
TBS Superstation	57.0
The Travel Channel	17.5
Turner Network TV	55.0
Univision	6.0**
USA Network	57.9
VH-1/Video Hits One	43.2
The Weather Channel	49.8

*includes broadcast TV as well as cable TV home

**number of Spanish-speaking homes only

filiation. The fee is based on the number of subscribers the cable system has. A typical basic network would charge a cable system somewhere between 3 and 25 cents per month per subscriber, depending on how popular it is. (ESPN can charge more than Nostalgia Television.) Sometimes a bit of barter comes into play in that some services allow the cable systems to insert their own commercials. But, as mentioned previously, these commercials are not a major source of income for cable systems.

The systems take on the burden of recouping the expense (and hopefully making a profit) by selling the cable TV service to consumer households. Most cable systems offer a "basic service" to their subscribers. This includes all the local origination and public access channels, all the local stations, and all the basic networks, for a cost of about $15 a month. Pay services are extra. Some cable systems will make two "tiers" out of these basic channels. They will offer local origination, access, local stations, and some of the public service networks and less glamorous basic networks (C-SPAN, The Silent Network) for a very inexpensive price of about $5. Then the second, and more expensive, tier will include MTV, ESPN, USA, A&E, and other more entertainment-oriented basic networks.

Advertising. Most of the basic networks sell advertising. Because of this they theoretically have the potential for much more income than the four commercial networks. They have two sources of income—cable system fees and advertising—and they do not need to pay affiliate fees. What, of course, has kept any one of them from being more powerful and permeating than NBC, CBS, or ABC is the fact that their audiences are much smaller than those of the commercial networks and, therefore, they cannot sell commercial time for as high a rate as the traditional networks can. Most cable systems are elated if they obtain a rating of 4, whereas commercial network program ratings tend to be in the 11–15 range. But if commercial network audiences continue to decline and cable audiences increase, the cable networks could outbid the networks for the best programming—and in a few instances they have already done just that.

However, the likelihood of the cable networks achieving really high ratings is low because most cable networks program to a very specific audience—Lifetime to women, ESPN to sports fans, Black Entertainment Television to people of color. They do not plan to attract large audiences. Some of the services, such as USA, do undertake broad programming for broad audiences, however, and consider themselves as competition to the commercial networks.

Superstations. Superstations, such as WTBS from Atlanta, WOR from Chicago, and WGN from New York, are a special type of basic service. They are not really networks. Rather they are regular TV stations that have been placed on the satellite for national distribution. The stations themselves do not earn direct money from becoming a superstation. They are placed on the satellite by other companies such as United Video. These companies collect the fees from the cable systems. The superstations make additional money because they can charge advertisers based on their national audience rather than just their local area.

Basic Variations. Of course, with 60+ basic channels, there are many variations on how they operate. C-SPAN, which features the proceedings of the House and Senate, is noncommercial and so lives totally on the amount of money paid to it by the cable systems. The home shopping networks, which make their money because people who watch call in and buy the products shown, usually do not charge the systems for carrying them. Sometimes networks will pay systems something similar to the traditional commercial TV affiliate fee, but usually this is just used as an enticement to get the systems to try the network. Once hooked, the systems must start paying the networks. At one time a Spanish-language network, SIN, paid systems for each Spanish surname on their subscriber rolls. The network was able to promote with the clever slogan, "SIN pays."

Regional networks abound in the cable world. Almost all of them are sports or news oriented (e.g., Home Team Sports, Prime Ticket, Madison Square Garden Network, and Orange County Now). Sports networks are active only when games are in progress, but most of the news services provide 24 hours a day of regional news information. Sometimes these news services are operated in conjunction with a local newspaper or local TV station. The regional networks, like the nationals, have had their ups and downs, consolidations, and demises, but generally they are able to attract enough advertising to subsist.

In addition to "moving picture" networks, there are other services offered to cable systems that they make part of their basic offering. These include digital sound services such as Home Music Store and electronic text services such as news bulletins from Associated Press and Reuters.

A word should be added about a new form of network-affiliate relationship that crosses the commercial and cable organizations. Fox network has recently affiliated with several cable systems that are now carrying the Fox

RATE CARD FOR

YOURTOWN CABLE TELEVISION SYSTEM

(monthly charges)

Local TV Stations Only	$5.00
Basic Cable (6 Local Stations and 25 Cable Services)	17.00
Pay Channels	
HBO	14.95
Showtime	14.95
The Movie Channel	14.95
Cinemax	12.95
The Disney Channel	10.95
Sports Action	11.95
Pay Channel Discounts	
HBO and Cinemax (save $5.00)	22.90
Showtime and The Movie Channel (save $5.00)	24.90
New Installation	45.00
New Installation if two Pay Channels are Purchased	30.00
Reconnect Current Line	20.00
Additional Outlets	10.00
Converter Box	5.00

City taxes are not included.
All rates are subject to change.

FIGURE 3.1
A typical cable system rate card.

FIGURE 3.2
C-SPAN, a non-commercial, non-profit basic cable service that features the proceedings of the House and Senate, receives its operating expense funds from the cable systems that subscribe to it. (Photo courtesy of C-SPAN.)

programming as a basic level service. This has been undertaken in areas of the country where independent stations are not available or are unwilling to carry Fox. In this manner, Fox can be extended to more of the population than it could by affiliating only with regular TV stations.

Pay Networks

Pay cable networks, such as HBO, The Movie Channel, Showtime, and The Disney Channel, do not sell advertising. They receive all of their income from the systems that carry them. The systems, in turn, charge the consumers for each pay network they subscribe to, usually at a rate of $10 to $20 per month per pay service. In other words, the pay services are on a more expensive tier than basic services. Sometimes a system will offer subscribers a discount if they subscribe to a number of pay services. The systems and the networks split the consumer fee money, usually about 50-50, but this split is subject to negotiation. Consumers who do not subscribe to the pay services see a scrambled signal on their TV sets when they tune in to the channel that the pay service occupies.

The major pay cable channels and the number of households subscribing to them (out of a total of 93.1 million households) are as follows:[3]

	In Millions
Cinemax	6.3
The Disney Channel	5.6
Home Box Office	17.6
The Movie Channel	10.0
Showtime	10.0

FIGURE 3.3
Prime Ticket, a regional network covering the Los Angeles area, covers local sports, such as events involving the Los Angeles Lakers. (Photo courtesy of Prime Ticket.)

To justify their additional monthly fees, pay channels were aware they had to offer subscribers true alternative services. No viewer is about to part with $10–$20 per month just to see the same entertainment already available on the traditional networks. The pay channels solved this problem by buying the rights to present feature films shortly after their theatrical runs and prior to their network availability. They showed these films uncut and without commercial interruptions—a plus worth paying for to many viewers. As the pay channels matured, they branched out into series production, but usually with shows that contained language or themes that the commercial networks couldn't, or wouldn't, schedule. In addition, they began to offer a menu of occasional specials such as sporting events, "made-for" movies, miniseries, stand-up comics, and musical concerts. The combination of you-can-only-get-it-here programming and the absence of commercials has allowed a number of pay channels to flourish.

Some of the regional sports networks are considered pay services even though they contain advertisements. Usually the decision whether to make such a sports channel basic or pay depends on the particular system. Cable operators have a habit, which the networks don't particularly like, of juggling regional sports back and forth between basic and pay. If they can get more money by offering a pay service, they do so. But if there aren't many bites from the consumers, they bump the network to basic.

Pay-per-view Networks

Of all the forms of cable networks, pay-per-view networks are the most unsettled because they are the newest form. Basically, subscribers pay only for those programs they actually watch. If they have not paid for a particular program, they see a scrambled signal. The network and the system split the subscriber fees, based on whatever percentage is negotiated. What the subscriber pays is based on what the market will bear. Movies can be seen for a few dollars, while major sports events have a price tag in the $20 to $50 range.

Most of the cable systems that offer pay-per-view programming employ technology (referred to as addressability) that allows for interaction. Viewers who want to see a particular program can press a button on a remote control device that sends a signal back through the wire to the cable system. The program is then unscrambled or in some other way made viewable to the consumer. A computer also notes that the subscriber should be billed for the program and this amount is added to the monthly amount the subscriber must pay. Systems without addressability can operate pay-per-view by having subscribers call an 800 number to order a particular program, but the instant access provided by the remote control works better.

The two main pay-per-view services, Request Television and Viewer's Choice, program 24 hours a day. Their staple programming is newly released hit movies, but they also program sports and entertainment specials. Both these services provide multiple channels that run the same movie (or the same ten or fifteen movies) over and over. The philosophy here is to compete with the video rental business. The pay-per-view services want to make it convenient for you to turn on the TV and see the movie when you want to see it without the hassle of driving to and from the video store. The jury is still out as to whether or not pay-per-view will achieve this goal.

Playboy at Night cablecasts each evening and is the oldest of the services that are now pay-per-view. It was originally a pay cable service, but because of its adult content, community groups objected to it. They pointed out that if parents subscribed to Playboy on a monthly basis and went out for the evening, their offspring could easily tune in the channel—accidentally or on purpose. So Playboy switched to pay-per-view, whereby each program must be specifically requested.

Other pay-per-view networks do not cablecast on a regular basis; they feature special events, primarily sports and concerts. TVKO, for example, programs a boxing event the second Friday of each month. Likewise, the Wrestling Federation supplies occasional wrestling matches, and Forum Boxing shows events from the Los Angeles Great Western Forum.

An interesting sidelight to the sports–electronic media relationship is what happens to the consumer along the way. At one time all

The pay-per-view services and the number of households subscribing to them are as follows:[4]

In Millions	
Action Pay Per View	2.0
Forum Boxing	Varies
Playboy at Night	5.0
Request Television	8.5
TVKO	Varies
Viewer's Choice	16.0
Wrestling Federation	Varies

FIGURE 3.4 ESPN cablecast this Pittsburgh Steelers National Football League game—a contest that at one time would have been seen only on ABC, CBS, or NBC. (Photo courtesy of ESPN. © 1988 Mike Fabus.)

football games available on TV were free—provided by the commercial networks. Then some of them went to basic cable (ESPN) and the viewer had to pay a small fee for the service. Some started appearing on HBO, so the subscriber was paying a heftier fee. If pay-per-view manages to corral the major sporting events, the cost could become prohibitive to many set owners. This potential has come to the attention of a few members of Congress/sports fans, but nothing specific has been done to prevent such a situation.

The Concept of Syndication in Cable

Syndication does occur in cable TV, but the syndicators sell not to stations but rather to networks. The basic networks, in particular, buy off-net and made-for programming. For example, USA Network has programmed "Murder, She Wrote," and Lifetime has used "Cagney and Lacey." Sometimes cable systems will pick up a discarded series ("Paper Chase," "The Days and Nights of Molly Dodd"), produce new episodes, and air it as a series. Fortunately for the syndicators, cable networks became interested in syndicated material about the same time the independent TV station market was drying up due to the advent of Fox.

On the other side of the coin, cable networks are starting to provide material for the syndication market. Nickelodeon is offering its original series "Double Dare" for syndication, and Showtime's "Brothers" has hit the syndication market. More interrelationships between cable and syndicators are likely to occur in the future.

Overall, the cable world is complex because it has so many players. New programming ideas are constantly occurring, some based on advances in technology and some on the human mind's ability to ferret out new ideas.

FIGURE 3.5 The cable distribution process.

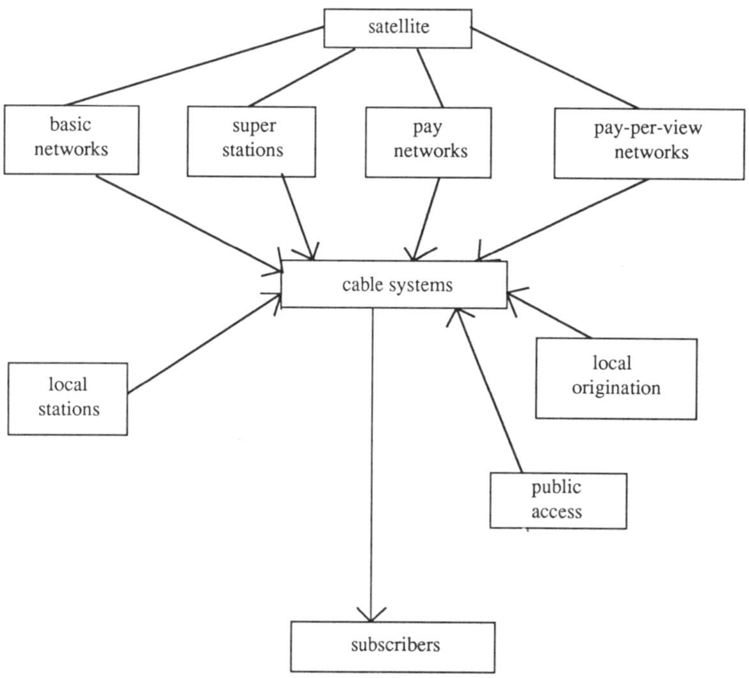

COMMERCIAL RADIO

The radio marketplace has changed greatly over the years. At one time it looked somewhat akin to the modern commercial TV marketplace in that networks supplied programs to affiliated stations and paid the stations an affiliate fee. However, times have changed, and although there still are radio networks and affiliated stations, they take on a very different function than they did during the "good old days" of radio.

Networks and Syndicators

In the radio business, it is hard to tell the difference between networks and syndicators. One distinction used to be that networks provided their material over phone lines and syndicators mailed tapes. But since the advent of satellites, both send most of their material over these birds. Some syndicators still use tape and, in fact, some networks augment their material with tapes (or compact discs).

During the 1960s and 1970s when radio networks were primarily purveyors of news, another differentiation between networks and syndicators was that networks provided programming that had a time factor while syndicators dealt with programming that had a long shelf life. But now both networks and syndicators provide music—a programming form that is not dependent on the day's happenings—and the feature material provided by some syndicators has very little shelf life.

It also used to be that networks provided more programming than syndicators. Networks would give daily or even hourly feeds while syndicators provided occasional, featurelike programs. But now an organization that has three weekly shows and some specials calls itself Global Satellite Network.

People and organizations within the radio business differ as to the definition of a network. Statistical Research, Inc., an organization that provides ratings for radio networks, has a four-part definition of a network that states that a programmer must have continuity of programming; written, contractual agreements with its affiliates; the capability of an instant feed to all affiliates; and a clearance system so it can determine which affiliates carry which programs. Bud Heck, president of the trade organization Radio Network Association, says, "If you hook two or three stations together, that's a network."[5] But his organization limits its membership to programmers that distribute to all of the 48 states in the continental U.S. If networks must distribute to all 48 states, then one of the most successful programs on national radio, "Rush Limbaugh," is not a network program even though it is heard on nearly 500 stations because the company that distributes it, EFM Media, does not cover enough states.

The fact that it is difficult to differentiate between a network and a syndicator is not really important. What matters is that there are about 500 companies out there in the jungle trying to supply program material to radio stations.

The largest network owners are ABC, CBS, Westwood One, and Unistar. All of these are actually several networks. Westwood One, for example, owns five radio networks called Westwood One Radio Network, Mutual

---

The following is a partial list of radio networks and syndicators, along with the main services they provide.[6]

- ABC Radio Networks—News, sports, music, and talk
- American Comedy Network—Short-form comedy and parody songs
- Associated Press Broadcast Services—News, sports, business, agriculture, features, and public affairs
- Bailey Broadcasting Services—Black-oriented programming including gospel and rap, celebrity interviews, and black history specials
- Dick Brescia Associates—Short-form programming from radio's past and satirical commentary by Stan Freberg
- CBS Radio Networks—News, sports, Hispanic programming, information features, and weekly music programming
- Christian Broadcasting Network—News, talk, information, and music targeted to Christian listeners
- CRN International—Ski/beach condition reports and Wimbledon Tennis updates
- DC Audio—Political satire vignettes
- EFM Media Management—Long- and short-form talk-oriented programs
- Finger Lakes Productions—Short-form informational programming including environmental tips
- Jones Satellite Audio—Easy listening, adult contemporary, soft hits, golden oldies, and country music
- Melcor — Big band and nostalgia music
- Charles Michelson, Inc.—Vintage radio from the 1930s through the 1950s
- MRN Radio—Live auto racing coverage and news and talk about auto racing
- Satellite Music Network—Oldies, hard rock, country, middle of the road, adult contemporary, contemporary hit radio, and urban adult contemporary
- Sports Entertainment Network—Sports-oriented features and talk programs
- Unistar—News, nostalgia, contemporary hit radio, classic rock, adult contemporary, oldies, country, adult rock
- Vanguard Audio Features, Inc.—Daily features on personal performance, parenting, and holidays
- Westwood One Radio Networks—News, information, entertainment, business, science, travel, talk, Top 40, classical, MTV News, rock, and country

Broadcasting, NBC Radio Network, NBC's Talknet, and NBC's The Source. Westwood One was formed in the early 1980s by an upstart entrepreneur, Norman Pattiz. He surprised the then semisomnolent radio network business by purchasing the Mutual Network in 1985. Then in 1987 David really slew Goliath when Pattiz purchased all the NBC networks. So NBC, the originator of network radio, is no longer in the business, although Westwood has kept the name NBC associated with the networks. The five networks broadcast different types of programming aimed at different demographics. Westwood One Radio Network is primarily individual music programs such as "Casey Kassem's Top 40 Musical Hits"; Mutual and NBC Radio Network present news, sports, and features; Talknet consists of four different hosts who each conduct three hours of talk programming; and The Source is news features, and rock concerts aimed at 18–34 year olds.

Similarly, ABC has eight different networks. Some are music but most are news—each aimed at a different type of station. CBS operates two news-oriented networks, and Unistar fields eight music services (Dick Clark is a mainstay) and two news networks. Don't expect either the number of network services or the content of them to be etched in stone. The radio network business ranges from flexible to fluid.

FIGURE 3.6
Eli Gold (left) and Barney Hall announce a Daytona race for MRN Radio, a syndicator that supplies solely auto races. (Photo courtesy of MRN Radio.)

Among the smaller network/syndicator services are MRN Radio, whose sole product is information about auto races that it broadcasts to some 400 stations; Dick Brescia Associates, whose main claim to fame is Stan Freberg and drama programming from the "golden age" of radio; and Finger Lake Productions, which offers short features primarily about the environment.

As with commercial and cable TV, regional networks are also plentiful. Kansas, for example, hosts four of them: one news, one farm reports, and two sports. Individual cities also have services; for example, Metro Traffic supplies traffic reports to many Los Angeles area stations.

The main method of payment out there in radioland is barter. However, some cash plus barter or cash only does exist. Occasionally a network will pay a station an affiliate fee. This happens most often in large markets or with top-rated stations. The network will be willing to pay for a particular station or market so that it can strengthen its pitch to advertisers. With 500 companies trying to eke a living out of radio, some unusual forms of compensation have been tried. One company that features ski reports gives these reports absolutely free to stations so that they can insert all the commercials. The company makes its money by charging the ski resorts for the mentions that they get in the reports. Short of illegal or unethical procedures, any method that earns a buck is fair game.

Stations

Radio does have affiliated and independent stations, but the difference is rather academic. Some stations undertake all of their own programming and don't use any of the networks or syndicators; they are truly independent. But a preponderance of stations affiliate with at least one network. In fact, many stations affiliate with several networks and also use syndicated material (perhaps Unistar for music, Mutual for news, and MRN for car races). This is totally unlike commercial TV, where stations are heavily defined by the network with which they affiliate.

Usually it is difficult to tell what network(s) a radio station has affiliated with because the station is trying to make everything sound local. Networks with name value (such as ABC, CBS, and NBC) flaunt their names rather frequently, but most networks are anonymous because that's the way stations want it.

The driving force in a contract between a network and an affiliate is the commercials. The contracts are written in such a way that stations are required to play the commercials supplied by the networks but not necessarily the programs. And that's what some stations do. Especially if they can negotiate an affiliation fee, they will play the commercials at the times designated by the network but air only some (or none) of the network programming. The rationale for this is that much of the network material is some form of talk, and the stations want to interrupt their music programming with as little talk as possible. But they do like the affiliation fee the networks will pay if the network commercials are aired.

On the other side of the coin, networks sometimes affiliate with several stations in one city. This is common for news and music programming but not common for sports where one station is usually given exclusive rights to air particular sporting events. Multiple affiliations in one city are particularly common among the networks that have a wide variety of services. For example, one of ABC's news networks might affiliate with the Top 40 station in town while a different ABC news network may be heard over the town's easy-listening station.

Syndicators sometimes offer features through what is called format exclusivity. The syndication company will not give its features to two stations in the same market with the same format, such as two contemporary hit stations, but it may offer them to stations with different formats (contemporary hits, country, all news).

Radio stations are much more likely to be individually owned than are commercial TV stations or cable TV systems. This is because running a radio station is less expensive than operating TV; it does not require the deep pockets of its sister medium. However, there are still many group owners in the radio business. But because stations are so individualistic and cater to such differentiated, fragmented audiences, group owners find it difficult to buy programming services to suit all their stations. Radio station programmers tend to have autonomy.

Following is the same list of group owners given as TV station owners in Chapter 2. However, this time what is listed is the radio stations that they own.[7]

Cox Enterprises
- WSB-AM/FM Atlanta, Georgia
- WSOC-AM/FM Charlotte, North Carolina
- WHIO-AM and WHKO-FM Dayton, Ohio
- WCKG-FM Elmwood Park, Illinois
- KFO-AM and KOST-FM Los Angeles, California
- WIOD-AM and WFLC-FM Miami, Florida
- WSUN-AM and WWRM-FM St. Petersburg, Florida

Dix Communications
- WMMZ-FM Ocala, Florida
- WTBO-AM and WKGO-FM Cumberland, Maryland
- WKVX-AM and WQKT-FM Wooster, Ohio
- WRAD-AM and WRIQ-FM Radford, Virginia

Gannett Broadcasting
- KIIS-AM/FM Los Angeles, California
- KSDO-AM/FM San Diego, California
- WDAE-AM and WUSA-FM Tampa, Florida
- WGCI-AM/FM Chicago, Illinois
- KCMO-AM/FM Kansas City, Missouri
- KUSA-AM and KSD-FM St. Louis, Missouri
- KOAI-FM Denton, Texas
- KKBQ-AM Houston, Texas
- KKBQ-FM Pasadena, Texas

Spartan Broadcasting Company
- WTWA-AM and WTHO-FM Thomson, Georgia
- WSPA-AM/FM Spartanburg, South Carolina

PUBLIC BROADCASTING

The main difference between public broadcasting and commercial broadcasting is that the former is not supported by advertising. It receives most of its financial support from one form or another of government money and from foundations, corporate underwriters, and members of the public who send in contributions. The result is that its programming mission can be different from that of commercial broadcasters.

This mission is to provide programming that is cultural and uplifting. As E. B. White said, public broadcasting "should arouse our dreams, satisfy our hunger for beauty, take us on journeys, enable us to participate in events, present great drama and music, explore the sea and sky and the woods and hills. It should be our Lyceum, our Chatauqua, our Minsky's, and our Camelot. It should restate and clarify the social dilemma and the political pickle."

Quite an order. Public broadcasting may not live up to all of that, but, in general, it presents programming that is intended to be intellectually elite. It does not reach a large audience; neither public radio nor public TV usually achieves an audience higher than 3 percent of the public, but it was established to be an alternative to the large-audience networks. In fact, it has raised the eyebrow of government, its audience members, and the commercial networks when it has ventured into more popular programming such as the transmission of tennis matches.

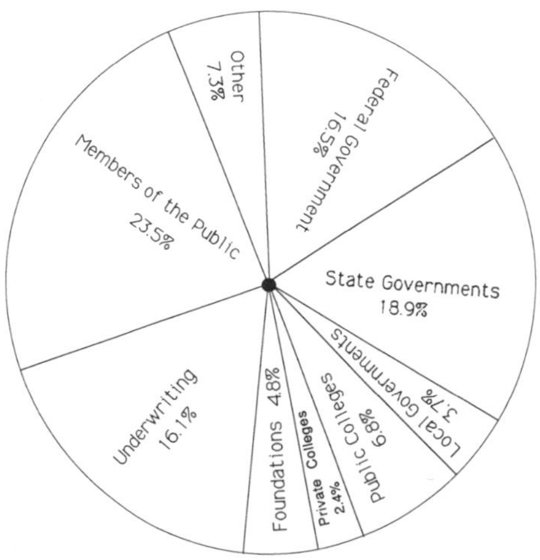

FIGURE 3.7
The percentage of its funding that public broadcasting receives from various sources. Source: *Facts About PBS* (Alexandria, VA: Public Broadcasting Service, 1990): 6.

Public Television Networks

The main public TV network is the Public Broadcasting Service (PBS). It was established in 1967 by an act of Congress. This act also established the Corporation for Public Broadcasting (CPB), the organization that receives money from the federal government to operate public broadcasting. The theory is that CPB is an independent body that insulates the broadcasters themselves from any influential strings that the government might want to attach to the money.

The Public Broadcasting Act of 1967 specifically prohibited CPB from operating a network, so it created PBS to acquire, schedule, and distribute programming. PBS itself does not produce any programs. It acquires most of them from public television stations, although some also come from outside agencies such as the British Broadcasting Corporation (BBC) and Children's Television Workshop (CTW).

The financial procedure for all of this is somewhat complex. CPB receives money from the government and gives some of it to PBS to run the network and the rest to local stations in the form of Community Service Grants (CSGs). The stations operate on the CSGs and the money they collect locally from viewers and from the state, local, or educational institutions that own the stations. However, they wind up giving most of that money to PBS in the form of a membership fee that covers the satellite distribution system and program costs. PBS, in turn, gives most of the money back to specified stations so that they can produce the programs that will air on PBS.

The money received from the federal government is not enough to fuel PBS and the stations. Additional money is raised when PBS or stations approach foundations such as the National Endowment for the Humanities or the John and Mary Markel Foundation to seek contributions toward specific programs or toward paying the operating costs of the stations or network.

Corporations such as Xerox, 3M, and

Exxon also contribute money to public broadcasting in return for a credit on the air. These "credits" have increased in length and scope over the years to the point that some say they look very much like commercials. But the corporations are not considered sponsors. They are called corporate underwriters, and their contributions to public television are tax deductible.

Most public TV stations also raise money by asking their viewers to contribute. They hold on-air fund-raising drives, sometimes in the form of auctions of unusual items, and they use the phone and mail to solicit money from the public.

The method by which PBS's programs are selected is an evolutionary complexity. (Public broadcasters seem to delight in making even the simplest of processes complex.) It used to be that programs were selected by a five-tiered voting process referred to as the Station Program Cooperative (SPC). (Public broadcasters also delight in an alphabet soup of abbreviations.) All stations interested in production presented their program ideas to the SPC. Then all stations voted on which of these submissions should receive a go-ahead. Stations paid for, and aired, those programs that they had approved.

The problem with the SPC was that every year it seemed to add a few complications. It became what one PBS executive referred to as "democracy bordering on anarchy." More importantly, hardly anything new ever got produced. For example, each year WETA in Washington would offer to produce "The MacNeil/Lehrer Newshour," and almost all the stations would indicate they wanted it. A fresh, new idea couldn't get enough station votes to be added to the schedule. The SPC became even more vulnerable when cable networks such as Arts and Entertainment and The Discovery Channel began seeking much of the same programming (particularly BBC material) as PBS. With the SPC in place, PBS could only acquire programming once a year. It needed to be able to act more quickly in order to counter the artistically oriented cable networks.

So the public television folks decided to hire a programming czar (her title is Vice President for National Programming and Promotion) who, along with her Washington-based staff, would decide what programs would be aired on PBS. In 1990 the SPC was abolished and the new plan was put into effect. Now the money that stations pay to PBS is not earmarked for specific programs. It is for whatever the PBS programmers decide should be aired. PBS still uses its member stations to produce most of its programming, so the money collected from all stations goes back to producing stations, just as under the SPC. This new system has created efficiencies and some new series, but a number of stations do not like being left out of the decision process. A committee of television station people has been appointed to advise the czar, and further refinements (with alphabet soup) will no doubt occur.

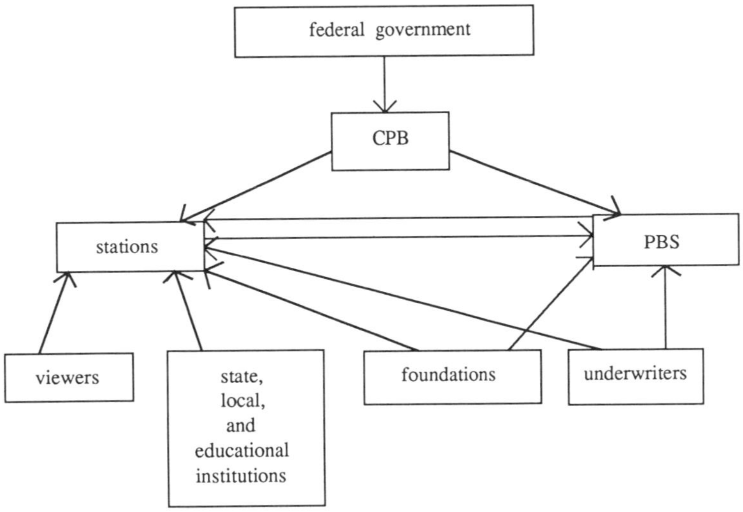

FIGURE 3.8 The rather convoluted process through which public television is funded.

PBS schedules daytime programs such as the children's show "Sesame Street" and self helper "This Old House." Its primetime programming is on six nights a week (not on Saturdays). Nothing in the agreements between PBS and the stations prevents stations from taping the satellite feeds and playing the programs whenever they want. Under the older system where stations voted for and aired different shows, hardly any public TV stations had exactly the same schedule. But under the new regime, concerted efforts are being made to convince the main public stations in each market to air the primetime programs at the same time, or at least on the same evening. In this way, shows can be promoted nationally much more accurately and efficiently.

PBS is by far the dominant network service for public television stations. However, there are four regional networks that were originally intended to serve the eastern, southern, central, and western parts of the nation. Some states, such as Pennsylvania and Wisconsin, have their own networks. Prior to the satellite age, regional networks were interconnected by wire that served only their particular areas. But now they place their signals on satellite and nothing prohibits a station in San Diego from airing something offered by the Eastern Educational Television Network—and, indeed, this does happen. The regional networks obtain their programs primarily from stations in their region, but they also acquire from foreign countries and independent producers.

Public Television Stations

All public TV stations are affiliates (usually called members) of PBS. There are no independent stations, but some stations are not as full-fledged as others. In markets where there are multiple public TV stations, one will be the dominant station and others will be more subsidiary. The dominant ones pay higher membership fees to PBS and use more of its programming. In fact, the subsidiaries, because of their limited-use discount, are not allowed to air more than 50 percent of PBS programs. The PBS programming they do air is usually shown after it has been seen on a primary station. While the primary stations consider the distribution of PBS programming to be their main mission, the other stations specialize in other missions such as providing classroom education for grade schools or college credit courses for adults. Both dominant and subsidiary stations have a fair number of hours to fill with programming other than that supplied by PBS.

FIGURE 3.9
Tony Saletan hosts this instructional program, "Let's All Sing," targeted for third and fourth graders. Public TV stations can obtain the thirty-two fifteen-minute music lessons through a Los Angeles–based educational program distributor. (Photo courtesy of Western Instructional Television.)

Syndication does not really exist in public broadcasting, although on occasion public stations will buy some vintage TV programs or films from the same syndicators who supply the commercial market. Specialized organizations, such as Western Instructional Television, supply instructional programs to be used in elementary and secondary schools or as college courses, and stations often buy from them, but these organizations do not refer to themselves as syndicators. Also, sometimes stations deal directly with British program suppliers. Stations exchange programs, use regional network programs, rerun PBS programs, and produce a fair number of their own local programs to fill their extra hours.

One element that sometimes affects a public TV station's role in the marketplace is its ownership. Some stations are owned by nonprofit community groups while others are owned by universities, public school systems, or a state government. A station owned by a public school system might be more likely to buy and schedule programs intended for in-school use than a station owned by a nonprofit community group.

Following is a selected list of public TV stations and their owners.[8]

- WAIQ Montgomery, Alabama; licensed to Alabama ETV Commission (a state government organization)
- KAID Boise, Idaho; licensed to Idaho State Board of Education
- KTWU Topeka, Kansas; licensed to Washburn University
- WGBH Boston, Massachusetts; licensed to WGBH Educational Foundation, a nonprofit community group
- KSPS Spokane, Washington; licensed to Spokane School District No. 81

Public Radio Networks

The same act that created PBS also created National Public Radio (NPR). In a manner similar to public television, the CPB gives grants to public radio stations that, in turn, use this money to buy programs from NPR. The public radio structure, like public TV, also receives money through foundations, corporate underwriters, and listener contributions. However, unlike PBS, NPR produces many of its programs itself in its Washington, D.C., studios. Also unlike PBS, the radio stations are not at all disciplined as to when they play what programs. They can cherry pick whatever they want from NPR's offerings and, for the most part, play their pickings whenever they wish. If a show contains a call-in element, the network will require that it be aired live, and if it contains dated information, the stations will be told they must play the program within a particular time frame.

A second public radio network, American Public Radio (APR), was formed in 1982 and operates out of Minneapolis, Minnesota. APR does not produce any of its programs; most are supplied by stations. The users of the programs pay an affiliation fee and are then able to use most of the programs aired by APR for free. Some that are expensive to produce cost the stations a small amount extra. As with NPR, stations are not obligated to use all of APR's programs nor are they obligated to play them at any particular time.

Some states have public radio networks that supply state-oriented programming (mostly news) to local stations.

Public Radio Stations

Public radio stations are free to affiliate with NPR, APR, or both, or they can affiliate with neither and remain independent. Stations that don't affiliate with a public radio network are often referred to as "noncommercial" radio stations because they are not truly within the "public" structure created by the Public Broadcasting Act of 1967. However, the words "public" and "noncommercial" are commonly used interchangeably to refer to any radio stations that are not within the commercial realm.

Most of the noncommercial independent stations are owned by universities and are used as training grounds for teaching students interested in entering broadcasting. The schools want to give the students on-air production and talent experience; they do not want to simply pipe through network shows.

FIGURE 3.10 "Mountain Stage" is a two-hour musical variety program produced by West Virginia Public Radio and aired on APR. It is recorded live at the West Virginian Cultural Center Theatre in Charleston and is hosted by Larry Groce (center). It often features the Fabulous Twister Sisters, fiddler Deni Bonnett (left) and guitarist Julie Adams. (Photo courtesy of American Public Radio. © Steve Payne.)

In order for stations to affiliate with either NPR or APR, they must have at least five full-time employees and must program at least 18 hours a day every day of the year. This eliminates many of the university stations that do not pay the students and often shut down for holidays. However, other university stations have decided to affiliate and are, therefore, truly public radio stations that are operated with a professional staff. In these cases students may be used as interns or given small blocks of time to program.

Of the stations that are affiliates, most contract with both NPR and APR. Then, in addition, they produce much of their own programming, often classical music.

As already indicated, some noncommercial radio stations are owned by universities; most of the rest are owned by other educational organizations, religious groups, or nonprofit community groups.

All aspects of public broadcasting programming differ widely from those of commercial TV, cable TV, and radio. Without a dependence on advertising and with a mandate to be

---

Following is a selected list of public radio stations and their owners:[9]

- KLRE-FM Little Rock, Arkansas; licensed to Little Rock School District and University of Arkansas
- WSHU-FM Fairfield, Connecticut; licensed to Sacred Heart University
- WRFG-FM Atlanta, Georgia; licensed to Radio Free Georgia Broadcasting Foundation
- KSKF-FM Klamath Falls, Oregon; licensed to State Board of Education

---

uplifting, this form of broadcasting is unique within the otherwise capitalistic system of the American electronic media.

• • • • • • • • • • • • • • • •

As we have seen in this and the preceding chapter, each of the electronic media has a singular way of doing business. Commercial networks are virtually autocratic in the selection of shows; syndicators must persuade every prospective buyer that their programs are rich in quality and potential; cable networks must convince cable systems to carry them; commercial radio networks and syndicators both supply stations with small program elements; public television retains a czar and staff to make program choices, but stations insist that their voices be heard; public radio stations can select individual network programs and air them whenever they choose. Regardless of these many different processes, all the media rely on a circle of suppliers to fill their creative needs. These programming sources are the subject of the next chapter.

• • • • • • • • • • • • • • • •

THE NINE LIVES (GIVE OR TAKE A FEW) OF "THE QUIZ KIDS"

Sometimes a program will appear on several of the different electronic media forms during its lifetime. Commercial network programs are regularly rerun on cable channels such as Nostalgia Television, The Family Channel, and USA. "Paper Chase," a dramatic series set in a law school, was produced for CBS's 1978–79 season and was then canceled by the network, but Showtime picked it up and aired it from 1983 to 1989.[10]

But the show that has probably been reincarnated more often and over a longer span of time than any other is "The Quiz Kids." In its many existences it has appeared on, or at least flirted with, radio, commercial network television, syndication, group-owned commercial stations, public broadcasting, basic cable, pay cable, and foreign production. The adjustments in its format as it moved from decade to decade and medium to medium provide a continuous history of programmers' attitudes toward their audiences.

Developed for radio by producer Lou Cowan in 1940, the show featured a panel of five exceptionally bright youngsters, ranging in age from 6 to 16, who attempted to answer questions sent in by listeners. The panelists, recruited from local schools through newspaper advertising, competed for points, and the three highest scorers returned the following week. However, the competition was designed to be incidental. Mr. Cowan's primary goal was to celebrate the richness of America's resources by showing that kids from any area were capable of great intellectual accomplishment.

Even the selection of the host, Joe Kelly, played to this objective. An unpretentious man, he frequently expressed awe and amazement at the knowledge of his youthful colleagues. Not only did the panelists have to be brainy, they also had to be affable, outgoing, and down-to-earth. Mr. Cowan wanted them to come across as normal in every respect except for their extraordinary abilities with instant recall. The audience found the format and the youngsters appealing, and the series was a radio network fixture for thirteen years.

In July 1949, "The Quiz Kids" made its TV debut. In recognition of the different needs of the new medium, the producers introduced visual elements—questions about nations were accompanied by flags; a math query about fluid ounces was demonstrated with milk bottles. Celebrities were invited to guest star and their appearances frequently were supported with production values. For example, Milton Berle, then the dominant performer on TV, was seen at a library (actually a separate set on stage) boning up for his joust with the kids. Joe Kelly hosted the early shows but was replaced by Clifton Fadiman, a more widely known personality as a result of his quizmaster duties on the perennial top ten radio series, "Information Please."

The TV version of the series introduced a new concern for the producers. Not only did the panelists have to be young, they had to look young. A bright, likeable 16-year-old with an emerging beard just wasn't going to look innocent enough to stir the audience's soul.

Despite the many modifications for television, the show didn't click. After a two-year, four-month run on NBC, followed a few years later by a nine-month appearance on CBS, the series left the air in September 1956. Shortly thereafter, Lou Cowan became an executive at the CBS network and any effort to sell the show would have represented a conflict of interest. After he left CBS, Mr. Cowan directed his energies elsewhere and the show remained dormant for twenty years.

In 1976 one of the original quiz kids, Harve Bennett, by now a leading TV producer ("Rich Man, Poor Man," "The Six Million Dollar Man") approached Geoffrey Cowan, son of the show's founder, who was now representing the project, and asked if there would be any interest in doing a syndicated version of "The Quiz Kids." Mr. Cowan liked the idea, and a pilot was made.

Once again, the format was revitalized. In deference to the perceived change in audience tastes and the medium's increased competitiveness, the show elements were jazzed up. Celebrities quizzed the kids, flash demonstrations such as fire jugglers were used to punch up the questions, the set pulsed with blinking lights, and a bells-and-whistles environment prevailed. The show was taken to NATPE (see Chapter 5) and supported with a major sales effort by Columbia Pictures Television, but to no avail. Station interest was weak and the show was never launched.

In its next incarnation in the late 1970s, the series popped up in Canada under the production auspices of the Global TV Network, based in Toronto. The format was simplified, sort of a back-to-basics approach. Canadian viewers were thought to require less glitz than their American counterparts, and a more restricted budget imposed restraints on production excesses. One lesson learned, or, more properly, relearned, in the two-year Canadian venture was the value of a rooting interest in quiz shows. One of the panelists was a brilliant youngster who was a smart-alecky know-it-all whom the audience quickly learned to loathe. They tuned in every week in the hopes that someone—anyone—would knock him off. Eventually a cute young lady did him in and a nation cheered. This was the same type of audience identification that made the American quiz shows of the late 1950s so popular and caused the rigging scandals.

When RKO found itself facing possible license loss for its radio and TV stations in 1980, it decided to class up its program schedule with quality programs. The company hoped this surge of responsible programming might favorably influence the licensing decision. Toward this end, the firm's executives contacted Geoffrey Cowan and his colleagues and made a deal to present "The Quiz Kids" in its original pristine form on all its owned stations. In Boston it was scheduled ahead of "60 Minutes" and enjoyed excellent ratings. But the ploy failed. RKO lost its license fight—and all further interest in the show.

CBS developed an artsy cable channel in the early 1980s and signed producer Norman Lear to host yet another version of "The Quiz Kids." To make the show appealing to the upscale audience, a leisurely pace was adopted with plenty of time available for the youngsters to show the range of their analytical skills and worldly awareness as well as their gift for instantaneous recall. Lear created a warm and folksy atmosphere, and the show radiated charm and fun. Unfortunately CBS Cable folded and took "The Quiz Kids" with it.

Although the show might seem like a natural for PBS, Geoffrey Cowan had no luck in his sales efforts. He was unable to obtain a corporate sponsor to underwrite the production costs and didn't succeed in persuading the Station Program Cooperative to take it on as a PBS venture. One year it did rank as the first program below those voted acceptance by the SPC, but that's as close as it came to being placed on public television.

At The Disney Channel, it met a similar plight. In 1989 Disney executives commissioned a run-through (a demonstration of the format without cameras or a full set) and requested a harder-hitting format. Game-show features were added, such as "Who Am I?" in which a celebrity pretends to be a figure in history (Dick Van Patten as Albert Einstein). The face of the historic person is gradually revealed as the panelists answer questions correctly. Disney thought it over and passed.

In its most recent life, but not necessarily its last, the Guber-Peters Entertainment Company, Inc., thought that the series might have a chance in TV syndication, but only if it branched out to reach adults as well as kids. To do this, the format required yet another spin. The result was "The Quiz Kids Challenge," a show that pitted three youngsters against three adults in genuine competition. Only the winners kept the prize money. The host, Jonathan Prince, had just come off a two-year run on the sitcom "Throb" and was seen as bridging the gap between the young and the adult audiences. Launched in 1990, the show lasted thirteen weeks and was canceled, as were all the other syndicated game shows that were introduced that season.

What next for "The Quiz Kids"? Surely with the introduction of game show channels on cable, it will surface again. But with what format? The marketing staff will analyze the potential audience, the programmers will respond by tinkering with the structure, and who knows, maybe a hit will once again emerge.

NOTES

1. *The Broadcasting Yearbook* (Washington, DC: Broadcasting Publications, Inc., 1991): D-40.
2. "National Cable Programming Services," *Daily Variety* (November 15, 1991): C-10–C-12; and "The Cable Network Programming Universe," *Broadcasting* (January 7, 1991): 94.
3. Ibid.
4. Ibid.
5. "So What, Exactly, Is a Radio Network?" *Broadcasting* (July 13, 1992): 28.
6. "Program Suppliers and Their Offerings," *Broadcasting* (May 18, 1992): 33 and 34.
7. *The Broadcasting Yearbook*, A-42–A-44 and A-55.
8. Ibid., C-4, C-19, C-25, C-29, and C-63.
9. Ibid., B-24, B-57, B-80, and B-270.
10. Alex McNeil, *Total Television* (New York: Penguin Books, 1991): 584.

4 Sources of Programming

No matter what the electronic media form, the person in charge of programming must figure out how to fill the hours of the day with specific shows, a daunting task. For example, an independent station that is on the air 20 hours a day, seven days a week must schedule *7280* hours of programming a year. Where does it all come from?

The answer is very complex. Ideas for programs can originate from just about anywhere. For example, legend has it that one night Mrs. Merv Griffin said to her husband, "Why don't you do a game where the contestants get the answers and have to give you the questions?" Merv liked the idea, and "Jeopardy" recently passed its 30th anniversary.

A big-name independent producer driving to work may hear a radio story about surrogate mothers and think of a heart-tugging idea for a movie-of-the-week. A writer may formulate an idea for a children's program while reading bedtime stories to her children. Early morning disc jockeys have been known to plan skits based on the dreams they had the night before. A soap opera writer has confessed to gaining ideas by going to bars and encouraging people to dump their problems on him. A group of students developed sketches about college dorm life for a public access cable series.

But an idea does not a TV series make. Most ideas must be funneled through a fairly well organized structure of suppliers. These companies provide the money and the technical and production know-how to make programming a reality.

COMMERCIAL TELEVISION

Despite the massive need for product at networks and stations, the number of sources of supply is surprisingly limited. The financial risks of program-making are high and the business is intensely competitive. Not too many firms can suffer reverses and still survive. Buyers tend to rely on companies that have delivered successes in the past, a practice that makes it difficult for newcomers to get on the scoreboard. Program concepts primarily originate with one of the following sources.

Major Production Companies

Large companies that specialize in the production of programs (e.g., MCA Universal, Paramount, Disney) have a number of creative people under exclusive contract (dubbed "umbrella deals") whose job is to develop new ideas for shows.

Networks, station groups, and stations like to do business with these firms, not only because they have a history of success as well as access to some of the best producers, writers, performers, and craftspeople, but also because they are financially sound and won't have to default on a commitment if unforeseen and expensive hurdles are suddenly confronted. Buyers know that if a program isn't coming together well, the "majors" have the resources to do whatever is necessary to fix it. These large solid firms account for the lion's share of network primetime programming.

Feature films, produced by major companies, are also a source of programming for networks and stations. After a movie has finished its theatrical run, it is released to television. However, networks and stations do not get first crack at it. The distribution is undertaken through a series of "windows"—the amount of time that transpires as the film is released to different media forms. Often a movie will have a six-month window between the time it finishes its theatrical run and the time it is available to cable TV pay-

FIGURE 4.1
"The Ropers," starring Norman Fell and Audra Lindley (center) and a cast that also included Jeffrey Tambor and Patricia McCormack (back) and Evan Cohen (front), was a series that spun out of the hit sitcom "Three's Company." Unfortunately, it never approached the popularity of its parent program. (Photo courtesy of and © DLT Entertainment Ltd.)

per-view systems. Then it has another window of a month or two before it is distributed in video stores. Following this is a window that leads to the pay cable services such as HBO and Showtime. Only after that are films made available to commercial networks. Even further down the pecking order are local TV stations and basic cable networks such as USA and Lifetime. The length of time of the various windows differs from film to film. On rare occasions, the order of the releases varies and commercial TV may obtain a movie before it is shown on cable. Sometimes the networks and stations buy the rights to air the films from the major production companies and sometimes they buy them through syndicators.

Also, feature films have been the basis for many TV series developed by the same major production firms. Before the picture is made, the company will usually negotiate the right to produce a TV version if it seems to lend itself to that medium. In recent years, we have seen such network primetime series as "In the Heat of the Night," "Uncle Buck," "Ferris Bueller," "9 to 5," and, of course, the granddaddy of them all, "M*A*S*H."

Both the majors and the independents (see below) borrow from themselves to develop new shows. Many successful programs contain subsidiary characters who have the potential to carry a completely new program. Called "spin-offs," these shows are frequently scheduled in the time period immediately following the parent program to maintain continuity with the established audience. "Laverne and Shirley" was spun off from "Happy Days," "Different World" from "The Cosby Show," and "The Ropers" from "Three's Company" (well, they don't all work). Probably the most fecund queen bee in spin-off history was "The Mary Tyler Moore Show." It begat "Phyllis," "Rhoda," and "Lou Grant," all featured characters in the original half-hour comedy.

Independent Production Companies

An important source of shows is a group known as "indies," small firms within which the owners are oftentimes the chief creative contributors (Stephen J. Cannell Productions; Steven Bochco Productions; Witt/Thomas/Harris Productions; Miller/Boyett Productions). The epitome of successful independent production companies is MTM. Formed in 1970 by actress Mary Tyler Moore and her then-husband, Grant A. Tinker, the company quickly built a reputation for quality. Included among its many hits are: "The Mary Tyler Moore Show," "Hill Street Blues," "St. Elsewhere," "WKRP in Cincinnati," "Phyllis," "Rhoda," and "The Bob Newhart Show." The company's most recent success is "Evening Shade," starring Burt Reynolds.

Two independents who have been prolific

in the 1980s and 1990s are Witt/Thomas Productions (or, on some projects, Witt/Thomas/Harris Productions) and Miller/Boyett Productions. Paul Junger Witt and Tony Thomas (son of comedian Danny Thomas) came from a production background, and Susan Harris was and is a writer. At the start of the 1991–92 season their network productions were "Good and Evil," "Blossom," "Empty Nest," "The Golden Girls," "Nurses," and "Herman's Head." Tom Miller, a veteran producer, and Bob Boyett, a former network executive, were similarly active in the fall of 1991. Among their productions were "Family Matters," "Perfect Strangers," and "Step by Step."

Indies are lean organizations with an aversion to overhead and a close personal attachment to the few productions they present each season. Some of them are started by artists who have abilities beyond their on-camera performance. Merv Griffin, Monty Hall, Dick Clark, and Henry Winkler are just a few personalities who have expanded into production and now run highly respected independent companies. One cautionary note to programmers: stars frequently develop shows that feature themselves in roles they have long harbored a yen to play but are not right for. For the program executive, burying the project without alienating the artist is one of life's more demanding challenges.

Some of the indies are so successful that their profitability equals or outstrips some of the major companies. The prime example is the Carsey-Werner Company. These two former colleagues at ABC, Marcy Carsey and Tom Werner, formed their own firm and within a few years were producing three major hits on network television: "The Cosby Show," "A Different World," and "Roseanne." Although all have been immensely profitable in their network runs, it is the income from syndication that is truly staggering. The revenue from the cash sale of the "The Cosby Show's" initial offering was over $500,000,000. Add to that the barter income and subsequent cash sales for renewals and the total income is estimated at well over $1 billion. The profit helped Mr. Werner become the principal owner of the San Diego major league baseball team.

FIGURE 4.2
Mary Tyler Moore was not only one of TV's most popular and beloved performers, but she was also co-founder of the enormously successful production company MTM Productions. (Photo courtesy of the Academy of Television Arts and Sciences.)

For the most part, however, indies struggle each year to get a pickup or two from some network that will keep the enterprise alive. Because of rising costs and the need for deep pockets, many people who would like to be their own bosses in independent companies find, instead, they must affiliate with one of the major companies. Dealing with the larger bureaucracies cuts down on the creativity that can spring from the lively imaginations of independent producers.

Grant Tinker, independent producer and former president of MTM, worries about the independent production business. "I don't know if the business even exists anymore . . . in a way I would like to be in it," he says in reference to the receding network audience, the proportionate lower license fees to producers, and the rising cost of production.

Specialized Production Houses

Some producers do not attempt to be all things to all media. They are content primarily to develop one form of programming and are the suppliers to whom buyers turn when an opening occurs for their particular specialty. Some examples are: Mark Goodson Productions (game shows), Hanna-Barbera (animated programs), Danny Wilson Productions (after school specials), Bill Moyers Productions (documentaries), and Len Hill Productions (made-for-TV movies).

Sometimes these entrepreneurs are so successful at what they do, they branch out and become equally effective at other program forms. At the beginning of his career, David Wolper was exclusively a producer of quality documentaries. He eventually went on to showcase some of TV's greatest works ("Roots," "Roots II," and "The Thornbirds"). Mostly, however, these boutiques work their special fields and fill an important programming need.

Foreign Production Sources

American companies are ever alert to the possibility of adapting a foreign success to the tastes of American viewers. In fact, two of the biggest TV hits of the 1970s were modifications of British comedies: "All in the Family" and "Sanford and Son." Australia has provided America with a number of series, and in 1992 Group W Productions attempted to syndicate an American version of the Italian game show "That's Amore."

It is rare for a show from another country to come over as is and play on American commercial television, even if it is in the English language. The sense of pace, the foreign accent of the actors, and the subtle language differences are difficult for American audiences to accept. America is the largest exporter of programs, but it imports less programming than any other country. However, this country's producers are very willing to buy good ideas from foreign creators and "Americanize" them for U.S. viewers.

A popular approach of recent years is to develop "co-productions." In previous decades this most often meant a foreign company contributed money to an American production in exchange for certain distribution rights overseas. "Not anymore," says David Gerber, chair of MGM/UA Television in charge of worldwide production. "From now on if we do something together it's going to have to be a global partnership." By this he means the foreign contributors will participate in story development, casting, and distribution strategy, as well as division of profits.

The appeal of a co-production is that a much larger pool of money is available to the production. A movie-of-the-week that might normally be budgeted at $2 to $2.5 million could go twice that high with foreign investments. These additional funds can give the movie a much larger and richer look and make it more attractive to viewers all over the world. Furthermore, most countries outside the United States have limitations (quotas) to the amount of foreign production that can be imported each year. The goal, of course, is to protect their actors, producers, directors, and production companies. However, any program that is partially owned by a company of that country has a much better chance to come in under the quota rules.

There are some drawbacks to co-productions. Not all types of shows are successful overseas. Comedies, in particular, don't travel well. Frequently the language and situations are too uniquely American to interest or amuse foreign audiences. Also, there are times when concessions are made in the story line to appeal to home-country tastes and these accommodations detract from the show's effectiveness in the United States. But as the financial crunch continues to hurt American producers, the search for foreign partners in co-productions will accelerate.

Networks

As mentioned in Chapter 2, commercial networks are a primary supplier of material for the stations that affiliate with them. In addition, regional, ad hoc, and specialized networks provide material to stations that have particular needs.

However, sometimes networks are the source of programming for other networks. This is a revolutionary change, brought about by the sinking fortunes of the networks. Old antipathies have been superseded by fiscal realities. Gone are the days when a show on one network wouldn't be allowed to have a guest performer from (gasp!) another network.

In 1989 Capital Cities formed the ABC Production Company under the leadership of its former entertainment division president, Brandon Stoddard. Of his eighteen initial projects, Stoddard announced two were designed for NBC and three for Home Box Office, both unrelenting competitors of ABC.

In 1991 ABC sold "Fugitive Among Us," starring Peter Strauss, to CBS. At about the same time the Nickelodeon cable service reported that it was producing a series, "Hi Honey, I'm Home" for broadcast on both Nickelodeon and ABC. Shortly thereafter there were more of these mergers between strange bedfellows:

- The USA cable network and CBS began airing of the Stephen J. Cannell production, "Silk Stalkings."
- NBC and PBS agreed to cooperate on the coverage of the 1992 political conventions. For the first two hours of the evening NBC permitted commentators Tom Brokaw and John Chancellor to appear on PBS while NBC presented its more commercially rewarding primetime schedule. Then for the remainder of the evening NBC covered the event with Brokaw and Chancellor providing the commentary.
- HBO announced that it was preparing three pilots for the Fox network and one for ABC.
- MTV and Nickelodeon reported they were developing pilots for ABC.

All of this intermixture is a manifestation of TV's changing scene, and not too surprising when one remembers that ABC owns ESPN and NBC owns CNBC. Then there is the Disney organization, which produces programs for all the networks and its own cable channel plus its owned station in Los Angeles, KCAL. In the near term we can look for a continuation of this cross-fertilization as the various distribution systems struggle to find combinations that will keep them solvent.

Stations

Some stations supply programming to other stations. This is particularly common among group owned stations. We have already seen how Group W's five owned stations interchanged material on "PM Magazine." Frequently a station group will try out a program on one of its stations to test its appeal. If it shows promise it will be extended to the other stations in the group. Finally, if it scores well in these owned markets it will be offered to other stations around the country. "Studs" is an example of this gradual rollout. The program was initially broadcast on Fox' Los Angeles outlet, KTTV. When it enjoyed early success, it was sold to a limited number of additional stations. By 1992 it was available for sale to all stations.

The industry has also come up with variations on this process. In 1991 Group W and the NBC O&Os agreed to cooperate on joint development ventures which, if successful, would be distributed nationally by Group W. In fact, the incredibly successful Oprah Winfrey show began as a local production of ABC's Chicago-owned station, WLS-TV. The station licensed King World to distribute the show nationally, and it now claims a station lineup of over 200.

A unique but growing station-to-station trend involves affiliated stations producing news programs for independents in the same area. For example, WPEC, a CBS affiliate in West Palm Beach, Florida, produces the 10:00 p.m. newscast for independent station WFLX in the same market. WFLX pays WPEC for the program, enabling WPEC to increase its staff by fifteen people. The situation is viewed as win-win by both stations. WFLX's general manager, Murray Green, says, "If our station started a newscast from scratch, it would take three years to get credibility."[1]

On a more ambitious scale, Tribune Broadcasting Company and Turner Broadcasting entered into a ten-year collaborative agreement involving news, documentaries, specials, and original long-form programming. By the terms of the arrangement the Tribune stations are permitted access to the Turner-owned CNN Newsource and Newsbeam. In turn, CNN will have access to Tribune stations' news material.

In general, commercial stations hardly ever supply programming to networks. However, in the news arena, the relationship between stations and networks is changing. Stations used to be almost totally dependent on the networks for national news because they did not have the technical capability to cover fast-breaking stories beyond the 30-mile range of the microwave dishes that they used to send stories back to the station.

However, with the advent of satellite news gathering (SNG), stations can now send local reporters anywhere in the nation or the world to cover stories that are then sent back to the home station by satellite. This has changed the relationship between networks and stations. Stations often bring their viewers national stories during the local newscasts that precede the network newscasts. Although affiliates are not rejecting the network evening newscasts, they have less need for them than in the past. In addition, as the networks cut back news budgets, they look more and more to the affiliated stations to send them stories about local events that have national significance. Although the networks are still the main purveyors of national news, the balance of power is shifting.

The Buyers Themselves

The networks and the large station groups are staffed with programmers whose main assignment is to evaluate the submissions of outside producers. However, they frequently come up with ideas for specials or series that they are unequipped or unwilling to produce in-house. In these cases, they contact a producer or writer who they feel is particularly qualified to develop the premise. Production companies are always eager to cooperate. Buyers tend to be most receptive to their own conceptions.

In his book *Three Blind Mice*, author Ken Auletta states that in the early eighties Brandon Tartikoff, president of NBC Entertainment, jotted down the phrase "MTV cops" and passed it on to writer Tony Yerkovich and executive producer Michael Mann. That thought blossomed into "Miami Vice." Another time Tartikoff visited "an aunt in Miami and came back with the germ for "The Golden Girls." Auletta goes on to quote Tartikoff as saying, "Ten years ago 90 percent of the [program] ideas came from the creative community. Now [1987] it's only 20 percent."[2]

The "American Music Awards" is a good example of how the buyer-to-producer process works. In 1973, ABC's five-year contract to present the Grammy Awards expired. In the judgment of network executives, the rights fees and other requirements for a renewal were too demanding and they elected not to meet the conditions. Instead, they decided to compete with an awards show of their own, but with a format more suitable for the viewer and less encumbered by the rituals and political necessities of NARAS, (the Grammys' parent organization).

ABC programmers asked Dick Clark to develop a format and structure that would fulfill the goals of the show. Although he had never previously produced a primetime special, Clark was selected because of his success with "American Bandstand" and his familiarity with the contemporary music scene. He worked up a tight, fast-moving format, persuaded virtually every major pop music performer to attend the event, and scored an enormous rating for the premiere production. On January 25, 1993, "American Music Awards" celebrated its 20th anniversary with a "top five" rating performance.

Syndicators

As mentioned in Chapter 2, syndicators supply a great deal of program material for local commercial stations. Some of this material is produced by the syndicators themselves (first run), some consists of programs that have already run on the commercial networks (off net), and the rest is movie packages.

First-run syndicated shows must meet the

same creative criteria as programs for any of the other electronic media. The viewers are the same, and they make no concessions just because the budgets of syndicated shows may be less than their network counterparts.

The bellwether forms for Monday–Friday syndication have been talk, talk-variety, games, service, and tabloid news. Aside from those shows that contain a news element, e.g., "Entertainment Tonight," the forms are all capable of multiple productions per day. It is axiomatic that the more shows that can be shot in a studio-day, the less the cost per program. Game shows, whether network or syndication, are routinely produced in blocks of three to five per day, and one-day shooting schedules of twelve shows are not unknown. Hour-long talk shows are shot at least on a two-a-day schedule and sometimes more. The point is to keep the costs down so that the series can be competitive in the rough-and-tumble syndicated marketplace.

Programmers in syndication show a disturbing tendency towards the herd instinct. Somehow the word gets around that the marketplace is looking for a certain type of program, and scores of syndicators immediately announce development plans to meet that need. Who the oracle is who makes these pronouncements, nobody knows. But programmers hate to appear unhip or out of touch and strive to show that they're always on the cutting edge.

In the past this has led to such absurdities as five production companies introducing strips solely devoted to news about soap operas. A second's deliberation would have revealed that the concept was too limited to have even a prayer of success. Furthermore, the strips would be promoting many series that were appearing on rival stations, not a compelling selling point to a potential buyer.

In one season in the mid-1980s, the syndicators went overboard on game shows; in another it was a superabundance of kids' cartoon programs. In the early 1990s, talk shows were the rage. In addition to Phil Donahue, Oprah Winfrey, Geraldo Rivera, Joan Rivers, Arsenio Hall, and Sally Jessy Raphael, buyers were offered Chuck Woolery, Jenny Jones, Ron Reagan, Maury Povich, Vicki Lawrence, and Dennis Miller.

FIGURE 4.3 ABC executives asked producer Dick Clark to develop a music award show that would be tailored for television. He created the "American Music Awards," which celebrated its 20th anniversary in 1993. The program always attracts music's top superstars such as Stevie Wonder (left) and Neil Diamond. (Photo courtesy of dick clark presentations, Inc.)

Students who intend to enter the syndication field are encouraged to make their own assessment of the market's needs and stick to their convictions. The competitors aren't necessarily any smarter or more tapped in; they're just more reactive to the same rumors.

Advertisers

As we have seen, advertiser-produced programming was the foundation of network programming from the start of radio, through the beginnings of television, to the early 1960s when cost considerations made ownership of a single program impractical. Only a few advertisers continued the practice throughout the years, most notably Hallmark Cards, whose quality programs have enriched the medium since the early 1950s.

Lately, interest in advertiser-produced programming seems to be on the upswing. Although limited mostly to one-time-only specials, a sole-sponsor show might soon become a primetime series. Reportedly King World came close in 1991 on behalf of a client of BBDO Worldwide (an advertising agency), but the project aborted over the issue of the cost of time. In addition to the Hallmark series, some other client-supplied shows include "AT&T Presents" (showing a made-for-TV movie), Procter & Gamble's "People's Choice Awards" and "Circus of the Stars," and "Chrysler Showcase" (made-for-TV movies). Procter & Gamble also underwrites "Cocina Crisco" ("Crisco Kitchen") for the Telemundo Spanish-language network.

In the hard-pressed network economy, advertiser-delivered programs are becoming increasingly attractive. To the struggling networks these shows represent time period sales at full rates with generally strong programming. The advertisers like to believe that there is the added advantage of "no clutter" (the phrase used by Werner Michel, senior vice president at the Bozell Agency). However, most viewers probably find it hard to distinguish between 24 minutes of one advertiser's products versus 24 minutes of many advertisers' products.

At the station level, a fair number of programming heads are accepting "infomercials." These are 30-minute advertisements masquerading as informative shows. They use interviews, demonstrations, and sometimes even little dramas to extol the virtues of a particular diet plan, a baldness treatment, or a particular brand of sunglasses. The advertiser pays the station for the air time and also provides the program, which is generally well produced.

In-house Production Units

The networks and most stations have their own studios and equipment that they use to produce programs—no intermediaries needed. The news departments make the most frequent use of these facilities. For the most part networks and stations have their own on-staff news personnel who see to it that the news is produced each day. Reporters hired by the station or network cover specific beats (city hall, Congress, the Middle East) and tape stories about significant events.

The news department also uses information that comes over wire services such as Associated Press and United Press International as well as tips or camcorder footage from individual citizens, stories in newspapers, and information from various data banks. The news business is also rather incestuous in that stations and networks obtain information by listening to each other. CNN has become a major source of news ideas for both networks and stations.

News directors and producers decide which stories should be placed on the air in what order. The buildup to a newscast is intense, and even during the news presentation changes are made to update stories—"Cancel the robbery story; we just got an update on the hostage crisis."

Often the news department is totally autonomous from the programming department. In these cases, the program director has no say over what happens in the news department, the news director makes the decisions. In other organizations, news falls under the aegis of the program director.

The other types of programs most likely to be produced in-house are news-oriented shows such as sports events, documentaries, and public affairs. The news director, the program director, or a committee of network or station management makes the decision as to what will and will not be produced. The number of programs, the specific content of the programs, and the budget for each must be decided in-house.

Live sports events are quite expensive for networks and local stations because they must pay for the rights to air the games and cover the costs for the people and equipment needed to televise the events. For many years, sports was an unqualified moneymaker. Networks and stations bid against each other for the rights in what were often bitter battles. But recently the advertising payments have not been sufficient to cover the costs, and the audience, too, seems to be tiring of the huge amount of sports available on TV. Programmers and managers are in the process of reviewing their philosophies about sports programming.

House-produced documentaries and public affairs shows can also lead to their own brand of problems. For example, after CBS produced and aired "The Uncounted Enemy" in 1982 it found itself engaged in a libel lawsuit against General William Westmoreland. The documentary had accused Westmoreland of purposely deceiving President Lyndon Johnson by estimating that the enemy troop strength in Vietnam was much lower than it really was. Although the case was settled out of court with no clear victor, it demonstrated the difficulties documentaries can cause.

Ironically, this is one of the reasons networks prefer to produce their own documentaries rather than buy them from outside sources. At least the networks know the quality of research and the source of the ideas and can defend themselves. If they accepted work from outsiders, they would still be liable but would have less knowledge of the process of production. Documentaries and public affairs programs are broadcast to serve the public interest. Rarely do they recover their costs. Advertisers do not like to sponsor what might be controversial and viewers usually prefer to watch entertainment shows.

Networks and stations also produce some of their own children's programs. The extent of in-house production is somewhat a function of the FCC's mood regarding the content of children's programs. When deregulation is in vogue and programmers are not required to consider the educational content of children's programs, many buy standard animated material from production companies and syndicators. When regulations require that children's programs contain certain educational or social content, the networks and stations often prefer to produce their own so that they can guarantee the needed elements. A new ruling from the FCC, effective January 1992, makes station license renewals dependent in part on the quality and frequency of children's programs. As a result, we may anticipate an increase in house-developed shows for the 2 to 12 year olds.

Although limited by FCC fin-syn regulations on the amount of programming they can produce (see Box), the networks have always had a fair number of house-owned programs scheduled throughout the broadcast day. All the early morning series ("Good Morning America," "Today," "This Morning") are network produced. Many daytime serials, such as "General Hospital," "All My Children," and "One Life to Live" are done in-house, and the same for the late night series such as "Tonight" and "Nightline." In primetime such series as "Moonlighting," "20/20," "48 Hours," "Fresh Prince of Bel-Air," and "60 Minutes" are all owned by their respective networks. As the FCC rule continues to relax, more of these types of production can be anticipated.

FINANCIAL INTEREST–DOMESTIC SYNDICATION

Back in the 1960s, the three networks, ABC, CBS and NBC, funded and produced a number of their own programs. In addition they held a "financial interest" in programs that were produced for them by production companies. Furthermore, they not only handled the domestic and international syndication of their house-produced shows, but also quite a few of the ones in which they maintained a financial interest.

All of this, plus their grasp on 90 percent of the audience, made the networks very powerful. In 1970, the FCC took a hard look at the domination of the networks and decided that the old adage "power corrupts and absolute power corrupts absolutely" applied. As a result the FCC instituted the financial interest–domestic syndication (fin-syn) rules.

The original 1970 rules stated that networks could produce and own only three hours per week out of their twenty-two primetime hours. In addition, they could not syndicate any of these three hours to stations or other buyers in the United States, but they could act as syndicators if they wanted to sell these programs overseas. The rules applied mainly to series.

The networks were free to produce, own, and syndicate news, made-for-TV movies, and miniseries.

The result of the fin-syn rules was that the networks had to pull way back on production. In general, they did less than the FCC had mandated. (In 1987, for example, the only network-owned and -produced primetime series was ABC's "Moonlighting.") Why bother to set up the structure (and overhead) needed to produce and syndicate material when all you had to work with was two or three series? Besides, the domestic syndication market wasn't that great in 1970 because independent stations hadn't yet come to the fore. The fin-syn rules were resented by the networks, but they caused only a minor blip on the overstuffed balance sheets.

The ones who profited from fin-syn were the Hollywood production companies. The networks had to come to them for almost all of their programming. They received money from the networks, which then had the right to air the programs twice. But once the programs finished their network runs, the production companies, who held the rights, were in a position to capture the profits from the syndication market.

Times changed. Syndication became very important; in fact, most series did not turn a profit until they were sold in syndication. The networks' fortunes (and power and audience) plummeted. They looked longingly at the profits being made in the growing syndication market.

And along came cable TV. Born in a period of deregulation, the cable companies could own programs they produced, networks that distributed those programs, and even the cable systems that delivered the programs to the subscribers. There were no sanctions whatsoever against syndication.

"Why," asked the network executives, "are we saddled with these archaic 1970 rules?" In the early 1980s, the networks began asking this question of the FCC. The FCC set up hearings on the subject. The Hollywood community showed up en masse to protest. Production companies did not want the networks back in the production and syndication business. Led by their flamboyant spokesman, Jack Valenti, president of the Motion Picture Association of America, the Hollywood producers managed to stall any decision, thus maintaining the status quo. In 1988, the FCC did increase the hours a network could produce per week from three to five, but this bone did little to satisfy the executives' appetites.

Then along came Fox. Fox appeared to be both a network and a production company. 20th Century Fox had been producing movies and television series for many years. When the Fox Broadcasting Company was formed to distribute programs to independent stations, the Fox production company provided some of the programming, as did other production companies such as Columbia ("Married . . . With Children"). In no way did Fox want to lose out on the money it was making from syndicating its TV series, which included, among others, the highly profitable "M*A*S*H." So Fox Broadcasting Company said it wasn't really a network. According to the FCC rules a network had to "deliver at least 15 hours of programming a week to at least 25 affiliates in 10 or more states." For several years, Fox kept its program offerings below 15 hours. Then in 1990, Fox decided it would like to program 18.5 hours, so it petitioned the FCC for a waiver from the fin-syn rules. In a spirit of collegiality, it also asked that NBC, CBS, and ABC be given a waiver. Valenti and company saw through this maneuver and quickly objected. The FCC gave the waiver to Fox but not to the other three.

"Wait a minute," said the networks. "This doesn't seem fair." The FCC, tired of the bickering, told the networks and the Hollywood community to hold meetings on their own and come up with a solution. "If you can't agree," said the commissioners, "we'll get back into the fray and come up with something no one likes." Truer words were never spoken. The two sides stalemated, and, in April 1991, the FCC came up with new rules—and no one liked them.

The new fin-syn rules were quite complicated, but in essence they allowed the networks to acquire financial interest in, and domestic and foreign syndication rights for, 40 percent of their primetime schedule. "Not enough," said the networks. "Too much," said the Hollywood producers. Both sides appealed the new rules to the courts. And there they stand.

Meanwhile, the networks, in anticipation of being able to produce at least 40 percent of their schedule, are beefing up their in-house production units. Obviously, opinions on the effect of changes in the fin-syn rules on the status of the programming structure run the gamut. Networks say little would change except that they would have a much-needed and much-deserved new stream of income from syndication. Production companies say they would be put out of business because the networks wouldn't buy from them. Crystal-ball watchers see mergers in everyone's future. Networks will buy established production companies—or vice versa. In that way one corporate entity can profit from both production and distribution. Needless to say, programmers are keeping a close eye on the continuing battle.

Members of the Public

On occasion, bona fide community groups can design a program concept and receive the sanction of a local station. For example, the local school district might propose an idea for a series highlighting the work of outstanding teachers. The station would agree to the idea and would probably even donate its studio and crew for the production. A series proposed by the Red Cross, the Alcohol Rehabilitation Center, or the American Heart Association might be similarly treated.

But generally, commercial networks and stations do not accept any programming concepts that originate from members of the public. This is not because the talent isn't there or the ideas are atrocious (although many of them are), but for legal reasons. Programmers have discovered that the simplest way to avoid litigation for plagiarism is to refuse any submission from an unaccredited source.

Generally, an unacceptable contributor is defined as one who doesn't have an agent who is recognized within the industry. When unsolicited program ideas arrive in the mail, network and station employees are instructed to send back the package unopened. If the envelope is not detectable as a program submission and the seal is broken, the company will return the contents along with a release form that assigns virtually all the rights to the production company. The terms are so burdensome that the sender is usually never heard from again.

According to The Disney Channel's vice president of business affairs, Fred Kuperberg, "Plagiarism is based on three criteria: 1) Access: did the alleged plagiarist have the means to appropriate the concept? 2) Substantial similarity: does the produced work bear a close resemblance in its essentials to the original proposal? 3) A protectable idea: is the basic concept significantly different from other programs to make it uniquely one's own?"

Although networks and production companies take extensive precautions to avoid plagiarism cases, they are troubled with suits year after year. Many are of the nuisance variety and can be quickly disposed of. Others are more serious and require the expensive efforts of a battery of lawyers to refute.

Even as experienced a producer as Aaron Spelling is subject to the accusation of plagiarism. In February 1992 a merchandise marketer, H. W. Broido, Jr., claimed that Spelling modeled the "Beverly Hills 90210" series after an idea of his titled "HolliWoods." Holli, an uprooted Midwestern teenager, moves through a variety of adventures in California and "discovers a world of sunshine, swimming pools, and temptations to grow up too quickly." According to Broido, he met Spelling when he was retained to manage the product licensing program on "Dynasty," a Spelling production. He alleged that Spelling was enthusiastic about the concept, especially when he mentioned that it would be an excellent starring vehicle for Spelling's actress-daughter, Victoria (a role she eventually played in "Beverly Hills 90210"). Spelling denied the charges, but Broido's lawyer stated that by the time the case is over "it will dwarf the Buchwald case[3] in its impact on Hollywood."[4] The Spelling case was still unsettled when this book went to publication. The warning is clear to programmers: don't expose yourselves to program submissions unless you have airtight protection. Even if the submitter is a relative or very close friend, you should have them sign a release document "just as a formality."

Sources of the Sources

Many, if not most, program ideas originate in the minds of network executives, producers, and their staff people. They may spring spontaneously or they may be the result of hours of research on what's currently working followed by countless more hours of searching for an inventive variation that will sell. No matter how ideas are arrived at, they will be the product of the imagination of one or more persons.

During the last decade, movies-of-the-week based on actual events have dominated. This may be because the fiction well has run dry or because buyers find true stories easier to promote ("Torn from the pages of today's headlines!"). Whatever the reason, it is a fact and producers are responding to the traumas of real life.

Although ideas can come from anywhere, some of the most common sources are the following:

1. Newspapers

The daily news is a fertile breeding ground for program ideas. Producers read the papers with an eye for the article that suggests a story. Unfortunately, this occasionally leads to an unseemly competition for production rights. A vicious or salacious crime will frequently touch off a battle among production companies that will become news in itself. The invasion of grieving people's privacy is one of the less admirable characteristics of the industry.

Newspapers, by nature, are a common source for news programs. Most network and station news departments subscribe to a host of newspapers and cull them carefully for both hard news and feature material. Ideas in the news are also used for talk show topics, documentaries, public affairs programs, and magazine shows.

2. Magazines

Many magazines run in-depth stories that deal with personal problems—a legal fight between a husband and wife for the custody of a child, a domestic dispute over the desirability of an abortion, the unpredictable life of a multiple personality. These are conflicts that make strong movies-of-the-week and talk show topics, so programmers avidly scan the magazines for possible productions.

3. Books

The blockbuster books (*The Godfather, Bonfire of the Vanities, Presumed Innocent*) are usually sold to movies for enormous fees. But books that have not enjoyed such popularity can also be the basis for good TV dramas. The competition for these volumes is vigorous, and the major talent agencies have staffers who specialize in knowing who is writing what, and when it will be released. Through a network of friends, contacts, and folks who "owe me one," the more nimble of these agency executives are able to read the book in galley form (or even photocopied pages of the author's own master manuscript) in order to make the first bid on a hot project.

CABLE TELEVISION

When cable television first began, its supporters were predicting that it would create many new types of programming and, therefore, opportunity for many new sources to contribute to television. The forecast has only partly come true. To be sure, the rise of ESPN, Nickelodeon, the home shopping channels, and other specialized fare has brought new forms to the screen. But the explosion of creativity that had been hoped for has not materialized. One reason for this was that the cable TV structure did not really have the money to engage in massive program experimentation. With so many channels cablecasting so many programs the audience became fractionalized. The piece of the pie was so small for most channels that revenue was hard to come by. But with income at a low level the amount of money available for productions

FIGURE 4.4 Producers of children's programming now have more potential customers than in precable days. Two networks that could be approached are Nickelodeon and Black Entertainment Television, which focuses on programs of interest to African Americans, including children. Shown here is BET's "Storyporch" (a), which is designed to foster an interest in reading and literature by having celebrity storytellers, such as Ossie Davis, read to children. Nickelodeon is totally devoted to children's programs such as "Eureka Castle." (b) (Photos courtesy of MTV Networks .)

was too skimpy to do anything ambitious. Therefore, the shows were often no-frills, bare-bone clones of network programs. Because these shows were skeletal, they didn't attract much of an audience, which in turn meant little revenue and just pennies in the pot for development of other productions.

However, in recent years the picture has become much more encouraging. HBO in particular has made substantial progress in the number and quality of its productions. In 1993 HBO received fifty-five nominations for Emmy awards. Most were for telefilms, but eight were for the comedy series "The Larry Sanders Show." This exceptional performance could herald the arrival of cable's long-awaited artistic revolution.

Production Companies

When a cable television network (be it basic, pay, or pay-per-view) wants a made-for movie or a series produced specially for it, it deals with the same major, independent, or specialized production companies used by the commercial networks. If it wants pre-produced materials, it contacts the established syndicators (or, more likely, the syndicators contact the cable networks).

However, because of the variety of specialized audience targets cable has, a larger range of ideas can be created by production companies. Pro-social children's programming can be developed even during periods when the FCC is in a deregulatory mood because Nickelodeon might be interested. Documentaries can appear on The Discovery Channel. Programming ideas geared toward blacks can be taken to Black Entertainment Television.

The pay-cable and pay-per-view movie services have intense relationships with the major motion picture production companies because they cablecast so many of their movies. These services have been able to sign some production companies to contracts that prohibit their films from appearing on competing channels. For example, HBO has exclusive deals with Paramount that keep that studio's movies off of Showtime and The Movie Channel. Movies from MCA Universal, 20th Century Fox, and Paramount appear on Request Television but not on the rival pay-per-view service Viewer's Choice.

The movie studios have also devised some unusual ways to supply cable channels. At one point Universal representatives invited three cable companies—The Family Channel, Arts and Entertainment, and Lifetime—to come into the same room and cherry pick over 300 films. The three services have different needs and different target audiences, so for the most part they were not interested in the same films. Representatives of the three companies went through the titles one by one and divvied them up among themselves. Universal walked away with $28 million, but each of the cable networks paid less than it would have had it negotiated for individual films.

Foreign Production Sources

Some of the cable channels (particularly Arts and Entertainment, The Discovery Channel, Family Channel, Turner Network Television, and Bravo) have much more extensive interaction with foreign sources than any of the commercial networks. These cable channels rely very heavily on material from the British Broadcasting Corporation (BBC) and the Canadian Broadcasting Corporation (CBC). They also undertake co-productions with production companies in various countries.

FIGURE 4.5
The Family Channel co-produces "Maniac Mansion," a comedy that features Joe Flaherty as an amiably incompetent inventor, with several Canadian companies. (Photo courtesy of Atlantis Films Limited, The Family Channel, YTV Canada, Inc., and Lucasfilm Ltd. Television. Photo by Marni Grossman.)

Networks

As already mentioned in Chapter 3, the main source of programming for cable systems is cable networks. The two have a tightly entwined symbiotic relationship.

Also as mentioned, many co-productions have been undertaken between cable networks and broadcast networks. But cablecasters have interacted in other ways with the commercial networks. NBC experimented (albeit somewhat unsuccessfully) with providing many of the sporting events of the 1992 summer Olympics to pay-per-view. In addition some of the cable networks that are owned in whole or in part by the broadcast networks interchange source material. Because NBC owns CNBC (Cable News and Business Channel), news collected through the traditional NBC sources winds up being used on CNBC also. When ABC bought a large film-footage library that contained a great deal of old news footage of a military nature, A&E (which is partially owned by ABC) produced a number of different series ("Heroes," "The Eagle and the Bear") that made extensive use of the library footage. Peter Hansen, A&E's vice president of programming, said, "We were able to milk ABC's library in order to produce some very fine, informative series."

Advertisers

On rare occasion a cable network will obtain a program directly from an advertiser. For example, General Motors supplies "American Playwrights Theatre" to A&E, which has script approval but does not become involved in other ways. Then, of course, there are the home shopping networks that are essentially nonstop commercials. Their program material is totally dependent on the advertising community. Infomercials also appear on various cable channels. In fact, there are vague plans afoot to create a cable channel that is nothing but infomercials.

In-house Production Units

Some of the cable networks produce almost entirely in-house. ESPN and the regional cable networks oversee most of their sportscasts. CNN has charge of the content of its news, and the same is true of the Weather Channel. MTV produces its video jock programs itself, but the videos are provided free of charge by the record companies. C-SPAN produces its own political material. Other networks produce some of their own material. When you see a stand-up comic on any of the channels, that program has probably been produced in-house. Public affairs and talk shows are also likely to be undertaken by a resident production crew.

Cable systems also use in-house production for local origination shows. A local crew will cover news, much as a local TV station does, and cable system equipment and studios will be used for public affairs programs. In-house production is also used to produce inserts for some of the cable networks. CNN and The Weather Channel, for example, leave time for systems to provide information about local news and weather if they so desire. The same cable system staff people who produce local origination produce these inserts.

FIGURE 4.6 The advertiser General Motors provides Arts and Entertainment with a series of dramas titled "American Playwrights Theatre." This episode, called "3rd and Oak: The Pool Hall," starred Mario Van Peebles (left) and James Earl Jones. (Photo courtesy of Arts and Entertainment/ Linda Alaniz.)

Members of the Public

Cable TV networks make very little room for submissions by individuals. Most of the major cable networks have adopted the same "return the envelope unanswered" policy as the commercial networks, but some of the more information-oriented networks are open to ideas from the general citizenry.

However, within the cable system public access channels, the individual reigns supreme. Of all areas of the electronic media, access is the most receptive to ideas stemming from the minds of members of the public. Access is truly a democratized concept. Anyone with an idea who can pull together cast, crew, a little money, and enough time to produce a show may do so. Some people (including any number of college students) have started in access, learned a great deal and proven themselves, and then moved on to paying jobs in the broadcast or cable worlds.

DOG DAZE

One of the more inventive ideas for a cable TV public access show came from an Agnes von Emery, who claimed she could read dogs' minds. Agnes, along with other people from the community, signed up to take a TV production course offered by staff members of her local cable company. The deal was that once someone finished the class, he or she could check out the cable company's TV equipment to tape in the field or could use the cable company's small studio to produce a show.

FIGURE 4.7
A large number of cable networks make use of in-house production units. A regional sports channel, Prime Ticket, tapes Los Angeles Kings hockey games (a). The anchors of The Weather Channel (b) announce from that facility's studio. VJ Karyn Bryant (c) intros music videos for MTV. And a C-SPAN crew sets up to tape in Washington (d). (Photos courtesy of Prime Ticket; The Weather Channel; MTV Networks; C-SPAN.)

a

b

c

d

Agnes showed little interest in the field equipment. She had her mind zeroed in on a studio production. Five minutes into the first class she interrupted the instructor and burst forth with her idea. She was going to invite people from the community to bring their dogs into the studio where she would read the dogs' minds. The instructor flinched a bit at the idea, but Agnes didn't notice. With exuberance she expounded for 10 minutes about her dog–mind-reading abilities and the importance of this to humankind.

Agnes passed the class and, as required by the cable system for all public access shows, submitted a one-page proposal for the program she wanted to do. To no one's surprise, it was the mind-reading show. The cable TV company program director had no alternative but to accept her proposal. He did make it clear that the dogs had to be housebroken and that she would be responsible for any damage they might do to the studio. No problem. Because she could read their minds, she would know when the dogs were thinking of mischief.

The production day arrived and Agnes appeared at the door with three dogs. She gave the pretense that the dogs belonged to people in the neighborhood, but everyone at the cable system suspected they were her pets. But the dogs were well behaved—not a bark or a puddle.

Agnes sat in the studio with the three dogs in a semicircle around her. And, for half an hour, she read their minds. "Fluffy would like to go for a walk." "Inky is starting to think about dinner." "Max is wishing his owner would come for him." Occasionally one of the dogs yawned in agreement. At the end of the program, she gave her home phone number and invited her audience members to call and make an appointment to bring their dogs onto next week's show so that they could learn the wonders of the workings of their dogs' minds.

After the show she was in a state of euphoria. She kept exclaiming about how well she and the dogs had performed. She was sure she was on the way to stardom and world salvation. On the way out, she scheduled studio time for the next dog–mind-reading show. She was never heard from again.

The lesson in this for a programmer involved in public access: assume nothing, expect anything, and plan for the worst.

COMMERCIAL RADIO

Most radio programming is produced by local radio stations that do not deal with a great number of outside sources. Also, the two main products on radio are music and news, so there is little need for the structure that accompanies drama and comedy programming.

Music

The main product on radio is music. Whether this music is being supplied by a network, a syndicator, or the station itself, the source for the music is record companies that very willingly donate copies of their new releases to those stations and services that are likely to program them. In fact, sometimes they are too willing and a practice known as payola creeps in: record and CD distributors give money or special favors to those in radio programming in exchange for airing of the music. Radio exposure is such an important factor in a record's success that distributors are willing to do almost anything in order to receive favorable, frequent airings. Payola was made illegal in the 1950s after it was discovered to be rampant in the radio business. It occurs infrequently now, but on occasion it rears its head at one or more stations.

Although the records are provided free, stations must pay license fees in order to play music. These fees are paid to ASCAP, BMI, and SESAC, music licensing organizations that collect money from stations and pay it to music composers and publishers. The amount collected is an annual fee based on the station's overall revenue—about 1.4 percent. Each of these licensing organizations has jurisdiction over different music. Most stations find they must pay all three so that they can air whatever music they want. The licensing agencies periodically collect play lists from a representative sample of stations and feed into a computer the entries of musical selections from each play list. Based on these, the agencies decide how much of the money collected should go to each composer and publisher.

Most station program directors prefer to make their own decisions regarding the music they play, so the disc jockey shows are produced in-house. "I'm known for being anti-syndication," says Phil Wilson, program director of KGGO in Des Moines. "I want to create my own sound. Most syndicators are too one-artist intensive for my taste." Program directors and/or disc jockeys select the music that will be aired, and the disc jockey provides the in-between chatter, usually in an extemporaneous manner. No production companies, advertisers, syndicators, or networks are involved in determining the programming of the typical disc jockey show. About the only music these types of stations are interested in obtaining from networks are star-studded concerts or perhaps special music host personalities such as Casey Kassem or Dick Clark.

However, some stations like to have their music provided by a network or syndicator who delivers the music either over satellite or through tapes sent in the mail. Stations that program oldies often have trouble finding records, and a syndicated collection that includes Andy Williams, Perry Como, and Peggy Lee is much appreciated. Automated stations also subscribe to prerecorded music. They can just turn on the satellite signal or the tape recorders and leave the station virtually unattended. Commercials are triggered to run by cue tones from the satellite or tape, and occasional announcements about the music are either supplied by the music service or a local announcer who has been prerecorded.

News

All-news radio stations have the same elaborate source structure as major TV news outlets. They subscribe to wire services, networks, and data banks, and hire a large number of reporters to roam the local streets and even travel nationally and internationally to cover stories. They must have a well-organized but flexible infrastructure to incorporate and update the latest happenings.

However, on many music-oriented stations, news is downplayed. Since the deregulation of the 1980s, radio stations are not required to program news, so many of them do only a minimal amount. They do not have their own full-blown news staffs and choose instead to subscribe to a network (or several) to receive news. Sometimes the stations use the network newscasts just as they are and sometimes they use the network actualities (interviews with people in the news) but surround them with copy read by a local announcer.

The sources of news for the networks are very similar to those for TV and all-news radio stations. ABC's radio networks, for example, have reporters and stringers around the world but also subscribe to AP and UPI.

Other Programming Forms

Play-by-play sports can be supplied locally, regionally, or nationally. Stations that program all-sports use all three, but most stations, if they do sports at all, either produce the game coverage in-house or use the services of one particular network.

Networks or local stations can be a source for talk shows, too. If they contain a call-in element, then stations must broadcast the network feed live so that their listeners can interact with the host.

Stations that like to present features have many, many sources to choose from. For example, National Black Network provides specials on Martin Luther King and Black History month. Syndicom offers comedy editorials; Morrie Trumble and Associates delivers leisure activity reports. CBS has "Walter Cronkite's Twentieth Century" and Westwood One presents "America on the Road."

Individuals can occasionally get program ideas accepted by radio stations or networks, but individuals can play a much greater role in radio programming when they call in to express opinions for a talk show or request music for a disc jockey program.

PUBLIC BROADCASTING

Most of the programming sources for public broadcasting are "all in the family." Stations supply networks and networks supply stations. Other entities do find their way into the system, but the network-station relationship in public broadcasting is tight.

Stations

As noted in Chapter 3, the majority of PBS programs are produced by affiliated stations, originally through the Station Program Cooperative and now through a more centralized structure. When the programmers located at PBS's headquarters in Alexandria, Virginia, decide which programs should be shown over the network, they ask one of the public stations to produce it and give the station the financial resources to do so.

A small core of public TV stations led by WGBH in Boston and WNET in New York produce most of the material for the other stations because they are best equipped for elaborate production. Also, as Dan Everett, director of broadcasting at WGBH, points out, "From the very beginning WGBH understood that ideas well produced that are of interest to a local community will be of value nationally. It also helps that we are located in Boston because this city is a rich cultural resource." But other stations also participate. For example, WQED in Pittsburgh recently produced a limited series, "Space Age," for PBS airing.

Public radio stations also figure as a major

supplier for public radio network programs. About 75 percent of APR's programming comes from this source. Stations offer programs to APR, and APR makes the decision whether or not to make these offerings part of the network feed. Sometimes the programs suggested have been aired locally and sometimes they are produced by the stations specifically with the intent of offering them to the network.

NPR's situation is different in that it often produces in conjunction with stations. Peter Pennycamp, NPR's vice president for cultural programming, says, "We do whatever works best. 'Car Talk' is produced in Boston because the two hosts, Tom and Ray Magliozzi, live there. 'Wade in the Water' is produced here in Washington at NPR because it is a cooperative project with the Smithsonian." In addition, stations contribute stories to the NPR newscasts—usually at the rate of about 300 a month.

Networks

PBS, NPR, and APR are primary suppliers to public broadcasting stations, but some of the TV stations also make use of the four regional public TV networks.

Some public radio stations affiliate with a specialized commercial radio network, the Beethoven Satellite Network. This network, based in Chicago, programs classical music. Its offerings are available to both commercial and public stations, and many public stations find it fits well with their locally produced programs, many of which are classical "disc jockey" shows.

Public television has not been left out of the recent co-production phenomenon that has surfaced between commercial and cable networks. WNET and The Learning Channel are jointly financing an interview program called "Charlie Rose." It will be cablecast, with commercials, by The Learning Channel the day after it airs noncommercially on WNET.

Foreign Production Sources

Public broadcasting obtains a significant amount of its programming from foreign sources, especially in comparison with the paucity of this product on commercial TV. PBS telecasts so much from the BBC that people have quipped that "PBS" stands for "primarily British shows." Some of these British shows are actually co-productions between a PBS station and the BBC—part of the financing comes from the public broadcasting money hierarchy and part comes from the Brits.

APR regularly airs newscasts from both the BBC and the Canadian Broadcasting Corporation. Both public radio networks feature dramas produced in foreign countries.

Production Companies

The regular major, independent, and specialized production houses rarely deal with public broadcasting. Its structure and program needs are so different from the commercial and cable outlook that production companies don't really fit in.

There are some companies (most of them nonprofit) that do supply shows to public broadcasting. One of them, Children's Television Workshop, started in the late 1960s, is in a class by itself. This organization, which has produced such highly acclaimed children's series as "Sesame Street," "Electric Company," and "3-2-1 Contact," is separate from PBS but is so closely tied to it that it could not exist in its present form without the public TV structure.

FIGURE 4.8
"3-2-1 Contact," which includes segments like this one where cohosts Z Wright (left) and Stephanie Yu uncover facts about genetics, is produced by Children's Television Workshop and aired through PBS. (Photo courtesy of Children's Television Workshop/Kimberly Butler.)

FIGURE 4.9 Ken Burns, an independent producer, has supplied PBS with several highly regarded series including "The Civil War." (Photo courtesy of Florentine Films.)

A newer production company, Ken Burns Enterprises, has supplied public TV with several limited series including "The Civil War," the highest-rated PBS series ever. Burns' company, although it is technically independent, works in conjunction with WETA in Washington, D.C.

Other companies produce and/or distribute instructional material that stations broadcast for school use. These companies include Great Plains National in Nebraska, Agency for Instructional Technology (AIT) in Indiana, and Western Instructional Television in California. College credit courses are available through several consortia around the country—Dade County Community Colleges in Florida and the Southern California Community College Consortium in the Los Angeles area. Sometimes these instructional programs are paid for and selected by local school districts or colleges; the station merely transmits them.

Informational material can be purchased from the educational arms of companies such as Time Warner and Wolper TV, and some firms specialize in offering classic (or sometimes just old) movies or TV programs.

In-house Production Units

Public broadcasting stations also produce their own programs—less than they used to because of declining economics, but still a fair amount, especially in comparison to many of their commercial kin. WQED in Pittsburgh, for example, boasts of "Black Horizons," the longest-running black public affairs program on public television. It also airs a weekly program, "The Editors," produced in conjunction with local newspapers, and its special "Breast Test" was successful in encouraging many women to call in for free mammograms. KOCE in Huntington Beach, California, produces numerous documentaries about its Orange County coverage area and also tapes one telecourse a year. Radio station KCRW in Santa Monica, California, covers city council meetings, tapes local musical groups, and, on occasion, produces original dramas.

NPR produces all its own news—primarily "Morning Edition" and "All Things Considered," which runs in the afternoon. It uses essentially the same news sources as the commercial stations and has about fifty correspondents spread throughout the world. Because money does not flow abundantly in the public radio environment, NPR often uses as its reporters spouses of newspaper correspondents who are stationed in foreign countries. APR, at this time, does not produce any of its own material.

Members of the Public

During the 1970s individuals around the country complained that their ideas could never make it through the public TV pipeline. They noted that public TV was supported by taxes and was indeed supposed to be "public." It should not lock out the people and air only what the establishment (PBS and the affiliated stations) wanted. As a result of these complaints, Congress set up a fund called the Independent Television Service to aid independents in producing or completing projects. This money is administered each year by the CPB. Producers who win grants are not guaranteed that their programs will be shown on public television, but the productions are carefully considered by both PBS and individual stations.

Independent producers are welcomed by public radio but are likely to have the greatest chance of success if they are working in conjunction with some formal public radio entity. For example, "Echoes," a program about new age music, is produced by Steve Rolin, an independent producer in Philadelphia, but he is working with the Pennsylvania Public Broadcasting system.

• • • • • • • • • • • • • • • •

Thus, the sources of programs are many and varied, and ideas are brought to buyers in a wide range of forms. Sometimes they are expressed in a single sentence; more frequently they are typed out in two or three pages. Occasionally, a program creator will present a few minutes of tape or film to communicate the core of the show.

When buyers like what they see or hear, they start a chain of events that is known as "development." This important process is addressed in the following chapter.

• • • • • • • • • • • • • • • •

FIGURE 4.10 "The Return of the Art Maker," an elementary school program starring Dan Mihuta, is distributed by Western Instructional Television, along with over twenty other series intended for in-school use. (Photo courtesy of Western Instructional Television.)

MAKING "PRIMARY COLORS": THE REAL WORLD OF PBS

Jeffrey Hayden, longtime producer and director of commercial TV shows, began collecting works of art by Sister Corita Kent in the late 1960s. Not only did he admire her bold colors and inventive forms, but as his collection grew he developed an increasing admiration for the woman herself. Sister Kent was a charismatic teacher of art at Immaculate Heart College in Los Angeles. She was not afraid to try new approaches to painting and serigraphs, nor was she fearful of suggesting new ways for nuns to become more involved in a changed society. Some of these ideas did not sit well with James Francis Cardinal McIntyre, archbishop of Los Angeles, and after thirty-three years of religious life Sister Kent left the convent. She settled in Boston and with difficulty learned how to make her way in the strange role of a single woman on her own. For the next eighteen years she continued to grow as an artist, and her spirit and genuineness enriched all who met her. In 1986 she succumbed to cancer.

Mr. Hayden felt that such a life deserved a memorial. Since his experience was in television, he decided that a one-hour documentary celebrating her art, her personal story, and her social consciousness would be a suitable tribute. In January 1987 he took his first step and obtained from the executor of the estate the rights to produce the film. He was not aware then that this would be the start of a four-year odyssey.

It didn't take long for Hayden to learn that the only hope for an airing of such a film would be PBS. Each of the commercial networks had instantly dubbed the project "esoteric" and summarily turned it down. A total neophyte in the world of public broadcasting, he contacted the only person he knew in the field, Henry Cauthen, president and CEO of SCETV (South Carolina Educational Television). Mr. Cauthen promptly acquainted him with the harsh realities of the medium: all production costs must be covered by the producer; some grants are available from PBS institutions but they must be matched dollar for dollar by outside sources. To help get the project under way, Mr. Cauthen advanced a modest amount for research and development.

With this seed money, Hayden worked up a promotional package consisting of a stylish brochure, a 30-page nuts-and-bolts presentation, and a 15-minute videotape that showed samples of Ms. Kent's art and scenes from her life. Armed with these materials, he crisscrossed the country making pitch after pitch to corporate leaders, private foundations, and government bodies. He drew up a list of every art dealer who had ever distributed Ms. Kent's works and asked them for contributions. He also contacted every collector known to have purchased a Kent painting or silk screen and requested their support. His efforts were tireless. They were also unavailing. Nobody wanted to make a move until they saw proof that someone else had given. It was a seemingly endless game of chicken and egg, egg and chicken. After eighteen nonstop months of telephoning, letter writing, and presentation giving, Hayden had virtually nothing to show for it. It was the "bleakest moment" of his life.

The break came from an unlikely source, a broadcast symposium. The topic, "PBS Documentaries," caught Hayden's eye and he decided to attend. One of the panelists, Ron Hull, was a specialist in raising funds for productions. At one point during his commentary, Mr. Hull stated that the PBS system was like a freeway: "You just have to know what access ramp to go up and then you get into the flow." Hayden buttonholed Hull after the session, described his project, and asked to be led to the ramp. Shortly thereafter, through the good auspices of Messrs. Hull and Cauthen, Hayden met in Washington with representatives of the National Education Association and the Corporation for Public Broadcasting. He eventually emerged with a grant of $100,000—an amount, of course, that he had to equal with outside funds. With the logjam broken, a number of corporate donors came through. So did many private owners of Kent artworks. One woman in San Luis Obispo, California, contributed $25,000 and then held a dinner party at which she raised another $15,000. When the spurt was over, Hayden found himself tantalizingly close to his goal, but not quite there. Further presentations were useless. All the wells were tapped dry. He was out of ideas, short of money, and exhausted. He needed one more miracle—and he got it.

The redoubtable Henry Cauthen called and reminded him that SCETV's facilities were first-rate and considerably cheaper than Hollywood, where Hayden had planned to do the postproduction work. Cauthen suggested he come to South Carolina and "see how many dollars we can save." They found a score of ways to be more cost-efficient and Hayden discovered that the lower rates would enable him to complete the film with the money he had collected. The hard part was over. Now all he had to do was make a picture.

On January 28, 1991, PBS presented "Primary Colors," a tribute not only to Corita Kent but to the tenacity of a determined producer. It had taken four years, almost to the day, for Hayden to see his dream realized. For a person steeped in the ample, if not downright profligate, ways of commercial television, the experience had been the most "draining, frustrating, and fulfilling" period in his long and successful career.

FIGURE 4.11 It was the courageous personal story of Sister Corita Kent as well as her artistic gifts that inspired producer Jeffrey Hayden to embark on his four-year effort to complete "Primary Colors." (Photo courtesy of The Saint/Hayden Co.)

NOTES

1. "More Indies Airing Newscasts Produced by Competitors," *Broadcasting* (September 23, 1991): 16.

2. Ken Auletta, *Three Blind Mice* (New York: Random House, 1991): 357.
3. Art Buchwald, syndicated columnist, sued Paramount Pictures on the basis that the company appropriated an idea of his to make the Eddie Murphy film *Coming to America*. Buchwald won, and the case caused the Hollywood community to expose its Byzantine method of calculating net profits.
4. "Spellings Sued over '90210' Idea," *The Hollywood Reporter* (February 14, 1992): 3.

5 Development

Development is the process by which a program progresses from the glimmer of an idea to a production commitment from a distribution system. Some employees at production companies phrase it slightly differently: "Development is taking an idea and getting it into such good shape that the network can't possibly say no." Whether the buyer is a network (broadcast or cable), a syndicator, or a station, the shaking down—or shaping up—of a concept is a mandatory procedure.

Very, very rarely will a buyer make an on-the-air commitment simply on the basis of a program description. Most buyers insist on seeing some evidence of the show's "doability" before thousands or millions of dollars are risked. This usually applies even to those few star producers who have firm air commitments. Although development changes from medium to medium regarding particulars, the process in general consists of escalating steps that finally reach a point at which the buyer believes the project has a chance for success and a green light is justified.

COMMERCIAL TELEVISION

The development for all forms of programs (primetime, daytime, specials, children's, even news and sports) in network and syndicated television is essentially the same. The time spans and the number of intervening steps may differ, but the course of the procedure is very similar. Primetime broadcast network series development is the most multiphased process, and for that reason it is used in the following schematic:

- The production company comes up with an idea and makes a date with the network to present the concept.
- At the "pitch" meeting the buyer either rejects the project or expresses an interest by commissioning further development such as: a page or two of "notions" (very brief plot ideas); "outlines" (more extended versions of sample stories); a "treatment" (detailed descriptions of characters plus some sample stories); or a "bible" (a major tome encompassing all of the above).[1]
- If the seller comes back with satisfactory material, the buyer's next step is to order a script for a "pilot" (a sample show on film or tape).
- Once the script is deemed acceptable, the producer is asked to present recommendations for cast and staff for buyer approval.
- When all the creative elements (key people who will work on or in the show) are approved and signed, the network authorizes a pilot production.
- After the pilot is completed and the producer has made delivery, the network begins extensive testing of sample audiences.
- Finally, key executives review the research findings, assess their scheduling needs, measure the show against other alternatives, and announce their decision.

Financing Development

In most cases, the development process from initial concept to pitch meeting is funded by the producer. Occasionally a writer or producer will receive up-front money to "go off and dream up something," but most buyers are loath to provide creative people with "thinking" money. They've discovered that too often they're just paying for something that already exists in a desk drawer somewhere.

Once a "next step" is authorized, the buyer underwrites the project all the way through the pilot. As we shall see, this does not always cover all costs. But the simple rule of development is: "You order it, you pay for it."

Fundamentals of "The Deal"

No money will ever be handed over by the buyer to the seller until a contract is in place.

A network or a syndicator is not about to part with substantial dollars only to find that the project is free to go elsewhere. Thus, when an executive says, "I like it, let's go forward," it means, "Let's go forward to the Business Affairs Department where your lawyer and our lawyer will discuss the terms and conditions." It is not within the province of this book to cover the range of program deals, some of which weigh slightly less than an air conditioner.

It's sufficient to say that the goal of buyers is to protect their investments. Their concerns include: license fee per episode, the length of time their hold is exclusive (can't be taken to another buyer), right of first refusal (the ability to match the best offer of a competing buyer when the initial term is completed), cost overruns (who pays), creative control, and so on. The sellers, of course, are equally interested in these matters and attempt to do their best to preserve their profit potential.

When the contract is essentially agreed upon, the project can proceed. In the real world, work gets under way long before papers are signed. If it didn't, few shows would ever get on the air. Companies that have dealt with each other many times in the past are sufficiently confident that a deal will eventually be made. Unless really substantive disagreements emerge, the parties proceed as though the documents will eventually be signed. At times the decibel levels go ballistic and the cage-rattling is deafening, but nobody makes any money unless there's a picture on the screen, a fact that both sides are aware of. Peace and reasonableness usually return in due course.

The Odds for Success

Development is a long and arduous journey for the production company and the odds against a pickup are very high. Consider this arithmetic:

In the course of a year network program development will receive in the neighborhood of 2000 primetime series submissions. These can range from an elaborate, multicolored, flip-card presentation, to a few sentences scratched on the back of an envelope. Whatever the form, the submission amounts to a seller trying to interest a buyer in a program concept. The vast majority of these notions are rejected on the spot. They either have no inherent merit or they do not address a need of the buyer.

Perhaps 10 percent strike a responsive chord, and the producer is encouraged to go further in development. This encouragement might be a commitment to fund a treatment or script, or it may just be a suggestion that the show holds some promise and is worthy of being fleshed out. Of these 200 positive responses, approximately thirty to forty will receive pilot commitments, or, as is currently fashionable among cost-conscious networks, minipilots (sample scenes and sequences).

Of these finalists, only five to eight will make the schedule in the average season, and, if two are strong enough to return for a second season, the program development team will feel it has had a good year. Unhappily for production company executives, the on-the-air commitment doesn't mean they should rush out and buy forty acres of Malibu beachfront. The company will still need at least three more annual pickups before there are enough episodes to release the series into syndication. If the renewals end before there are eighty to a hundred shows, the series has little chance to make the big syndication profits that keep the company viable.

The license fees paid by a network to a production company are almost always less than the cost of the program. In the 1990–91 season the average cost of a one-hour film show was $1,408,364. But the average fee paid by the network for the right to play the picture two times was $1,100,790, leaving the producer with a deficit of $307,574 *per episode.*[2] The figures for 30-minute series were equally sobering (see Box). If the production of a one-hour series experiences a loss of $307,574 on each of eighty episodes, it will accumulate a deficit of $24,605,920 before there is a chance of recoupment. The production business is not for the faint of heart. However, like the lottery, when you hit, you hit big. That's why suppliers keep playing the game.

The differences between production costs of a network half-hour program in 1985–86 and 1991–92 point out the growing gap for production companies.[3]

1985–86	
Cast	$40,000
Guest Cast	$3,000
Writer	$12,000
Director	$10,000
Below-line Costs	$150,000
Postproduction	$25,000
Producing Unit	$40,000
Agents	$45,000
Total	**$325,000**
License Fee	-$275,000
DEFICIT	**$50,000**

1991–92	
Cast	$125,000
Guest Cast	$10,000
Writer	$14,000
Director	$13,000
Below-line Costs	$275,000
Postproduction	$60,000
Producing Unit	$125,000
Agents	$78,000
Total	**$700,000**
License Fee	-$450,000
DEFICIT	**$250,000**

Development Departments: Networks and Production Companies

At each of the networks the entertainment division is divided into dayparts (different portions of the broadcast day) with separate staffs responsible for development in their areas of specialty. The reason for this is that each part of the program schedule, for example early morning, daytime, primetime, and late night, has a different audience composition and therefore requires a different program approach. Primetime, of course, is the most important and therefore has the largest number of development executives. Reporting to the vice president in charge of overall program development are executives for primetime comedy, drama, specials, movies-of-the-week and miniseries, and, most recently, "reality" series. Many of these department heads will have staffs of two or three people to aid in the screening of submissions. Usually the development executives for the other portions of the program day (daytime, children's, early morning, and late night) report to the vice president in charge of those areas. That person, in turn, reports to the president of the entertainment division.

The major production companies also have program development departments that consist of a chief and some support people. These program specialists have two key functions: 1) to be intimately familiar with the needs and interests of the buyers, and 2) to guide the efforts of the producers and writers under exclusive contract to the studio so that their product will meet the buyers' needs. They also screen submissions from outside creators who wish to develop their ideas with that particular company.

The Role of the Agent

Agents have been major figures in the entertainment business ever since vaudeville. Their historical function is to obtain work for their clients, originally just performers but in later years writers, producers, directors, and others involved in the creative process. They locate the job opportunities, negotiate the fees and conditions, and deduct 10 percent for their labors. With certain clients, their duties go far beyond the traditional and incorporate such functions as developing long-term career strategies, arranging accounting and investment guidance, overseeing travel schedules, and even serving as advisers in marital and romantic complexities.

Because agents represent the key creative elements in any programming enterprise, they play a vital role in the fulfillment of any project. They are keenly desirous of protecting the interests of their clients and for that reason usually attend meetings in which spheres of authority and influence are being determined. They are always the ones who travel to the networks or production companies to protest to management the slights, real or imagined, that their client has been receiving.

A few decades ago some of the more gifted and/or more powerful agents began to make programs possible by bringing together all the key creative elements. Projects originally thought to be unworthy of air commitments suddenly became viable because of the quality of the writer, producer, director, and stars that the agent was able to assemble. At first the agents were satisfied merely to collect commissions on those artists whom their company represented. But eventually they reasoned that the project never would have seen the light of day if they hadn't put all the parts together. That accomplishment deserved a larger take than just the sum of the commissions. Explains veteran agent Sy Fischer, "Without the agent there never would have been a project as a whole, so why shouldn't the agent commission the whole?"

This led to the concept of the "package commission," in which the agency receives 10 percent of the entire cost of the show. The difference in revenue is significant. The sum of income of a director, a writer, and performer on a current situation comedy might total $100,000 per episode. The agent who represents all three would collect 10 percent of that amount, $10,000. But the whole production may actually cost $500,000. The agency's commission then soars to $50,000, a cool $40,000 more per episode. The money comes out of the license fee a network pays to the production company.

In its original conception, the package commission was justifiable and generally unresented by production companies and distributors. But as time went on, agencies sometimes insisted on a package commission even when they represented only one creative element. They reasoned that their client (a major star, for example) made the project viable, and therefore they were entitled to a package commission. This was interpreted by buyers as a power play in which the agency used the appeal of a client to extract a higher commission for itself. The device still raises hackles when it is employed with marginal validity.

Starting the Process

When producers and writers come up with a notion that seems promising, they review it, sometimes exhaustively, with production company executives. (This is true for most types of programs, but, again, we will use primetime network series as an example.) Does the idea hold up for limitless episodes? Are the characters strong enough? How can we make it more salable? Does it need another element?

This last question can be the cause of much hand wringing. The attachment of a major performer may be just the addition necessary to push the project into the win column. Certain stars have a magic appeal to buyers and almost any submission that includes them will guarantee a pilot. But there are risks. The performer may not be right for the part. Should the entire concept be distorted to accommodate the style of the star? What if the network has cooled on the artist; then the project is dead in the water before it starts. Also, the demands of the star may make the project prohibitively expensive. Unless the show is expressly built around the talents of one performer, producers prefer to present the idea without a specific star and stay flexible to accept a casting suggestion from the buyer.

When the members of the production staff are confident that the network "can't possibly say no," they develop the verbal presentation that they will make to the buyer. Many experienced presenters rehearse this pitch in painstaking detail. Mock dialogues are played out: "If we're asked where the conflict is between the mother and daughter, Irv will respond by pointing out that. . . ." When commentaries have been developed for every anticipated objection and question, a call is made and a date set for the presentation.

The Pitch Meeting

At the appointed hour, the sellers travel to the office of the buyer to present their idea. The composition of the selling "team" may range from two to six people. Small production houses will usually be represented by just a major executive of the firm and the chief creative person (writer or producer) connected with the concept. Large companies will arrive with some combination involving a major executive, a development person, the executive producer, the line producer, the writer, and possibly an agent. The buyer may be represented by just one person or by an ensemble, depending usually on the importance of the producing company. If a superstar producer or writer is making the presentation, the audience will probably include the president of the entertainment division, the head of all development, and the entire staff of whichever department is appropriate—comedy, drama, reality. Networks are eager to be ingratiating with superstars, and a large turnout for a pitch is a way to demonstrate respect and affection.

The meeting never starts right in. No matter how gigantic the project, it is obligatory that the first few minutes be devoted to trivia. ("How 'bout those Lakers?" "Have you had Spago's new pizza yet?") Eventually one of the sellers, most likely the senior executive, will begin the presentation with a few scene-setting remarks such as the genius of the writer, the eternal appeal of whatever storyform they're about to present, or the time period needs of the network. The presentation is then turned over to the principal architect of the show who describes its concept, characters, and appeals.

The most effective presentations are those that cover all the main points in a taut, compelling manner. If the seller gets carried away and goes on interminably about nonessential matters, the buyer's attention drifts and the sale is all but lost. Most buyers wait for the presentation to be completed before they offer questions and observations. It is at this point that the seller's hours of preparation can pay off. The more buyer-doubt removed, the better the chance of a pickup. Any suggested improvement by the buyer should be treated with the utmost enthusiasm; the time to quibble over changes is *after* the deal has been secured.

Some sellers are brilliant at pitch meetings and can make any program sound like "M*A*S*H" reincarnated. Frequently the verbal virtuosity conceals conceptual weaknesses. The buyer is well advised to enjoy the performance but remain vigilant; viewers will be watching the flawed show, not the dazzling pitch. On the other hand, many program creators are not born sales people and their presentations fall short of the show's potential. Buyers must train themselves to pierce through the bungled rhetoric and find the essence of the idea.

Many sellers and some buyers do not like the pitch process. Even a production legend such as David Susskind, whose programs won so many Emmy awards he was elected to the Television Academy's Hall of Fame, loathed the pitch process. He called it "demeaning, nerve-wracking, and exhausting." Despite his revulsion, he was a very effective spokesperson for his projects and enjoyed over thirty years of prosperity and prestige.

Following the seller's presentation, the buyer will generally ask a number of questions and then make some sort of pronouncement. A few are very visceral and decisive: "I love it; let's do it," or just as frequently, "It's lousy; forget it. What else do you have?" More often the response will be cautious. If the buyer is convinced there are no fundamental flaws and that the production, with proper scripting, casting, and execution, has a chance, he or she will request additional information or material.

Further Development

The form that this next step will take depends on the nature of the show. If it is a dramatic program, it may be a long-term story projection or a script. If it is a comedy show, it may be notions for episodes or a script; if it is a reality program, it may be authorization to shoot some person-on-the-street interviews or a sample event just to get the flavor of the program. In all cases, this further development will be an intermediate step, a way to move the project along without a full financial commitment. The buyer's goal is to obtain further clues about the show's potential at a minimum cost.

Usually this stage will involve a series of telephone conversations and meetings in which the differences between the original concept and the buyer's desires become narrowed. Many times the gaps are never bridged and the project aborts. As was mentioned earlier, about 10 percent of the time the seller satisfies the buyer's objections, if any, and they proceed to the next step.

If the result of what is requested in the way of interim development is poor, the buyer will drop the project and save some development dollars for new submissions. If the development is impressive, the material won't be wasted. More than likely it will be used either in the pilot or somewhere during the series run.

The Pilot

When buyers are satisfied that the program is fundamentally sound, the staff and cast are equal to the creative challenge, and the pilot script is in shooting shape, they face a key decision: Does the idea go to pilot? At risk is an enormous outlay of money. Pilots cost more than the average episode of a series because there are no economies of scale and anything done for the first time is more time-consuming and expensive. Furthermore, if a pilot goes to series, the cost of the sets and props can be amortized over the length of the commitment. But if the show is not picked up for more episodes, the entire expense of the pilot must be laid against the program development budget.

It is not uncommon for the cost of the pilot to be greater than the network's financial contribution. The production company must either cut corners to deliver the pilot within budget or cover the additional costs. Usually they choose the latter. A downsized, on-the-cheap pilot might quickly become a dead pilot. Because this is the only sales opportunity the project will have, the producers, amidst much muttering, almost always spend the extra money.

Because costs have risen so drastically and network revenues have dropped so precipitously in recent years, network executives have increasingly ordered "demonstrations" or "minipilots" (selected scenes) instead of full episodes. The argument in favor of these truncated prototypes is that they reveal enough of the core idea and the staff's execution to make a sound judgment, at a substantially reduced price tag. The argument against is that the chosen sample may be "loaded" (with guest stars, action scenes, etc.) and not be indicative of the series. Minipilots sometimes become more useful as sales tools than as a basis for program decisions. "I'd much rather do a minipilot," says Saul Turteltaub, veteran writer-producer of network comedies. "It's much easier to be terrific for eight minutes than for thirty."

Even though the minipilot has its flaws and will from time to time lead to blundered program choices, it is here to stay because of the new economic realities in the industry.

Testing

Rarely does a buyer watch a pilot and instantly commit it to the schedule or the scrapheap. These days the screening of the prototype by network executives is just the start of a lengthy process of assessment through research. This phase of program selection is dealt with in detail in Chapter 6.

Why buyers require tons of statistical support before they make a program choice has been the subject of many conferences, panel discussions, and print interviews. The cynics say the buyers have no faith in their own judgments and need mounds of data to justify their decisions. They also claim buyers engage in research to avoid later criticism. If a program goes on the air and bombs, their excuse is that the research was faulty and misleading.

Not so, counter the defenders. The stakes are high and the criteria for success are inexact. Furthermore, as objective as network executives try to be, they are products of their backgrounds, which may not be representative of national tastes and interests. Why shouldn't they sound out the opinions of those who will be watching (or ignoring) the product. Every manufacturer in the country test-markets products before launching them nationally; why not networks? Perhaps the truth is somewhere in between.

The Decision

In most cases, before a final judgment is made, a number of company executives are consulted. These usually include, but are not limited to, people from programming, sales, research, promotion, production, finance, planning, and top management.

The need for each of these disciplines is obvious. The show is worthless if the sales department can't sell it and/or the promotion department has no hook on which to hang a campaign. There is no sense proceeding if the finance people say it is too expensive to ever be profitable or the research staff says there is only a small audience available in the intended time period. In some instances, experts in a specialized aspect of the company need to be consulted, e.g., the standards and practices division (censor) if the program contains borderline scenes and language.

But the process, as unwieldy as it may seem, is soundly constructed. Without input from all affected parties, a programmer could blunder fatally because of some overlooked item not inherent in the show itself. Experienced programmers know how to obtain useful information from marginal contributors without letting them get the impression they are programming legends-in-the-making. The final decision is rarely a consensus of all the viewpoints. After the program has been thoroughly reviewed, the two or three top members of the inner circle will usually go off for a private conclave and make the selections.

BIRTH OF A SERIES

When a commercial network begins its deliberations for the fall primetime schedule, the process is protracted and, in some cases, wondrous. Approximately a dozen executives are involved, and they meet daily from the early morning through dinnertime six days a week for as long as two weeks. The principal prop throughout the discussions is a large board, surfaced with Velcro, that presents a space for the 22 hours of each network's primetime schedule. Also available are cards for every show on the air, or known to be in development, for the four networks.

Most of the meetings for NBC, CBS, and ABC take place in New York, where the corporate managements are located. Fox meetings are located in Los Angeles at their corporate headquarters. Some potential producers elect to take up residence in New York for the entire period just to be on hand in case a network programmer wishes more information or wants to discuss a change that might put a show on the schedule. Since most of these people are away from home, there is a good deal of socializing, wining, dining, and gossiping each evening. In fact, the first hour or two of every day's deliberation is usually devoted to the rumors and tidbits that were picked up the previous night. A typical contribution will consist of an executive reporting with all solemnity, "I had a drink with Mark Johnson [producer] last night, and he told me that Lance Perkins [president of a rival network] personally—personally—put his card on the board in the Thursday at 9:00 slot and it stayed there for an hour and a half! Mark is sure he's a lock for the schedule." Other executives nod and ponder this critical bulletin. This morning-after reporting is a pointless exercise, but it has become a time-honored rite within the scheduling mystique.

For hours, day after day, the Dirty Dozen stare at the board. From time to time one will stroll up to it, pull down a few cards, insert a few others, step back, take a look, and wait for some crowd reaction. Eventually someone will offer an opinion, mostly negative (but polite) and the adventurer will return to his or her seat.

As the locked-in announcement date approaches, the mood becomes less congenial; old rivalries and turf battles surface, and soothing words from the senior official are often required to restore tranquility. When only a day or two remains and all the members have contributed the best of which they are capable, the majority are dismissed and only three or four are asked to stay behind. They will put the final cards in the final slots and decide the schedule.

The compression of a large group in a small room for a long period can lead to some strange behaviors. One of the rituals of the process is to "rate out" a schedule. Once the group has filled in all the time periods for all the networks, the research vice president is dispatched to his computer to determine the probable season-long rating performance of all the networks. Mind you, no network has yet announced anything, much less a complete schedule. But the traditions must be preserved and a competitive standing forecast.

One year ABC's deliberations were becoming strained. No matter how the executives manipulated the cards, the "rating out" always resulted in a third-place finish. Tempers were raw, moods were sullen, eyes were glazed. There were only a few days left before the announcement and the schedule was coming up short, mainly because of a hole on Tuesday at 8:00 p.m. A strong half- hour comedy was needed and the cupboard was bare.

From out of the gloom a voice said, "Damn! If only 'Temperatures Rising' [the previous year's loser in the time period] had done better. Hospital comedies can be great." Long pause. Finally another voice, "Yeah, and if only Paul Lynde had worked in 'Sedgwick Hawksides' [a sitcom modeled after Sherlock Holmes]." Another lengthy pause and then a third voice. "Maybe we shudda put 'em together." The room took a beat and then began to pulsate:

"Hey, maybe you got something. How about we make Paul the general manager?" "And the owner is his mother, a real battle-ax." "Paul plays great against somebody like that." "That's good. And let's add a surgeon who is a real con artist. Writes phony books and doesn't know a semicolon from a hemorrhoid."

And so the frenzy of creativity went on, fueled by days of frustration and incarceration. An executive rushed from the meeting to call Bill Asher, the producer of both shows, to see if he'd go along with the plan. The executive returned moments later and breathlessly announced, "Billy will do it!"

Before the cheers died down, the research vice president raced from the room to rate it out. For 45 minutes the Dirty Dozen paced nervously and waited for the verdict. Finally, the door opened. The researcher juggled a few sheets of paper, put on his professorial face, and quietly stated: "NBC has a 30.3, CBS a 31.9, and ABC a 32.1. We win the year."

Bonhomie ran amok! Much back-slapping and shoulder-punching, a chorus of "Aww-riights," a few more hugs, and the happy warriors dispersed in triumph.

Such bursts of irrationality are not uncommon during these pressurized sessions. Sanity usually returns the next day, the madness is erased, and the ship is righted. Not so that year. The network went to the post with its hybrid sitcom in the saddle at 8:00 p.m. Tuesday. It staggered through the fall and was mercifully replaced early in the spring. As for the network, it finished the season in its then-customary third place, about four share points short of the researcher's rosy prediction.

Syndication

As we saw in Chapter 2, the development process in syndication differs substantially from that in the network because of the manner in which the two systems deliver their product. To summarize, and oversimplify somewhat, when a station affiliates with a network, it automatically agrees to present the programs that the network designates for broadcast. Programs are sent at a specific time via satellite to all affiliated stations, who, in turn, transmit them to the homes within their coverage area. When producers sell a show to a network, they don't have to worry about obtaining national distribution; it comes with the sale. The network pushes the button and 99 percent of the people in the nation are able to see the show.

The process is far more complicated in syndication. There is no delivery button. Syndicators obtain commitments from stations on a market-by-market basis. The process can be somewhat shortened by selling the program to a station group (Tribune, Taft, Gannett, etc.), and they will place the program on all their stations. However, that still leaves a lot of hard slogging ahead to achieve sufficient national coverage (70 percent is generally considered to be the minimum) to justify a production commitment.

For production companies, all of this involves much patience and occasional heartbreak. They must go through the same steps listed in Chapter 2 to develop a show and either fund it all themselves if they are a production/distribution firm, or try to persuade a syndicator to underwrite the project. A prototype for a one-hour talk show may run \$750,000–\$1,000,000 and a game show may cost in the \$500,000–\$750,000 range. But the pilot for a syndicated series is a commitment for far more than just the cost of the production.

It also means the company, or its distributor, is prepared to undertake the laborious and expensive process of promoting, clearing (signing stations to carry the show locally), and selling the barter (obtaining national advertisers). These supplementary costs can quite easily soar past the \$2–3 million mark. (It has been reported that King World spent

$45,000,000 to promote the launching of the 1990 version of "Candid Camera.") Even for major production houses, these are substantial investments.

Once the prototype is completed and the decision has been made to go to market, the show is basically in the hands of the marketing specialists. From time to time the programmers will be asked to participate in the selling effort, particularly with executives of station groups or stations in large cities. These buyers like to hear directly from those responsible for the production what their series plans will be. And the programmers will certainly have to attend the NATPE convention.

NATPE

NATPE (National Association of Television Program Executives) was founded in 1963 to bring together station program directors for an exchange of ideas and viewpoints. Those who had developed successful concepts in their markets were encouraged to share them with lodge brothers and sisters from around the country. Also, those looking to improve their jobs were eager to let the program world know of their ingenuity in case a station in a larger, more affluent market might be searching for a hot young programmer. The gatherings were an immediate success and attendance grew from 71 people in 1964 to 8915 in 1991.[4] Many stations send not only their program director, but also the general manager, the sales manager, and the station owner.

It was inevitable that with so many station buyers under the same roof, sellers would seize the opportunity to present their wares. By the mid-1970s, the NATPE convention had become the most important event in the business life of syndication producers and distributors. All pilots, demonstrators, and promotion campaigns were geared for the big push at the convention.

With such spirited competition among highly creative and aggressive personalities, the convention gradually took on a carnival atmosphere. One couldn't walk thirty paces without being invited to a seller's suite by a trio of midgets, models in nuns' habits, or a pair of gorillas. The taste level plunged to unacceptable depths, and in the 1980s NATPE officials took steps to curb some of the more outrageous promotions.

Syndicators spend hundreds of thousands, even millions, to construct exhibition areas designed to attract buyers into screening rooms to view pilots. A syndicator's principal objective is to build a sales momentum that will stampede the convention. If word gets around the convention floor that "The Gloria Gloman Show" is blazing hot, station buyers will rush to the booth to lock up the program before competitors in their market can make a bid. To develop this effect, syndicators make frequent and imaginative use of hyperbole. Every show is "snowballing," all station lineups and coverages are "growing too fast to calculate," every "I'll-think-about-it" becomes "It's a virtual wrap." Announcements at NATPE can be politely referred to as an excess of exuberance.

The bazaar goes on for five or six mind-numbing days, and at the conclusion sellers generally have a good idea of their program's chances. In recent years, the convention has become somewhat less pivotal to a syndicator's survival. Many distributors discovered that if they waited for the convention (usually held in January or February) to launch their major sales effort, they would be too late. Competitors would have contacted key markets months earlier and picked up all the best clearances.

Now NATPE is used more for muscling the wafflers and establishing relationships. For in-studio producers who are attending their first NATPE, it can all become quite puzzling. When they hear that the syndicator cleared Chicago, they become euphoric. Then they notice that the sales staff is somber.

"Why aren't you excited?"

"Because it's a tier deal."

"What's that?"

"It means we've got prices for six different time periods."

"So?"

"So, if they run it at 2:00 a.m., you won't get any viewers and we won't make a dime."

Station Development

When stations decide to program a local show, they do not go through the elaborate development process engaged in by the networks. The ideas usually come from within the programming department and are in response to some need in the community. As the amount of local programming has dwindled in the last few years, so has the amount of development.

For example, Bob Brooks, program director of KFOR, Oklahoma City, has largely abandoned the development of daily local series and instead concentrated on weekend specials and kids' shows. The Fourth of July is exuberantly celebrated in the market, and Brooks has catered to this interest with a number of holiday specials. He believes local production in the future will be mostly event programming keyed to area needs and interests.

Another reason for the downgrading of local development is that many stations now belong to large groups such as Fox, Tribune, and Westinghouse. The creation of new programs is the responsibility of a corporate production unit rather than the program directors at the individual stations. As Don Tillman, vice president of programming and production at KTTV, Los Angeles, points out: "The corporate people develop the shows and use one or more of the stations as a launching pad. 'Studs' was first aired here, 'Foreign Affairs' in New York, 'America's Most Wanted' in Washington, and so on. These shows are always made with a national rollout in mind. We experiment with them in our markets, and when they're ready we widen distribution. But it's all done with little or no creative involvement from the local P.D."

As noted in Chapter 2, most station development is limited to news programs. This is partly because of the reasons offered above, but to a larger degree because news programs are the best method for a station to establish a strong local identity. "It has become more and more obvious that the future of independent stations depends on the way they reach their local market," states Steve Bell, general manager of KTLA, Los Angeles. "In cable homes we have to compete with fifty stations, and the only way to stand out and to continue to be the station that viewers want to turn to is through your local programming that they can't get elsewhere."[5] Accordingly, Bell hired a staff of over twenty people and launched a daily, two-hour newscast to compete directly against the "Today" show and "Good Morning America" from 7:00 to 9:00 a.m. While four other Los Angeles stations produce newscasts between 6:00 and 7:00 a.m., KTLA is the only station in the market to go against the entrenched network powerhouses. After six months on the air, the show's intensely local orientation has succeeded in making it fully competitive with its veteran rivals.

One variable that could cause a resurgence in local nonnews production in the near future is the changing face of station compensation. As the trend continues towards the reduction or elimination of the network's financial contributions, the possibility of station defections from certain time periods and dayparts grows. Some stations may conclude that a locally produced program in which the station keeps all the revenue may be a better risk than a network-supplied show for which there is no compensation and just a few spots in and around the program for local sale. If this were to happen, the program director could quickly become a very busy person in the creative field.

CABLE TELEVISION

The first step of development for a cable network is for somebody to figure out what the network is all about. Commercial TV networks never had to go through that phase, mainly because they took over the broad-based "something for everyone" concept developed by the original radio networks back in the 1930s.

Developing a Network

From the very early 1970s days of HBO, cable networks had more specific aims. Either their programming was narrow (all-movies, all-sports, all-news) or their audience was narrow (children, women, Hispanics). Unfortunately for many of the now-defunct or struggling networks, this first stage of development was rather haphazard. Many networks were formed during the cable frenzy of the late 1970s and early 1980s that did not benefit from a great deal of research into the possibilities for success. People with vague ideas started networks, not wanting to be left out of their part of the pot of gold. Some of these ideas worked and some didn't, as can be seen by a quick overview of a sample of cable networks and the convoluted history they encountered.

CABLE WAR STORIES

Here's what happened to a few of the networks during the early 1980s.

Probably the most flamboyant of the channels was CBS Cable, founded in August 1982. It was the commercial network's attempt to enter the cable market, which at that time was far less glamorous and prosperous than the mighty CBS. The network developed a critically appraised cultural channel that featured primarily original programming that cost on average $100,000 an hour to produce. With the deep pockets of its parent organization, CBS Cable wined and dined cable system operators, trying to convince them to make the service available to their subscribers. It got few takers, in part because the high-class programming appealed to a very narrow range of people, but perhaps more because the CBS Cable personnel, most of whom had come from the commercial network business, exuded a "we're CBS, we can do anything" attitude. The service didn't fare much better with the advertisers, who were supposed to pay the bills. Again, in commercial network fashion, CBS Cable people wouldn't even talk to advertisers who were not willing to spend at least $1 million in commercial time. With a minimum of advertising and with very few systems carrying the service, it soon languished. CBS decided to pull the plug in September 1983, after spending an estimated $50 million.[6]

Another "big-guys" disaster involved ABC and Westinghouse teamed up against Ted Turner. Turner had founded two 24-hour-a-day news channels, CNN and CNN Headline News, in June 1980 and a year later was making slow but steady progress toward black ink when ABC Video Enterprises and Group W Satellite Communications announced a competing service, Satellite News Channel (SNC), that would take advantage of both their established news-gathering organizations. Ted Turner did not take this lying down. He filed suit against the Goliaths claiming antitrust violations, and ABC and Group W, in turn, filed a suit accusing CNN of anticompetitive business practices. While the case was wending its way through the courts, ABC and Group W discovered that the cable business wasn't as lucrative as they had predicted. Westinghouse's Dan Ritchie explained, "We were on track with the costs and subscribers, but we were not on track with our revenues." What eventually happened was that, in an out-of-court settlement, Turner bought out SNC. He paid Group W and ABC $12.5 million to shut down their service and leave him with a cable news channel monopoly. Said Turner, "It was a real bloody war where they came in and tried to knock us out." But he felt he prevailed because "we started first and we had excellent services."[7]

On a less contentious note, two services, Daytime and Cable Health Network, merged in 1984 to form Lifetime. Daytime had been started in 1982 by Hearst and ABC Video Services as a program service for women. Cable Health Network was launched the same year by Viacom, Dr. Art Ulene, and Jeff Reiss to provide information about health, science, and better living. By 1983, both services found themselves drowning in red ink and arranged a friendly merger that was called "a very good move that will permit reductions in programming and personnel costs while creating a basic cable network on the way to attaining that critical mass of subscribers needed to attract advertisers."[8] So far, so good. As of the present, Lifetime is still in business.

Organizations entering the cable network business these days do undertake more orderly research into the possibilities for success. They study the cable networks past and present and try to come up with some idea that has not already existed. Usually they can't think of much because just about every conceivable idea has been tried.

But they can also look at existing channels, see what is popular, and attempt to present a variation that would tap into a similar vein of audience interest. For example, after MTV proved to be a great success, many other companies decided they could cash in on the music video craze. A few, such as Country Music TV and VH-1 (owned by the same company that owns MTV), have succeeded, albeit in less grandiose style than MTV. But others failed, some mercifully before they actually started. Ted Turner forayed into this field then pulled the plug in less than a month. More recently, several companies have noted the success of Univision with Spanish-language programming and are in the process of developing competing channels.

Most importantly, a potential cable network must convince cable systems to carry it. With most cable systems already filled to their channel-capacity brim, finding willing takers is no easy chore. Better for the organizers to learn that there is no room at the inn while they are still in the concept stage than after they have invested money in programming and satellite time. Smart would-be network organizers have found that one way to assure that systems will carry their channel is to offer system operators part ownership in the network channel.

Program Development

Once a cable network has developed its general philosophy, it begins program development. When networks are looking for original programming, their development process can parallel that of the commercial networks—pitches by independent producers, pilots, etc. Gary Nardino Productions in association with Paramount sold "Brothers" to Showtime after CBS had rejected it. Finnegan and Pinchuk, an independent production company that produced "Gore Vidal's Lincoln" for NBC, also produced "Quicksand: No Escape" for USA.

The width of the door opening for production companies varies from one network to another. Showtime is more likely to accept pitches than HBO. The people at HBO usually come up with their own ideas of what they want and then approach production companies that they feel can carry the ideas through to fruition.

Because most cable networks are so specific, companies wishing to sell to them must study their needs carefully. There's no sense pitching a children's program to Playboy. Some networks have narrow desires within their narrow programming. In 1991 Nickelodeon was looking for original nonviolent animation, but in future years it may prefer live-action drama. Bravo had a year when it was seeking high-quality American independent films that were not well known. Producers with these types of films in their pockets were more successful with their pitches than were people promoting opera.

Of course, many cable networks don't accept pitches from anyone. They develop their own ideas internally, much like an independent TV station. They buy from syndicators and produce their own material.

Cable System Development

The main development decision-making for a cable system operator comes when the system is first established and must decide what, from the many networks and other services available, to select for its channels. Sometimes these decisions are made on a monetary basis. Basic networks do not all charge the same amount. Theoretically, a cable system could load up on all the cheapest channels, but a system that does not carry the more popular (and more expensive) networks such as ESPN, WTBS, and CNN has the potential of losing income because its lineup will not attract subscribers. A system will certainly make more money using one of its channels for Cinemax than it will using that same channel for public access, but community members may want access more than they want a fourth pay movie channel. Conscientious cable programmers will think of the needs and demographics of the community when developing a channel lineup. A system servicing a retirement community could well skip MTV but would want to include Nostalgia Television.

Development of locally originated cable programming is fairly simple but assumes an involvement in the community, either for local origination or access. Some cable systems make a practice of encouraging members of the local community to produce material for public access; others just accept what comes in the door.

Overall, development in the cable world is

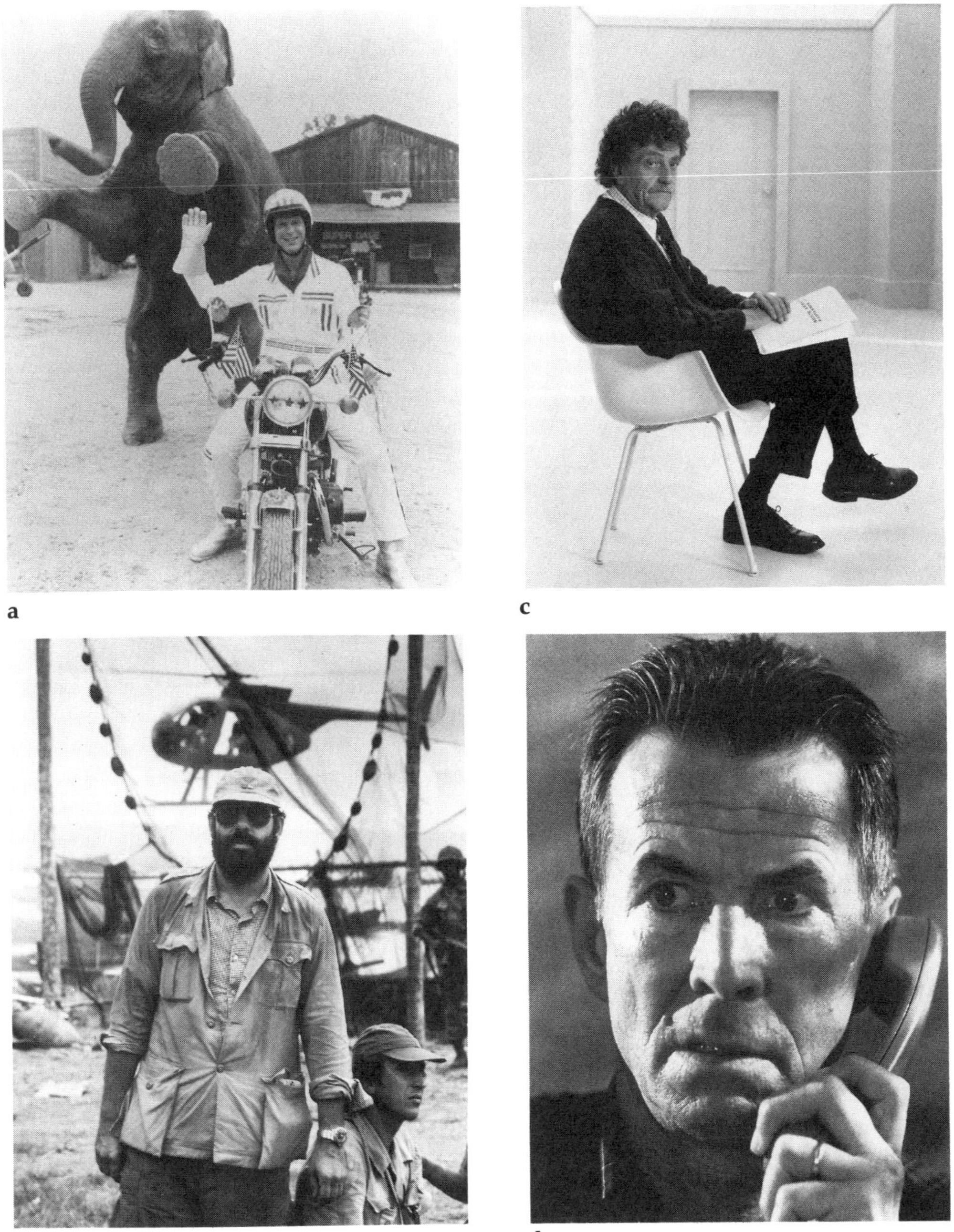

FIGURE 5.1
Various methods of development occur within cable TV. For example, Showtime acquired the four programs shown here in different ways. "Super Dave" (a) was originally a skit within another Showtime program called "Bizarre." When "Bizarre" went off the air, the network negotiated with the skit's creator, Bob Einstein, for a show that featured him as the star. "Hearts of Darkness: A Filmmaker's Apocalypse" (b) is a documentary shot by Eleanor Coppola about her husband, Francis Ford Coppola, while he was directing the movie *Apocalypse Now.* The Coppolas brought the finished product to Showtime, which purchased it to premiere exclusively on Showtime. Kurt Vonnegut's "Monkey House" (c) consists of several half-hour specials based on Vonnegut's short stories. *Psycho IV: The Beginning* is an original motion picture that aired exclusively on Showtime. The idea for the movie, which starred Anthony Perkins (d) as Norman Bates, the character he had immortalized thirty years earlier in Alfred Hitchcock's *Psycho,* was developed in-house by Showtime's staff.

b

a

FIGURE 5.2 The Family Channel develops some of its own programs such as "Big Brother Jake" (a), a series about a stuntman and his extended family. Others come from outside producers, such as "Rin Tin Tin K-9 Cop" (b), a series from the production company Katts and Dog. (Photo (a) courtesy of The Family Channel; photo (b) courtesy of Katts and Dog and The Family Channel.)

less frenetic than it is for network commercial television. The stakes are not as high because the audience levels are lower. But more and more cable networks are undertaking original programming as opposed to movies and syndicated material. Some are hoping to develop a megahit such as "The Cosby Show" or "Cheers." If this happens, the development process for cable will be a wilder game

COMMERCIAL RADIO

Development in radio deals more with formats than with individual shows. Much of the programming develops itself once the format is decided. For most radio stations, the format is a particular type of music, but talk formats such as all-news, all-sports, and all-talk can be chosen. Selecting a format involves thinking about the type of peripheries, such as news and features, that will be used along with the basic music or talk.

Developing a Format

One of the problems with selecting (or maintaining) a music format is that music is constantly evolving. What might be referred to as "new wave" doesn't stay "new" for very long. When a particular type of music, such as rock, becomes popular, it tends to be subdivided into categories (hard rock, soft rock, classic rock, album-oriented rock) so that stations can have a unique sound.

Nevertheless, format names do exist, and format changes are made with considerable frequency, in part to take into account the newer forms of music. The job of determining format is one usually undertaken by top management that may or may not include the head of programming. Often the program director is hired after the format decision is made so that someone with knowledge about that particular type of music can be chosen. In fact, in some existing stations, a format change is made right under the nose of the program director without that person being involved. If, for example, a station switches from easy listening to country, it simply fires the old program director (and the disc jockeys) and hires new ones.

When a station first comes on the air, it must, of course, select a format. Not very many open frequencies are left in the United States, however, so very few new stations come from scratch. Format changes are common among established stations that are not doing very well in the ratings and among stations that have just been purchased by new

The major music formats and the approximate number of stations playing them, as listed in *Broadcasting Yearbook*, in 1985 and again in 1990 demonstrate the changing world of radio formats.[9]

1985	
Adult Contemporary	1690
Beautiful Music	450
Big Band	130
Black	270
Classical	340
Country and Western	2440
Golden Oldies	300
Jazz	270
Middle-of-the-road	1430
Progressive	620
Top 40	960
Urban Contemporary	100

1990	
Adult Contemporary	2310
Beautiful Music	370
Big Band	170
Black	170
CHR/Top 40	970
Classic Rock	160
Classical	370
Country	2500
Gospel	190
Jazz	300
Middle-of-the-road	600
New Age	50
Nostalgia	80
Oldies	720
Progressive	140
Rock (AOR)	490
Urban Contemporary	210

management (often because they were not doing very well in the ratings and were about to go under financially).

Radio Consultants

A booming business of radio doctors (consultants) has arisen to help stations make the decision regarding a format change. Many radio station owners or general managers do their own research to determine what format would be best, but some hire one of the consulting firms (Drake-Chenault Enterprises is the grandaddy of them all) to at least assist. These companies have run research studies in many other cities and have developed a basic procedure that usually works.

For one thing, they have acquired a knowledge about the radio business that includes:

1. The appeals of various formats. They know from other studies they have conducted and from general industry reading the pros and cons of the various formats and the type of people each appeals to.
2. Cost effectiveness of the various formats. Again through experience, they can fairly easily cost out differing types of programming for specific markets. For example, all-news programming is more expensive than middle-of-the-road music and can only be undertaken in markets that are large enough and rich enough to be able to support it with advertising.

When the consultants come to town, they check out the area, looking mainly at:

1. Demographics. If the community contains many young people, their presence might be a determinant in planning for a format. If a large ethnic group is within the station's reach, this presents the possibility for a particular type of programming aimed at those people. The consultant uses census data, chamber of commerce data, and personal tours of the community to determine demographic information.
2. Life-styles. If the schools are on split shift and, as a result, young people are available as listeners during the morning, this could impact on the format and its accompanying features. If a city has massive traffic jams, this could mean more time needs to be devoted to traffic information than might otherwise be planned. Interviews with station management and with citizens of the community and trips to shopping malls, restaurants, and parks can give insight into life-styles.
3. Sales of records and CDs. Consultants visit music stores and talk to store employees and customers to determine who is buying what type of music when and where. The age and sex of different types of music purchasers is noted.

Consultants also do a thorough examination of the competition by noting:

1. What is aired on all the competing stations. Consultants usually spend at least one full day listening to each station in the market. They log everything they hear—station identifications, disc jockey comments, news stories, editorials, features. They make careful note of all the formats and features currently on the air.
2. The ratings of all stations. This consists of a very thorough breakout of ratings data for each station, giving not only its overall rating number but its rating according to specific demographics and specific times of day.
3. The reach of all the stations. Some stations are more technically powerful because of FCC requirements and therefore reach a larger geographic area. A weak station with a popular format might not reach the whole market, so a more powerful station with the same format could attract listeners on the outskirts.

And the consultants must consider the opinions of the top station management including:

1. Programming biases. If the station owner abhors country-western music and has stated in no uncertain terms that he or she does not want to be associated with a country music station, that alternative is ruled out, even if it seems to be the most economically viable one.
2. General objectives. Sometimes managers have specific personal or station goals. If someone wants the station to earn as much money as possible, a different approach can be used by the consultant than if someone wants to make a reasonable profit but make sure the station participates fully within the community.

Once the members of the consulting team have gathered and analyzed the information, they discuss it with management and make recommendations. Perhaps they feel the station should program easy listening with 10-minute news updates on the half hour. The reasoning for this might include a large number of middle-aged people who spend a great deal of time in their cars traveling between office and factory locations of a large employer whose business is spread out over town. They want information mixed with relaxation. Although two other stations in town program easy listening, one of them is very low-powered and can't be heard at two of the company locations. The other is the second highest-rated station in the market, but it has news for only 2 minutes on the hour, so some members of its audience could be attracted away by the longer news.

Management is, of course, free to ignore the advice of the consultants, and sometimes the chemistry between consultant and management is such that the two never come to an understanding. But usually management will listen to the advice it has paid dearly for and go with what is suggested. If it doesn't work—well, there are always other consultants.

Developing Programming

Once management and/or a consultant have determined a format and an overall programming philosophy, individual programs must be developed. For many music stations, the switch from one program to another is virtually seamless. The disc jockey may change, but the sound of the music remains the same. The programming is not really developed; it is supplied. Someone does have to select the music to be played from all the new submissions and old records or CDs available. At some stations the program director takes total charge of this; at other stations the disc jockeys have a large say as to what gets played. Stations that specialize in the newest of the new may have a person who does nothing all day but listen to submissions and winnow them down to a smaller stack that is passed on to the program director.

For all-news stations, the programming is also supplied, in this case by daily events. A news director must decide which news services to use, which stories to cover, which stories to put on the air, and how much time to allot to each story.

Talk stations do give thought to the development of specific programs. Hosts and their topics of conversation are thought through, usually by the stations' programmers. On occasion potential talent will pitch an idea to a talk station and then be hired to conduct the show.

Networks and syndicators often develop concepts in conjunction with their affiliated stations and their advertisers. In order for their material to be successful, stations must use it, and advertisers must want to sponsor it. Sometimes the two are at odds. Unistar, for example, had a successful weekly show that featured big bands of the 1950s. It was carried by many stations, but advertisers didn't like the older audience it attracted. After much discussion, Unistar compromised and shortened the show. Most of Westwood One's decisions are made at meetings attended by company programming, advertising sales, and affiliate relations executives. National Black Network found it had to provide a music bed for all of its features because its affiliates wanted to emphasize music and downplay talk.

Overall, radio development is quite different from that of television. The great degree of content fragmentation and the less expensive nature of radio make it a separate animal from its visual cousin.

A TANGLED FORMAT TALE

Jim Gabbert never dreamed that a series of radio station format decisions would bring his life full circle. As a child he spent most of his time in Central America, where his father was U.S. Ambassador to Costa Rica. His parents always sent him to schools where only Spanish was spoken, so when he returned to the United States to study at Stanford University, he "didn't speak English so good." His Spanish was excellent, however, so he found a job as a disc jockey on a Spanish-speaking radio station. His English soon improved and he lost his interest in Spanish, but the radio bug had smitten him. Throughout college and for a number of years thereafter, he was responsible for the start-up of a number of FM stations, and eventually he owned FMs in the San Francisco area and Honolulu.

Then in 1980, he entered what he refers to as "midlife crisis number one." He sold all his radio stations and bought TV station KOFY (pronounced "coffee"), Channel 20 in San Francisco. At the time he purchased it, KOFY was a Spanish-language station, but he changed it to a typical independent, showing mostly American reruns.

The radio bug still lingered, however, and in 1986, he bought a 1000-watt San Francisco daytimer station that he christened KOFY-AM and turned into an oldies station. Ironically, it, too, sported a Spanish-language format when he bought it.

His oldies format was based on music of the 1950s and 1960s, a period in history that was undergoing a nostalgic revival in the mid-1980s. Other stations in town were programming the same oldies, but he had a new twist. He programmed the station tongue-in-cheek, poking fun at the 1950s, the nostalgic craze, and the station itself. The disc jockeys were named after defunct automobiles—Henry Hudson, Norman Nash. Vintage, and not-so-vintage, cars were given away in contests.

The first year the format was a phenomenal success. But then other stations began imitating KOFY, and since KOFY beamed a mere 1000 watts and competing stations powered signals as high as 50,000 watts, the more powerful stations began eroding KOFY's audience. Gabbert petitioned the FCC to be allowed to increase the power of the station to 50,000 watts.

While that was in progress, Gabbert decided to renew a courtship with his old love, FM radio. He tried to buy a San Francisco station known as K-City. It had a unique eclectic format that could best be expressed as little-known songs by well-known artists. The station had a small but loyal following, and a few of the hard-core listeners sensed that Gabbert would be altering the format. They formed a coalition and petitioned the FCC to prevent the sale. Although the FCC had previously ruled that a change in format was not a valid reason for holding up change of ownership, the coalition managed to stall the sale process for eighteen months. During that time Gabbert's bank became $250,000 richer on the interest he was paying on a loan he had acquired in order to buy the station. Through compromise, Gabbert did eventually acquire the station, but he had to grant the coalition the right to select the program director and participate in programming decisions.

"It didn't work," says a wiser but more weary Gabbert. "Even the members of the coalition couldn't agree. Someone would suggest, 'Let's play "Jealous Man," by John Lennon.' Five people would say yes and five would say no. I had absolutely no control. After a very brief period of ownership, I sold the station to Viacom."

Gabbert had other problems that were ransacking his wallet and adding to his desire to sell the FM and gain some cash. His AM upgrade to 50,000 watts had been approved, so he needed to build a new antenna. Several years earlier he had purchased some land in Hayward (near San Francisco) that had once been used as a landfill but was covered over in 1954. He figured that would be the perfect place for his brand-new antenna. But because the land had once been a dump, he found himself dealing with 26 government agencies—everything from the Water Quality Board to the Army Corps of Engineers.

"We had to dig 35 wells and run tests. They made us hermetically seal the earth over the 1954 garbage and then put four feet of dirt on top of that. But I couldn't destroy any plants on the endangered species list. There was pickerelweed on the property, and we had to dig it up before we laid down the four feet of dirt and then plant it back on the top. I had to make provisions for the hairless popcorn flower, and I had to set traps to see if we could catch any red belly marsh rats, because, had there been any, we would have had to protect their habitat. All these studies were well intentioned, but they were taken to an extreme. I spent $250,000 on environmental impact studies and another $35,000 for a bird study. We had to make the land around the towers into a park for the city of Hayward. It took two years for the studies and only four weeks to put the towers up. All this so we could broadcast our oldies format at a higher, more competitive power."

But by now the format was floundering. Gabbert realizes in retrospect that he was so consumed by the FM legal problems and the AM tower environmental problems that he wasn't paying attention to the main aspect of his business—programming. The 1950s nostalgia that was in style in 1986 had petered out by the late 1980s, and KOFY's ratings were petering out with it. The preoccupied Gabbert tried to bandage the station with a satellite-supplied oldies service.

Then, while on a business trip, he was having lunch across from a store selling 1970s memorabilia. The place was packed. It occurred to him that the 1950s–1960s craze was over, and a 1970s craze had started. He slid his station into 1970s music, with a noticeable improvement. But the station, despite its 50,000 watts and environmentally proper antenna location, was not performing to Gabbert's liking, and, in 1991, he decided to sell it.

None of the bids was high enough, but he did notice that everyone interested in buying the station was planning to turn it into a Spanish-language station. He paused, investigated the market, and found that, indeed, a Spanish-language station made sense.

He didn't hire a consultant. "With my 37 years in the business, I don't need a consultant. What has happened is that the financial types—the bean counters—have taken over radio. They need consultants because they don't know anything about the business."

Gabbert took the station off the market and in November 1991 launched KOFY with a Spanish format that featured primarily popular music. He didn't change the call letters. "KOFY is such a good set of call letters, I didn't want to give them up. Besides, it's my philosophy that you should position call letters, not a format. In that way you can slide your format around without being obvious about it like we did when we went from 1950s to 1970s oldies."

Within six months, KOFY was the top Spanish-language radio station in San Francisco. Gabbert attributes this to a unique approach. "The other Spanish-language stations sound like Mexican stations. We program like American stations, except we're Spanish. We don't talk over records, and we only program ten commercials an hour, not eighteen or twenty. Spanish-speaking people who live in San Francisco want to be Americans, not Mexicans."

Gabbert has taken the Spanish influence to his TV station, too. Although it still programs primarily American reruns, it provides Spanish translation on the second audio program channel (SAP) so that the programs can be heard in either Spanish or English.

It wasn't until he was several months into Spanish programming that Gabbert realized the irony of his situation. "I started as a Spanish-language deejay, but twice I eliminated Spanish formats. Now here I am, back in the Spanish-language broadcasting business."

Lessons for the student of formatting:

1. Ideas for formats can come from anywhere—from people trying to buy your station, from a lunch counter view in a strange city.
2. Life goes on, even if your format doesn't. Don't become so preoccupied that you lose touch with what is happening in your community and the world. If the 1950s craze is over, it's over.
3. Study all aspects of change before becoming committed. Such "hidden" factors as citizens groups, environmental studies, and government rulings can ruin profitability.
4. Don't think you always have to make changes. Sometimes the old ideas (or people) who are part of your background or the station's may still have merit.
5. The radio business isn't easy, but those like Jim Gabbert who truly love radio will find a way to make it work. Developing and programming a radio station is full of rewarding, challenging experiences.

PUBLIC BROADCASTING

The word "development" has a different connotation in the noncommercial structure. Most public broadcasting organizations have a department of development that is responsible for obtaining grant money. This department is often compared to the sales department of a commercial broadcaster because it is responsible for bringing in revenue. Theoretically, the organizations that donate money to public broadcasting do not influence the content of the programs, but sometimes whether or not a program reaches the airwaves at all is dependent on the financial resources. So in that way, the fund-raising undertaken by development departments does affect program decision-making.

Public broadcasting also develops programs and series in the more traditional way. As with commercial television, ideas surface and are selected. However, the procedures for determining what will and will not be aired are quite different.

Public Television

The methods of development within public television are in a state of transition because of the abolishment of the Station Program Cooperative and the establishment of a more centralized system. Under the SPC, a peer-review system existed—stations submitted proposals to their sister stations and these stations accepted or rejected them. Now, however, a group of executives at PBS makes programming decisions, taking into account the desires of the stations.

Stations as Developers. For the most part, public TV program ideas are developed by employees of public TV stations. The formal pitch process does not exist. Telephone dialogue often takes place between the PBS programmers and the station programmers, but eventually the station people put their ideas in writing (something similar to a treatment or "bible") and send them off to PBS. The PBS programmers then decide which of the proposals to fund. The winnowing process of commercial TV is not part of public TV development; pilots and testing are virtually unheard of.

Independent Producer Involvement. Independent producers can become part of the public TV development process—with difficulty. The best route for a person with an idea is through a station because the stations are the ones that submit ideas and produce programs. However, the people who work for the stations have their own ideas, so are not overly eager to involve independents who, in a way, represent competition. The best way for an independent producer to gain an audience at a station is to come in the door with money. Someone who has procured a grant to produce a program or series will be taken seriously.

Obtaining a grant from one of the government or private foundations (National Science Foundation, National Endowment for the Humanities, Markel Foundation) is a specialized type of development process. It involves filling out forms and writing detailed proposals. You are competing, not just with other TV program ideas, but with all types of proposals that relate to a specific subject matter. For example, someone wishing to produce a TV series about Leonardo da Vinci might be competing at the National Endowment for the Arts with a painter wanting to utilize new materials to convey colors and a choral group wishing to tour Europe.

Usually it is easier to obtain a grant if you have connections with a station and the grant folks can be sure your program will actually air. So a sort of vicious circle sets in—a person needs a grant in order to gain station interest, but station interest is necessary in order to win the grant.

The larger public TV stations are more likely to work with independent producers than are the smaller ones. WGBH in Boston, for example, encourages independents to pitch ideas or bring in finished work. If WGBH programmers like what they see, they will buy the finished product from the producer or encourage the person to seek funding so that the project can be produced. For really good ideas, WGBH will join with the independent producer to find financing and then the two will co-produce the idea. Most stations do not have the resources of WGBH, however, so they are not really interested in this type of development. As Roberta K. Smith of KOCE in Huntington Beach puts it, "Producers don't really stand a chance unless they come in with funding or a finished product that can be aired."

Public Radio

Both NPR and APR divide their programming into two departments, one dealing with news and the other dealing with culture. According to Peter Pennycamp, NPR's vice president for cultural programming, "Culture encompasses any material that deals with human beliefs and expressions." APR has divided its cultural programming into two streams: classical music and cultural-variety.

The Think Tank. NPR has a program development method for its cultural programming that is unique to broadcasting. It co-sponsors an ongoing think tank, a group of respected intellectuals who meet for four days once a year to brainstorm about the role of culture in the United States. These people try to break with the historical patterns of thinking, and in so doing directly and indirectly come up with program ideas for NPR. For example, think tank member Bernice Johnson Reagon of the Smithsonian formulated the idea for "Wade in the Water," a series about sacred music in America that traces both its roots and its influences on other forms of music such as Broadway musicals. This series is one tentacle of an overall project on sacred music that also includes a book and a museum exhibit. In this context radio is considered to be an exhibitor, showcasing aural information just as a museum is an exhibitor of pictures and artifacts.

Networks, Stations, and Individuals. Ideas for cultural programming also come from the NPR staff, from stations, and from members of the public. NPR welcomes outside submissions and pitches—no agents needed. Anyone is welcome to send NPR a tape (in essence, a pilot), and be assured that it will be listened to by network personnel. If some submission sounds like a good idea, the programming staff may try to arrange a "marriage" between the submitter and other people within the public radio family (another independent producer, someone in a local station) who may have similar interests. (The financial stakes are low enough that a plagiarism suit is not likely.)

Because NPR's programming is purchased on a cherry-picking basis by the stations, station manager opinions sometimes figure into the development process. Not only will a program lose money if not many stations air it, but it will not have any impact. But many programs enter the production stage without assurances that they will be adequately funded. NPR foots the bill (sometimes up to $400,000—small potatoes when compared to television) and then goes looking for corporate or foundation underwriting.

APR's cultural programming comes from

public radio stations and, to a limited extent, from individuals. A low-key pitch process does take place in that both stations and independent producers need to present their ideas to the network programmers, who then accept or reject them. Because APR does not produce any programs, individuals have the best chance if they have their own funding and production facilities. Their best bet is to work in conjunction with a public radio station. Again, having grant money in hand is the best way to get a foot in the door.

In the area of news, APR and NPR differ widely. NPR develops all its own material while APR uses news produced by outside sources—BBC, CBC, and Christian Science Monitor. NPR's news organization is divided, much as newspapers are, by desks—national, foreign, Washington, and science. The science desk has an advisory board, citizens that meet once a year to give direction and then communicate with the desk at other times. (In general, public broadcasting is awash with advisory boards.)

Bill Buzenberg, NPR's vice president for news and information, characterizes factors that affect the overall approach of NPR's news: "We give more depth and analysis to stories than the commercial networks. Our coverage ranges from things everyone reports to things no one refers to. We also believe in using sound production creatively. For example, often we place music 'buttons' at the end of stories."

Formats. Public radio stations have development elements that are similar to commercial radio stations. Very often they will be format-oriented—jazz, classical music, foreign music, public affairs. In many instances the format is chosen because no other station in the market (including other public radio stations) is using it. The stations then select APR and NPR programs that fit most closely with their formats and program other material of their own. Public radio music-oriented stations are much more receptive to news and features than their commercial counterparts, however. Most of them include a fair number of talk-oriented information programs.

Problems often occur when a university station that has been operated as an independent station to teach students decides to change its format and become an APR/NPR affiliate. Invariably the students are removed from the station staff and professionals are brought in. No longer is there room for the mistakes that naturally occur with learning. Program quality improves, but the students feel disenfranchised and often approach the college administrators to appeal the decision. In most instances they lose. Universities feel they gain more prestige and community support as an affiliated station than as a student learning vehicle.

Development in the public radio area is able to have its unique characteristics in part because the cost of radio is much less than that of television. Then, too, with its cultural mission and its lack of commercials, it can create a different sound. Often it is a place where fun and challenge reign supreme.

FIGURE 5.3 "Echoes," an APR music program that brings a wide variety of styles together, is produced by independent Pennsylvania producers John Diliberto and Kimberly Haas, who work through Pennsylvania public radio stations. (Photo courtesy of American Public Radio.)

• • • • • • • • • • • • • • • • • •

Early in this chapter, we alluded to the testing process to which programs in development are subjected. It is a widely practiced, but somewhat controversial, procedure. Although the merits are disputed, programmers should be familiar with the objectives, techniques, and uses of these tests. That takes us next to the role of testing.

NOTES

1. When buyers deal with producers with whom they have worked successfully in the past, they may short-circuit the process and immediately commit to a script, pilot, or even series. The assumption in the schematic is that no such relationship exists between the two parties.
2. "Networks Hold Line on License Fees," *Broadcasting* (February 10, 1992): 3–4.
3. "The Rise and Rise of Program Prices," *Broadcasting* (September 23, 1991): 44.
4. Interview with office of Phil Corvo, President and COO, NATPE, February 19, 1992.
5. "KTLA Newscast to Vie with Network Morning Shows," *Los Angeles Times* (January 24, 1991): F-5.
6. "CBS to Drop Its Cultural Cable-TV Service After Failing to Draw Needed Ad Support," *Wall Street Journal* (September 9, 1983): 7.
7. "Turner the Victor in Cable News Battle," *Broadcasting* (October 17, 1983): 27–28.
8. "Daytime + CHN: Joining Together to Stay Afloat," *Broadcasting* (June 20, 1983): 37.
9. *Broadcasting/Cablecasting Yearbook 1985* (Washington, DC: Broadcasting Publications, Inc., 1985): F-65–F-92; and *The Broadcasting Yearbook* (Washington, DC: Broadcasting Publications, Inc., 1990): F-72.

6 The Role of Testing

Research is a heavily resorted-to device in the program process. Programmers are reluctant to commit huge amounts of money to productions without some evidence from the public that the project has a reasonable chance of success. That would seem to be unassailable prudence. But it is routinely assailed. Doubters ask the questions: Who's being researched? How reliable are these potential viewers? Who's doing the interpreting? What's the track record for all this statistical stuff?

According to some critics, the history of predictive research in television is abysmal. Dan Enright, a producer of programs ever since the medium's inception, states: "Nearly every network program that sees the light of television has been tested by focus groups and other mass devices. And each has passed its test with flying colors. However, once they hit the air, more programs fail than succeed, contradicting the test findings. Why do we continue to resort to TV program testing when its failure is so monumental? Testing [is] the bane of programming. It's amazing how we still cling to it."[1]

Supporters of research may well argue that failure cannot always be ascribed to faulty testing. Many cancellations are the result of poor execution, inept scheduling, misdirected promotion, or murderous competition. And so it goes, attackers vs. defenders. In this chapter we examine the various techniques used by researchers and explore the strengths, weaknesses, and interpretations. No matter what side of the argument a programmer comes down on, a knowledge of the testing systems and procedures is advisable.

COMMERCIAL TELEVISION

All network programs and most syndicated programs, regardless of the daypart, are subjected to some testing. However, because the stakes are so high, primetime shows go through a constant testing process from initial concept through the first year on the air and sometimes even later. There are a number of firms that do this type of research, the most active of which is Audience Studies Institute (ASI), a company that claims to be able to test virtually anything. In its sales information packet, ASI reports that it is "the largest communications-specialist data research organization in the world." It lists the following client services:

- Television pilots and programs
- Continuing series
- Title, concept, and script appeal
- Movies made for television
- Children's programs
- Target audiences
- Program scheduling
- Personalities: news anchors, program hosts
- Competitive environment
- Station image
- Program and channel awareness
- Tune-in campaigns

In addition, the company can provide "customized" research. For example, it will conduct studies for a client to determine an audience's awareness of a program's existence. Then it will probe to see if those who know it is on have actually sampled it, and, if so, whether they are watching more or less than they used to.[2]

A research design can be developed to investigate any problem confronting a programmer. Its value and reliability are subject to a host of variables, first among which is the quality of the test sample.

Sampling

Program testing requires the recruitment of groups who represent the universe to be measured. For example, if a producer wishes to test the effectiveness of a program with the 18–34-year-old female audience, the testing service must locate a sufficient number of viewers in this category to draw reliable conclusions. In addition, the survey group must offer a balance of economic, social, ethnic, and geographical considerations in order to prevent tilted or biased results.

According to Paul S. Lendburg, executive vice president of ASI Market Research, Inc., "Sampling isn't the most important thing in research today; it's the only thing." The reason is apparent. Even the most carefully constructed questionnaire or most skillfully conducted discussion session will yield worthless results if the respondents are not representative of the whole. Examining forty 18- to 25-year-old females all of whom are the same color and come from the same church group in a community of 300,000 homes is not going to help a producer develop a format for a television show intended for a broad national audience.

Respondents are usually selected by phone. Data available from many sources enable a recruiter to establish the general socioeconomic level of each home to be called. The respondent is taken through a detailed questionnaire, and if the answers fit the needs of the study, the person is asked to participate. Of course, there are financial and practical limitations to assembling a perfectly balanced sample group. Survey participants cannot be flown from Maine to California just to provide a Northeast presence in a test. For that reason all test results must be interpreted by experienced analysts who are able to take into account the skews and tilts caused by sampling imperfections.

After the sample is selected the research is conducted. There are four primary methods of conducting program research: focus groups, mini-theater tests, cable-based studies, and telephone research.

Focus Groups

For this method, the research firm brings together a small group (usually ten to twelve participants) to discuss a programming matter under the guidance of a trained moderator. The subject of the focus group might range from the appeal of a variety of potential hosts for a game show, to the acceptability of the type of language used in a sitcom pilot.

In some cases, the session will begin with a screening of the program to be studied, although focus groups have been used to test program concepts that have not even been committed to paper, let alone videotape. The conversation typically lasts about two hours. The moderator will have a carefully prepared list of topics but will not slavishly follow it. The goal of the session is to stimulate candid comments that will allow a client to hear what the audience really thinks. For later reference the sessions are generally videotaped by an unobtrusive camera.

One of the virtues of focus groups is that they can be conducted in any city with a minimum of equipment and expense. TV sets and VCR units are available everywhere, and the moderator need only bring a tape and pad to be in business. The chief value of a focus session is that the producer or programmer can gain reactions directly from the consumer; the words are not filtered through a computer or hidden in a welter of statistics. The passion, or the tepidness, of the consumers' reactions can serve as guidelines for format adjustment and improvement. Because there are only about twelve participants, the findings are never quantified. However, any unanimously negative expression would be a clear signal that remedial action is called for. A focus session is a hunt for clues and should be used as such.

There are two major concerns about this form of research: the limited size of the group, and the potential for a strong-willed participant to unduly influence the opinions of others. The size of any one group cannot be enlarged without risking a loss of control and candor. However, clients who wish a greater range of responses can lessen the effect of the small sample by ordering additional sessions. The second concern can be more damaging. Dominant personalities do not always reveal themselves in a preinterview and can burst out during the session to the surprise and chagrin of all. Experienced moderators are frequently able to cope with this dismaying situation by repeated appeals "to let everyone have a fair chance to speak." On those occasions when the consumer simply will not cooperate, the session has to be invalidated and a new one scheduled—a waste of time and money to all involved.

More than any other method of research, the focus group must be interpreted by a seasoned analyst. After conducting hundreds of these sessions, moderators become sensitized to group dynamics and are able to detect the most subtle undercurrents. Many times the subtext of the commentaries will be more valuable than the boldly stated opinions. Programmers are well advised to listen carefully to the savvy moderator to learn what the group is really saying.

Mini-theater Research

For mini-theater research, a sample group of people from the desired universe are brought to a small theater where they record their likes and dislikes. Within commercial television, the main use of the mini-theater is to study series pilots. These tests allow the programmer and the production staff to see if the concept is sound and the execution satisfactory. Although many of the directorial flaws, casting mistakes, and other production missteps become apparent when the pilot is privately screened, the executives and producers still have no way of knowing if the public will find the show appealing. For that, some expression from potential viewers is required.

ASI attempts to fill this need with their mini-theater research service. Twenty-five to thirty carefully selected target consumers are invited to a specially equipped facility. On the arm of each seat is a small electronic device that contains a dial and five markings: very good, good, normal, dull, very dull. All of these units are connected to a central computer that records the consumers' choices. These responses can be broken out into virtually any demographic configuration desired by the client.

After participants have been comfortably seated and a moderator has explained the operation of the dialing device, the pilot is presented on a large screen. All the dials have been preset on "normal." As the show unfolds, consumers twist the dial to signify their pleasure or disinterest in the events on the screen. The sum of their reactions is instantaneously calculated by the computer and is converted into a running graph not unlike an EKG readout. The people who have commissioned the study are seated in client booths watching the same film on a special monitor on which the graph plays along the bottom of the screen. Clients are immediately able to see the joke that worked, the heart-tug scene that failed. The entire picture, including the graph, is taped and made available for subsequent, second-by-second study.

Following the screening, consumers are asked to fill out questionnaires that contain personal information and supplementary inquiries about the program. The questionnaires are coded by seat number and can be linked up with the computer readouts of the dial responses. Thus, the client can obtain the reaction of any demographic subgroup in the audience.

Sometimes, approximately ten to twelve consumers are asked to stay behind to participate in a follow-up focus session. The purpose of this final phase is to probe the feelings and reasoning behind the reactions of the respondents. The moderator, of course, has seen the graph and can explore highs and lows in greater detail.

For decades, ASI conducted its theater research on a much larger scale. As many as 300 to 350 consumers were invited to react to programs in a large theater. Approximately fifty of the participants were hooked up to special sensors that measured their pulse and perspiration as the show evolved. The greater these emotional responses, the more effective the show was judged to be. However, the results were ambiguous and difficult to interpret, and the technique was subsequently shelved. The minitheater replaced the large house because virtually the same reliability could be achieved for substantially less cost.

Also, the smaller sample helped to eliminate a misuse of the service. With 300 consumers in the survey, there was a great temptation on the part of clients to seek one number, the average score of all the dials over the course of the screening. ASI has always cautioned against such a simplification of the system, but programmers and producers found it a short, convenient way to interpret the findings. Producers quickly learned that decision-makers were placing great emphasis on that one figure and hit upon a technique to make it work for them. The trick was to insert a high-powered scene very early in the show. The earlier the viewer could be moved off the "normal" reading and into "good" or "very good," the better were the chances that the overall rating would be impressive. Consequently, veteran producers inserted rousing action or sex scenes at the beginning of programs, many times with little or no relevance to the drama, just "to bounce the needle."

They would test the show, edit (or reshoot) scenes, test it again, edit again, and keep up the process until they had engineered a lofty overall score. Then the pilot would be turned over to the buyers for evaluation. The first thing they did was test it at ASI. Sure enough, there it was, a whopping number! A hit was assured; everyone was delighted. Unfortunately, many of those "guaranteed" smashes quickly wound up in the Nielsen graveyard. ASI ended the game by reducing the sample and withholding any composite number. Clients were encouraged to use the service for its original intent, spotting flaws and discovering targets or opportunity.

Cable-based Research

One of the major drawbacks of theater research, at least as perceived by many clients, is the unnatural setting of the test. Television is usually a solitary or family experience, most frequently occurring in the comfort of one's home. Yet the testing takes place in a public facility with the ever-present possibility that crowd response and other distractions might influence judgments. This dissatisfaction led to a search for a method by which people's opinions could be solicited in the normal viewing environment. The solution was cable-based research.

Firms that offer this service develop relationships with cable operators throughout the country. Through these connections, they are able to recruit the desired number of subscribers who are willing to participate in a survey. The consumers are asked to watch the program that will be fed on a certain channel number at a specific time. A reminder call is usually made on the day of the test. Upon the conclusion of the program, an interviewer phones the consumer and administers a detailed verbal questionnaire to obtain the information most desired by the client.

As in other forms of testing previously described, a consumer sample can be assembled to meet any demographic need. Without the distraction of other participants and within the environment of one's own living room, a participant is more likely to offer reliable responses.

Telephone Research

Phone research has limited value for determining reaction to proposed television programs because the consumers, obviously, cannot simultaneously watch a program and answer a questionnaire. However, phone research can be used effectively to find out information about programs that are currently on the air.

Interviewers call appropriate samples of people and ask their responses to set questionnaires. Sometimes the interviewers merely obtain the cooperation of the people and mail them the questionnaires. Of course, this method usually requires a large number of follow-up calls to remind the participants to fill out and mail in the questionnaires.

One of the major phone/mail research studies is TV-Q, a popularity evaluation service. These periodic studies, provided by Marketing Evaluations, Inc., purport to measure the audience's familiarity with programs and personalities and the intensity of their appeal. A nationwide panel of 1600 viewers completes a mailed questionnaire that lists up to 1200 performers and programs.

Results are distributed to clients in the distilled form of two scores, "Fam" (which indicates the percentage of respondents who were familiar with the subject), and "Q" (the percentage who checked "one of my favorites"). Demographic breakouts according to age and sex are also offered in the report.

Subscribers primarily use the data to determine the intrinsic appeal of a show. When the Nielsen rating is low but the quality of the production seems to be first-rate, the programmer will frequently look to the TV-Q for glimpses of encouragement. If the Q score is high but familiarity is weak, it suggests that those who have seen it like it, but too few have seen it. An expansion of audience awareness through a heavy promotion campaign might result in success. If the ratings are low but the Q for a particular demographic unit is high, it could mean a change of time period is called for.

"Without the TV-Q ratings, 'Tour of Duty' and 'Wiseguy' would not have been on this season's schedule," said David Poltrack, CBS's senior vice president of planning and research, in December of 1988. "Their Q scores helped to get them renewed despite low Nielsen ratings."

The application of TV-Q findings to casting decisions is the cause of great concern among performers. The Screen Actors Guild has railed against the service for over a decade. Kathleen Nolan, former president of the Guild, calls it an infuriating catch-22: "If you're not on the air, you're not familiar. If you're not familiar, you're not on the air." Performers believe that a good Q score is a function of the part that is being played. The actor who portrays Saddam Hussein will be less liked than the one who plays General Schwarzkopf.

Program executives generally deny a reliance on TV-Q for casting decisions. Actors and other on-air people remain skeptical. "They shouldn't be," says Tony Barr, a former vice president of dramatic programs at CBS. "In my 11 1/2 years at CBS, I never once even heard the phrase 'TV-Q' mentioned. It simply was not a factor in any talent decisions."

HOME FREE

"Hour Magazine," Group W's long-running (1980–89) women's service show, is a classic example of how research was used for a syndicated series. The program was conceived in testing and monitored by research throughout its ten-year history.

Group W's five owned stations informed their sister division, Group W Productions, that they had a need for a daily 60-minute program devoted to useful information for over forty topics (medicine, diets, mental health, beauty, fitness, etc). The most popular of these subjects served as the core format for the series.

Two pilots were shot, each with a different host. Both were tested and the version that featured Gary Collins scored higher by a wide margin. He won the job. Because Collins had no previous experience as a talk show performer, he traveled to several Group W stations to host their local daily interview programs. All of these shows were taped, edited, and tested to discover Collins' strengths and weaknesses.

After the show went on the air, maintenance studies were conducted every season. This annual tracking of viewer reaction enabled the production company to identify and repair any negative trends before they became fatal. One of the criticisms leveled against the show for its entire history was that it was too slow. The deliberate pace always made general managers and agency executives, by nature hyperkinetic types, impatient. They always urged the producer to "tighten up those spots, move 'em faster." However, the company knew from its maintenance research that the pace was just right. Each year the question was asked, "Is the pace of the show too fast, too slow, or just right?" Every year the response was the same: 12 percent too slow, 6 percent too fast, and 82 percent just right. Viewers repeatedly told researchers they wanted all serious topics treated authoritatively and in depth.

Annually some competitor would launch a syndicated women's service show and triumphantly announce it would move with breathtaking speed, no segment more than 2 minutes, and twice as many topics each day as "Hour Magazine." "I used to love it when I'd read those statements," said Marty Berman, the show's executive producer. "I knew we were home free for another year."

Station Testing

Very little program testing is done by either affiliated or independent stations, primarily because there are so few locally originated shows. Research is confined almost exclusively to news productions. "We do quite a

FIGURE 6.1
A sample page of a TV-Q questionnaire. (Courtesy of Marketing Evaluations, Inc.)

0713-1291
PP286

DECEMBER PROGRAM POPULARITY POLL

a) Your Sex: Male ☐ Female ☐

b) Your Age Group:

6-11 ...	☐	35-49	☐
12-17 ...	☐	50-54	☐
18-24 ...	☐	55-64	☐
25-34 ...	☐	65 and over	☐

On this page, and on the top portion of the following page, we'd like to know **your overall opinion** of each of the television shows listed, as well as **the number of times**, if any, you have watched them in the past 4 weeks.

For each program, **FIRST** answer the shaded column under "Your Opinion":

- If you have **never** watched the program, write in an "N".
- If you have **definitely watched** the program, write in a 1, 2, 3, 4, or 5, depending on how you feel about it.

THEN, for each program you have ever seen, place an "X" in the appropriate box to indicate **how many times** in the past 4 weeks you have watched the program.

YOUR OPINION

Write In:	If the program is:
1	One of your favorites
2	Very good
3	Good
4	Fair
5	Poor
N	A program you have never seen

NUMBER OF TIMES WATCHED IN PAST 4 WEEKS

"X" One Box To Show If You:

0 = Didn't watch in past four weeks
1 = Watched 1 time in past four weeks
2 = Watched 2 times in past four weeks
3 = Watched 3 times in past four weeks
4 = Watched 4 times in past four weeks

	Your Opinion	0	1	2	3	4
MONDAY NIGHT						
Evening Shade	___	☐	☐	☐	☐	☐
Fresh Prince Of Bel-Air	___	☐	☐	☐	☐	☐
Major Dad	___	☐	☐	☐	☐	☐
Blossom	___	☐	☐	☐	☐	☐
MacGyver	___	☐	☐	☐	☐	☐
Murphy Brown	___	☐	☐	☐	☐	☐
Designing Women	___	☐	☐	☐	☐	☐
NBC Monday Night Movie	___	☐	☐	☐	☐	☐
Northern Exposure	___	☐	☐	☐	☐	☐
FBI: The Untold Stories	___	☐	☐	☐	☐	☐
American Detective	___	☐	☐	☐	☐	☐
ABC Monday Night Movie	___	☐	☐	☐	☐	☐
TUESDAY NIGHT						
Full House	___	☐	☐	☐	☐	☐
In The Heat Of The Night	___	☐	☐	☐	☐	☐
Home Improvement	___	☐	☐	☐	☐	☐
Rescue 911	___	☐	☐	☐	☐	☐
Roseanne	___	☐	☐	☐	☐	☐
CBS Tuesday Night Movie	___	☐	☐	☐	☐	☐
Law & Order	___	☐	☐	☐	☐	☐
Homefront	___	☐	☐	☐	☐	☐
Coach	___	☐	☐	☐	☐	☐
I'll Fly Away	___	☐	☐	☐	☐	☐
WEDNESDAY NIGHT						
Unsolved Mysteries	___	☐	☐	☐	☐	☐
Dinosaurs	___	☐	☐	☐	☐	☐
Night Court	___	☐	☐	☐	☐	☐
Doogie Howser, M.D.	___	☐	☐	☐	☐	☐
Brooklyn Bridge	___	☐	☐	☐	☐	☐
Jake And The Fatman	___	☐	☐	☐	☐	☐
Seinfeld	___	☐	☐	☐	☐	☐
Anything But Love	___	☐	☐	☐	☐	☐
48 Hours	___	☐	☐	☐	☐	☐

	Your Opinion	0	1	2	3	4
WED. NIGHT (cont.)						
Quantum Leap	___	☐	☐	☐	☐	☐
Civil Wars	___	☐	☐	☐	☐	☐
The Wonder Years	___	☐	☐	☐	☐	☐
Davis Rules	___	☐	☐	☐	☐	☐
THURSDAY NIGHT						
The Simpsons	___	☐	☐	☐	☐	☐
Pros & Cons	___	☐	☐	☐	☐	☐
Top Cops	___	☐	☐	☐	☐	☐
The Cosby Show	___	☐	☐	☐	☐	☐
Drexell's Class	___	☐	☐	☐	☐	☐
A Different World	___	☐	☐	☐	☐	☐
The Trials Of Rosie O'Neill	___	☐	☐	☐	☐	☐
Cheers	___	☐	☐	☐	☐	☐
Wings	___	☐	☐	☐	☐	☐
Beverly Hills, 90210	___	☐	☐	☐	☐	☐
Knots Landing	___	☐	☐	☐	☐	☐
PrimeTime Live	___	☐	☐	☐	☐	☐
L.A. Law	___	☐	☐	☐	☐	☐
Street Stories	___	☐	☐	☐	☐	☐
FRIDAY NIGHT						
Family Matters	___	☐	☐	☐	☐	☐
America's Most Wanted	___	☐	☐	☐	☐	☐
Step By Step	___	☐	☐	☐	☐	☐
Hidden Video	___	☐	☐	☐	☐	☐
Matlock	___	☐	☐	☐	☐	☐
Best Of The Worst	___	☐	☐	☐	☐	☐
The Carol Burnett Show	___	☐	☐	☐	☐	☐
Hearts Are Wild	___	☐	☐	☐	☐	☐
Perfect Strangers	___	☐	☐	☐	☐	☐
Dear John	___	☐	☐	☐	☐	☐
Baby Talk	___	☐	☐	☐	☐	☐
Pacific Station	___	☐	☐	☐	☐	☐
Reasonable Doubts	___	☐	☐	☐	☐	☐
20/20	___	☐	☐	☐	☐	☐

Please be sure you've written in your opinions

-1-

number of focus sessions over the course of a year on our news personalities," says Bob Brooks, program director of KFOR in Oklahoma City. "It helps us to see if they're in tune with our audience."

Newscasters historically have resented being submitted to focus group scrutiny. They feel they should be evaluated as reporters, not showpeople, and that focus research places too much emphasis on appearance and not enough on journalistic skill. In a perfect world, program executives might tend to agree. But many viewers do base their news selection on the clothes, hairstyle, mannerism, and personality of the anchors, and as long as that is so, decision-makers will take steps to be sure these qualities are present.

Almost all syndicators can produce research by the ton to show why their programs can't miss in any market. Station program directors see no need to conduct their own studies to confirm or deny the syndicator's data. They usually examine the figures with their station rep firm and come to their own conclusions.

FIGURE 6.2
Ted Danson, who played nondrinking bartender Sam Malone on "Cheers," demonstrates his technique to "Hour Magazine's" host, Gary Collins. Repeated audience testing revealed that viewers wanted the show's segments to be thorough and unhurried and that "secret weapon" helped the program enjoy a nine-year run. (Photo courtesy of Group W Productions.)

CABLE TELEVISION

Cable network shows are much less likely to be put through program research than are the commercial networks. The cable networks that produce original programming occasionally test with ASI or a similar company, but generally testing is not considered high priority. Cable origination was born with low budgets that did not include money for testing. As the stakes become higher and the budgets rise, mini-theater testing may become a more common phenomenon.

The type of testing that is most common in cable involves focus groups that discuss overall programming concepts. The pay cable services will gather a group of subscribers to discuss the movie lineup needed to keep the subscribers from canceling their subscriptions. The culturally oriented services will ask about the proper balance of music to drama to documentaries. Usually this type of research is conducted by an outside firm rather than the network itself.

If cable networks want to undertake cable-based research, they can do so easily because they have many systems at their disposal. With little trouble they could arrange for a local access channel to cablecast a program before it is shown nationally. But they don't do this very often.

MTV, however, is utilizing cable systems to test the idea of breaking MTV into three 24-hour channels. It gained the cooperation of five cable systems in four markets to test the structure of a number of different formats, such as all-rap or all–heavy metal.

Several companies are using cable systems to test various types of pay-per-view. For example, a United Artists system near Denver is being used to compare the desirability of scheduled pay-per-view vs. video-on-demand. For the scheduled PPV, every household wishing to watch the movie must begin and end at the same time. For the video-on-demand, up to 47 homes can watch on different schedules, stopping for breaks in the middle if they so desire. But, again, both the MTV and PPV experiments are involved with concept testing rather than individual program testing.

COMMERCIAL RADIO

Commercial radio makes extensive use of testing, especially in relation to the music played over the airwaves. A large number of companies (*Broadcasting Yearbook* lists over fifty of them[3]) exist to assist radio stations, networks, consultants, or station group owners with their specific needs. Some of these specialize in specific aspects of radio programming such as adult contemporary music, small markets, news, or FM, but most of them will conduct research studies into anything the client requests.

Sampling

Because radio is so narrowly focused, the companies that conduct radio research often have difficulty obtaining an appropriate sample. A radio program director trying to learn the type of news items that appeal to the station listeners, most of whom are 12- to 24-year-old males, would obtain little value from a research study if the research company used the general population as a base.

Research companies often call a large number of people and administer a prequestionnaire before they solicit the people for the actual radio testing. This prequestionnaire consists of life-style elements such as age, income, and general radio listening patterns. When more than enough people have been found to fulfill the needs of the particular research, a random group of these people are asked to participate in the actual research. Or, sometimes, when a person has been identified from the prequestionnaire as appropriate to the research, he or she is administered the actual research questionnaire immediately. Of course, if the research is being conducted for a local station as opposed to a network, the sample should be drawn from a particular market.

To solve the sampling problem, research companies or radio stations themselves sometimes use only known station listeners as the universe from which to select the sample. Stations keep lists of people who call in or write, and then contact them when research needs to be undertaken. This sample base is not very effective if the station is trying to gain new listeners, but it is effective if the goal is to please the current listeners.

FIGURE 6.3 Rap music, such as that by Fab 5 Freddy, is being tested by MTV to see if it can constitute its own cable network. (Photo courtesy of MTV Networks.)

Testing Methodology

Telephone research is the most common methodology for radio testing. Unlike video programs, audio musical selections can easily be played over the phone, so people do not need to be brought into a theater where the artificiality of the situation can corrupt the findings. Phone research is usually conducted from a research company's central office that can be located anywhere in the country. The people undertaking the phone interviews are all in one location where they can be carefully trained and supervised. This type of testing is relatively inexpensive.

However, people are not usually willing to spend much time on the phone answering questions, so the number of musical selections that can be tested at any one time is limited. Also, the low fidelity of the phone system can interfere with a listener's enjoyment of a song that might sound much more acceptable on stereo FM. As a result, minitheater testing (usually called auditorium testing in radio jargon) is also used for radio, particularly when a large number of musical selections need to be tested. Typically 75 to 125 people are brought into an auditorium to listen to music.

Focus groups of ten to twelve people are

also used, especially to discuss broad concepts such as proposed format alterations or changes in call letters. Often questionnaires are distributed to participants prior to the focus group to minimize the group psychology or the effects of a dominant personality. These questionnaires are used during the focus groups to bring up ideas that may have been expressed on paper by the more reticent members of the group.

A unique form of information gathering is used by KSRF, a Los Angeles area station that prides itself in breaking new hits. The station has two phone lines, one for requests and one for suggestions. Both are connected to answering machines, and the messages are listened to daily and weighed heavily. As programmer Rachael Donahue puts it, "We are like the king who said, 'My people are in the streets. I must go out and see which way they are going because I am their leader.'"

FIGURE 6.4 The cubicles used for telephoning by interviewers at Denver-based Paragon Research. (Photo courtesy of Paragon Research.)

Research Topics

Radio stations, because of their local and individualistic nature, have a wide variety of research needs. Some of the most commonly tested topics include:

- Format. When a station is contemplating a format change and wants to find out what new direction to take, it will commission research to explore community response to a variety of structures. In addition participants usually listen to tapes that contain sample material of various formats and are asked their opinion of each. Usually this type of research is conducted in focus groups, although a station that has narrowed the format choice down to three or four can use phone research to tabulate opinions.
- Music within format. Although a good program manager will have a feel for which new releases fit within the station format, his or her judgment should sometimes be checked or enhanced by playing the new releases for audience members and seeing which they feel belong. Usually this research is auditorium-based because entire selections are presented, although short samples of the music can be played over the phone.
- Music mix. The order in which various musical selections are offered is another element that lends itself to auditorium research. Do the listeners like three oldies in between each new release, or would they rather have a stronger emphasis on new recordings? Should an instrumental always be followed by a vocal? How often should there be commercial breaks? For this type of research, participants usually listen to several tapes demonstrating different music mixes and indicate a preference.
- Music call-outs. These are conducted to determine which songs should be taken off a station (or a music network such as Unistar) and which should be added. Even though people like to hear their favorite songs over and over, popular music tends to burn out after a short period. Listeners switch the dial when something they have heard too often comes on. By determining ahead of time which songs are about to burn, a station can engage in guardian maintenance and save itself lost listeners by removing the song before it chases away the audience. In a similar vein, listeners can choose the new releases they would most like to hear, and the station can quickly add these to the playlist. This type of research is usually conducted over the phone with small bits of the old and new songs being played and listeners being asked their opinion. For most formats, this research must be directed specifically at people who listen to the particular format

and must be conducted on a weekly basis. In fact, there are those who think music call-outs are useless because of fast burnout. Leon Cleveland, programming head for National Black Network, says, "Radio is so quick, you can't test. Songs are in and out too fast."

- Perceptual call-outs. The term "perceptual" is used to refer to the research of items other than music. For example, phone calls to people to determine the popularity of a new disc jockey, the effectiveness of a currently running contest, the likability of a talk show host, or the desirability of news on the half hour would be considered perceptual call-outs. All of the respondents must be people who listen to the station.
- Life-style analysis. Wise radio station program managers will want to know something about how audience members live their lives when the radio is off in order to better serve them when the radio is on. Life-style research explores such characteristics as education, income, hobbies, opinions on social issues, use of various media other than radio, major purchases, and personal values. Through this type of research, program directors can not only learn those characteristics that set their audience members aside from rival stations, they can also discover particular segments of station listeners who differ on important attributes. For example, a sizeable group of people available on weekends might like sports while those available at 10:00 a.m. weekdays are neutral or negative toward sports. This knowledge might lead the station to adopt a block of weekend sports programming.

MUSIC CALL-OUT
(C) 1990
PARAGON RESEARCH
WXXX
01-01-1999

ALPHABETICAL BY ARTIST

(SCORE / BURN% / UNFAMILIAR% / RANK)

	TOTAL RESP.	TOTAL MEN	TOTAL WOMEN	TOTAL 18-24	TOTAL 25-34	TOTAL 35-44	WXXX CUMERS	WYYY CUMERS	WZZZ CUMERS	WXXX FANS	WYYY FANS	WZZZ FANS
B-52's ROAM	70.83 0.00 27.02 7	73.75 0.00 35.48 3	69.11 0.00 20.93 7	70.00 0.00 33.33 7	66.17 0.00 15.00 7	76.47 0.00 29.16 3	70.83 0.00 27.02 7	77.27 0.00 35.29 2	53.12 0.00 11.11 7	75.00 0.00 30.61 4	70.19 0.00 21.21 6	57.14 0.00 41.66 7
Bad English PRICE OF LOVE	73.61 5.40 27.02 3	71.05 6.45 38.70 4	75.00 4.65 18.60 4	80.00 3.33 33.33 1	66.66 15.00 25.00 6	72.36 0.00 20.83 6	73.61 5.40 27.02 3	68.75 17.64 29.41 5	78.57 0.00 22.22 3	73.12 6.12 18.36 6	73.80 3.03 36.36 3	68.75 16.66 33.33 6
Jackson, Janet ESCAPADE	79.05 8.10 0.00 1	76.61 9.67 0.00 1	80.81 6.97 0.00 1	78.33 6.66 0.00 2	77.50 10.00 0.00 2	81.25 8.33 0.00 1	79.05 8.10 0.00 1	80.88 11.76 0.00 1	86.11 0.00 0.00 1	81.63 8.16 0.00 1	77.27 9.09 0.00 1	75.00 8.33 0.00 3
Madonna KEEP IT TOGETHER	73.10 4.05 10.81 4	61.53 9.67 16.12 7	80.62 0.00 6.97 2	71.29 6.66 10.00 5	73.43 0.00 20.00 3	75.00 4.16 4.16 4	73.10 4.05 10.81 4	68.33 5.88 11.76 6	85.71 0.00 22.22 2	73.93 4.08 4.08 5	70.53 0.00 15.15 5	77.77 8.33 25.00 2
Motley Crue WITHOUT YOU	72.45 6.75 20.27 5	69.79 9.67 22.58 5	74.28 4.65 18.60 6	78.12 10.00 20.00 3	70.00 5.00 25.00 5	67.50 4.16 16.66 7	72.45 6.75 20.27 5	75.00 5.88 17.64 3	78.57 0.00 22.22 3	71.15 6.12 20.40 7	75.00 6.06 24.24 2	69.44 0.00 25.00 5
Myles, Alannah BLACK VELVET	77.04 6.75 17.56 2	76.04 6.45 22.58 2	77.70 6.97 13.95 3	72.00 13.33 16.66 4	82.81 5.00 20.00 1	78.75 0.00 16.66 2	77.04 6.75 17.56 2	71.42 11.76 17.64 4	75.00 0.00 11.11 5	80.48 8.16 16.32 2	71.15 6.06 21.21 4	79.16 8.33 50.00 1
Penn, Michael NO MYTH	72.41 2.70 21.62 6	69.00 3.22 19.35 6	75.00 2.32 23.25 4	70.45 0.00 26.66 6	73.21 10.00 30.00 4	73.86 0.00 8.33 5	72.41 2.70 21.62 6	65.38 0.00 23.52 7	66.66 11.11 33.33 6	76.19 2.04 14.28 3	67.39 6.06 30.30 7	75.00 8.33 41.66 3

FIGURE 6.5 A music call-out sheet prepared by Paragon Research. The sheet itself is color-coded. "Escapade" and "Black Velvet" are printed in green to represent that they are very strong (the numbers indicate that, too). "Roam" is in red to indicate that it is weak. The rest are printed in black, which means they are holding their own and doing all right, but not spectacular. (Courtesy of Paragon Research.)

Life-style analysis is easily done through telephone interviews.

In addition to these major forms of research, radio stations often want to explore other areas. They may desire research regarding new call letters, a new logo, the feasibility of a particular promotion campaign, or any other of a variety of subjects related to the overall programming.

Qualitative research is very important to the radio business, especially in markets where the competition is keen. Program managers who do not keep up with the latest thoughts and preferences of the listening audience will quickly lose touch—and their jobs.

PUBLIC BROADCASTING

Testing is not high priority in public broadcasting, in part because of its cost and in part because public broadcasting shows do not live or die by ratings. Sometimes the producers of PBS programs (primarily the public TV stations) will test shows just as commercial producers do, but many shows hit the airwaves without benefit of testing.

One major exception to this is the programming that comes from Children's Television Workshop. This organization undertakes major statistical pretesting to determine not only how well children like their programs but how much they learn by watching.

On occasion a public TV station will pretest a local concept. For example, in 1991 WGBH in Boston decided to program a talk show without a host. Since this was a rather revolutionary idea, the station tested the idea over and over all summer and made many changes before launching the show in the fall.

The Corporation for Public Broadcasting (CPB) holds focus groups around the country to discuss the overall impact of certain programs and how public broadcasting can better serve the public. Sometimes local stations piggyback onto these focus groups and arrange to have the participants queried about the local programming. Advisory boards, although they do not constitute a formal testing audience, are often used as a sounding board for ideas and programs.

APR conducts a limited number of focus groups, some before the program is aired and some after. It has separate focus groups for audience members and for station program directors. Often very different results emerge from these two groups. "We use these studies to convince program directors that they shouldn't program for their own taste," notes Dan Jensen, communication specialist at APR.

NPR has a policy of not testing programs until they are on the air. As NPR's Peter Pennycamp explains, "The program is given several months to develop and shake out, and then focus groups are brought in." This time factor constitutes a major difference between commercial and public broadcasting. Because public radio is not dependent on advertiser support and ratings, it has the luxury to let something grow on its own.

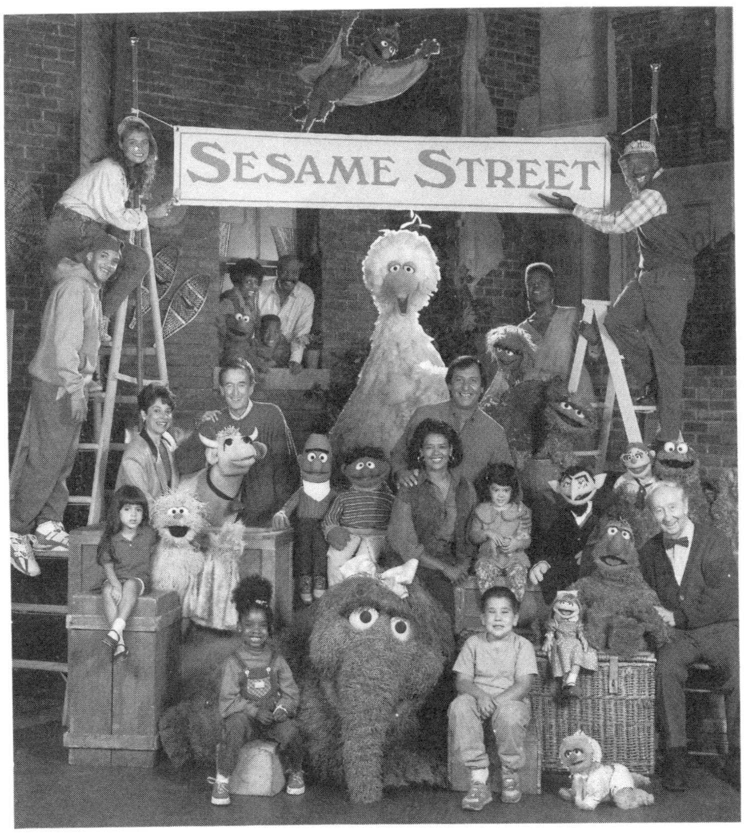

FIGURE 6.6 "Sesame Street" underwent intensive testing before its debut in 1969, and elements of it are still tested on a regular basis. (Photo courtesy of Children's Television Workshop.)

FIGURE 6.7
"All in the Family," with Carroll O'Connor playing the "lovable bigot" Archie Bunker, dominated the primetime ratings in the 1970s. But when the pilot was tested, the show's breakthrough language and story lines puzzled and annoyed respondents, who gave it a failing grade. (Photo courtesy of the Academy of Television Arts and Sciences.)

DOES THE RESEARCH WORK?

Attitudes toward whether or not research works differ widely, and sometimes passionately. Some producers consider qualitative research a valuable tool in the shaping of their productions. Others think it's a pestilence. They would rather put their faith in the opinion of one experienced staff member than a battery of audience tests. Gene Reynolds, producer of many popular shows including "M*A*S*H," a multi-Emmy award winner, refused to even read a packet of research material that a network had provided on a new series he was about to launch. He had his own vision of the series and was more confident in his own instincts than in the reactions of untrained critics.

It is a matter of record that the pilots of two of the biggest hits in the history of network television received moderate to low test scores. "Batman," a late-1960s weekly primetime series based on the cartoon character (later converted into a major motion picture success), was produced in a broad, somewhat tongue-in-cheek fashion. With its colorful villains and larger-than-life leads, it zoomed to the number one position in a matter of weeks. But prior to its premiere, it fared abominably in a large theater test. As one ABC program executive recalls, "The theater audience was puzzled about what they were looking at. Was it a comedy, an adventure, a combination of both, or what? So we ordered a second test and had the moderator come out before the screening to tell them it was a put-on, it was fun, it was OK to laugh. And they did. The second score was very good. We changed all our promos overnight to stress that the shows were fun. The spots positioned the audience properly and we had an instant hit."

"All in the Family" was a show unlike any other that had ever appeared on television. The protagonist, Archie Bunker, was a bigoted, racist male chauvinist who referred to minorities in derogatory phrases and told his wife to "stifle" herself. The reaction of the test audience was primarily bewilderment with touches of outrage, and the scores were decidedly subpar. Nowhere in the survey findings were there any indications that the show would become an enormous hit.

The experiences with "Batman" and "All in the Family" led ASI management to admit that their findings were most reliable when they could be measured against the norms of other programs of similar and familiar content. The service was least effective when attempting to evaluate unique program concepts for which no normative values had been established. Translation: the system can be unreliable if the show presents something completely innovative.

Programming is an inexact art. The process is filled with hunches, feelings, guesses, and prayers. Any hard facts that can be introduced into the mix can help to reduce the number of variables and lead to more reliable decision-making. Solid data should be encouraged. So-called "visceral" programmers should welcome information that will contribute to more enlightened judgments. But programmers would do well to heed the words of ASI's Paul Lendburg: "The purpose of qualitative research is to identify opportunities and help eliminate flaws, thereby reducing risks."

Research holds a potential danger: any number, no matter how questionably arrived at, may be considered more valid than a seasoned opinion that is unsupported by charts and graphs. All research data should be given their proper weight and added to the host of other considerations that enter into a programming decision. Many of these other factors are dealt with in the following two chapters, which discuss elements that help make for successful programming.

NOTES

1. "Let's Replace TV Audience Testing with Ingenuity and Daring," *Los Angeles Times* (March 23, 1992): F-3.
2. *ASI Entertainment Research Promotional Packet* (Burbank, CA: ASI Market Research, Inc., 1991): passim.
3. *The Broadcasting Yearbook* (Washington, DC: Broadcasting Publications, Inc., 1991): I-25–I-30.

7 Elements of Successful Programming: Part I

In one respect programming is no different from any other marketed product: success is measured by the achievement of an objective. In most cases, the goal of programs is to attract the largest possible audience. But not always. As we shall see, some shows have a different purpose. It is imperative that a programmer know what a production is to accomplish before the development process begins.

After the target is set, the programmer must decide the type of appeal that will best attract the audience. Can the viewers/listeners best be reached by laughter, or adventure, or calming thoughts? And once that motivation is selected, what elements should go into the show to give it the best chance of working?

Even if all three of these stages are carefully considered by the programmer, there is no guarantee of success. Audiences are notoriously unpredictable, their tastes mercurial. But one thing is certain: no show ever succeeded by ignoring its objective, arbitrarily selecting its appeal, and dismissing essential elements.

COMMERCIAL TELEVISION

Because commercial television must be responsive to advertisers, the overriding elements for success have to do with delivering an audience for the advertisers. But other, more altruistic, objectives can be present.

Programming Objectives

First, let's examine the various goals that may be established for a program.

Widest Possible Audience. Television is a mass medium. It is far and away the principal leisure-time activity of the nation. It is also a mass advertising device. Corporations spend hundreds of millions to stimulate desire for their products. Therefore, the objective of most shows is to attract the largest possible audience. The more viewers, the higher the advertising rate, the greater the gross revenue, the larger the profit. Conversely, no audience, no advertisers, no profit, no broadcaster.

A Specific Target Audience. Quite frequently a broadcaster (or advertiser) is more interested in reaching a particular segment of the audience rather than a large body count. For example, a luxury automobile company may have affluent males as its principal advertising target. This marketing goal will be better achieved through commercials in a golf tournament than in a broad-based situation comedy whose larger audience consists primarily of women and children of lesser means. However, broadcasters must be careful about scheduling programs tailored to fit the specialized needs of a sponsor. The "demographic show" with its highly focused but limited appeal may substantially reduce the audience size of the programs that follow. The consequent revenue loss may exceed the profit from the specialized presentations.

The Fox network was founded on demographic programming. In order to attract a commercially successful audience, Fox had to reach viewers who were not being satisfied by the three long-entrenched networks. The company focused on the 18- to 34-year-olds, the demographic group most desired by advertisers. Garth Ancier, former president of Fox Broadcast Entertainment, claims, "You no longer have to have a 30 share; you can have a narrow audience [and still succeed]. Young-appeal shows such as 'Married . . . With Children' and 'In Living Color' are two successful examples of the strategy."[1]

In fact, there is nothing new about this approach. Leonard Goldenson, former president of ABC Companies, Inc., says it reminds him of the early days of his network (the mid-1950s) when he began with fourteen stations, and both CBS and NBC had over seventy affiliates. "They're really copying the same thing we did when we started. It's a sound principle—going after young families, trying to counterprogram wherever you can, trying to come up with some kind of innovation—and take chances."[2]

Children, ethnic groups, and infrequent viewers are all part of a broadcaster's constituency. Programs designed for their interests must be represented on the schedule. Some services cater totally to minority audiences, and, although most of these are trying to make a profit, they also consider service a primary objective. A number of UHF independent stations program in one or more foreign languages to meet minority needs.

Prestige. On occasion a commercial broadcaster may wish to present a program or series whose sole objective is to gain prestige. A superb documentary such as "Blacks in America," or a quality drama such as "Antony and Cleopatra" appeal to relatively small audiences, but they are appreciatively reviewed and redound to the credit of the network and/or station. Some broadcasters present these shows out of a sense of obligation; others do so for more practical reasons such as a need for improved public relations or a difficult upcoming license renewal hearing. Regardless of the motivation, they are rare but welcome additions to the schedule.

Awards. Awards are conferred for programming excellence. Broadcasters who commission prestigious productions usually are also seeking recognition in the form of awards and honors. These tributes are an integral part of the goal to raise the stature of the network or station. More often awards are the by-product, not the objective, of excellence. Many of the Academy of Television Arts and Sciences Emmy awards are presented to avowedly commercial series or episodes that were produced with exceptional quality.

Although it is unusual, some series have been known to invest additional money in an episode in an attempt to make it worthy of award consideration. These efforts are not just for ego gratification. They can produce highly tangible results. "Hill Street Blues," a superbly produced and acted series, was faring poorly in the ratings in its first season. However, at Emmy time it walked off with eight awards (a single-season record for a series) and the subsequent excitement attracted many viewers. It went on to enjoy a much-saluted six-year run.

SAMPLE AWARDS

Here is a partial list of awards that electronic media practitioners can strive to attain.[3]

- Academy of Television Arts and Sciences Emmy Awards—outstanding achievement for nighttime programs shown nationally.
- Achievement in Children's Television Awards—significant contribution towards improving children's radio and television.
- American Women in Radio and Television Awards—excellence in portraying a positive and realistic image of women.
- Alfred I. duPont–Columbia University Awards—outstanding work in news and public affairs.
- Gabriel Awards—serving viewers and listeners through the positive, creative treatment of issues of concern to humankind.
- IRE Annual Awards for Investigative Reporting—initiative, originality, and persistence in breaking down secrecy in coverage of significant matters.

Iris Awards—outstanding local television programs.
Robert F. Kennedy Journalism Awards—outstanding coverage of the disadvantaged.
The Livingston Awards for Young Broadcast and Print Journalists—best coverage of news by journalists aged 34 or younger.
National Academy of Television Arts and Sciences Emmy Awards—outstanding achievement in daytime television.
The Ohio State Awards—excellence in educational, informational, and public affairs broadcasting.
George Foster Peabody Awards—achievement in news, entertainment, education, programs for children, documentaries, and public service.
RTNDA Edward R. Murrow Awards—overall excellence in news and documentary coverage.

Particular Local or National Purposes. Every market has regional concerns. The responsible station broadcaster will present programs that address area problems. Sometimes this is altruism and sometimes it is necessity. Under the Congressionally passed Communications Act, broadcast TV stations are required to show that they are acting in the "public convenience, interest, and necessity" in order to keep their licenses. A conscientious programmer will take this responsibility seriously and program something that is truly of value to the community rather than the first inexpensive idea that comes along.

When a disaster, such as a flood or earthquake, hits a local area, the media has an obligation to bring as much information and aid to its audience members as possible. This becomes a temporary, but overriding, objective. The same is true nationally when there is some major event such as the Kennedy assassination or the Persian Gulf conflict.

Fundamental Appeals to an Audience

After programmers have identified the objective, they must select the appeal through which the audience will be reached. There are two basic conduits from which to choose: information and entertainment. Information can be divided into a current up-to-date report of events (hard news) and a more general combination of facts, gossip, and opinion (soft news). Music, comedy, and drama are the three broad areas of entertainment.

Hard News. Year after year people report that TV is their main source of news and the medium they are most likely to trust if they hear conflicting stories. Networks and stations invest a large portion of their time and money to provide comprehensive news services. Although local news can be extremely profitable for a station, the news operations of the networks have traditionally been bathed in red ink. In recent years the new ownerships of the networks have been paring staffs and costs to bring their news departments in line with economic realities.

Sports events, which provide hard news and entertainment, are also undergoing major adjustments. The reduced audiences for network presentations no longer allow the networks to bid astronomical amounts for major events. Executives in free TV and cable predict that the manner in which sports are brought to the TV public will change dramatically in the next ten years.

Soft News. Human beings are inquisitive. They want to know what's going on with others and within themselves. Programmers can tap into this curiosity to obtain sizeable audi-

FIGURE 7.1
This chart shows the percentage of people who have indicated which media form they feel is most believable. Sources: *Public Attitudes Toward Television and Other Media in a Time of Change* (New York: Television Information Office, 1985): 5; *America's Watching* (New York: Television Information Office, 1987): 5; and *America's Watching 30th Anniversary, 1959–1989* (Washington, DC: Television Information Office, 1989): 15.

	1959	1961	1963	1965	1967	1968	1971	1972	1974
Most believable:	%	%	%	%	%	%	%	%	%
Television	29	39	36	41	41	44	49	48	51
Newspapers	32	24	24	23	24	21	20	21	20
Radio	12	12	12	8	7	8	10	8	8
Magazines	10	10	10	10	8	11	9	10	8
Don't know/ no answer	17	17	18	18	20	16	12	13	13

	1976	1978	1980	1982	1984	1986	1988
	%	%	%	%	%	%	%
Television	51	47	51	53	53	55	49
Newspapers	22	23	22	22	24	21	26
Radio	7	9	8	6	8	7	7
Magazines	9	9	9	8	7	6	5
Don't know/ no answer	11	12	10	11	9	12	13

ences. TV talk and magazine shows, such as "The Oprah Winfrey Show," "Donahue," "Geraldo," and "Entertainment Tonight," provide a wealth of information, or at least factoids, to satisfy this interest.

Music. Whether we need it to soothe a savage breast or just gentle away the time, music has always been a staple of the electronic media. It is deep, visceral, and fundamental.

Although radio is currently the dominant outlet for music, television is still a source of considerable exposure for artists. Some stations, following cable channel MTV's lead, present music videos. Such one-time events as the Grammy Awards and the American Music Awards always attract top ten ratings. However, in today's market, a performer, no matter how gifted, can no longer simply stand up and sing a hit number. It is visually too unsatisfying. The hundreds and thousands of music videos, some costing in the seven figures, have created a consumer demand for rich effects as well as artistic excellence. This reality has all but made the TV variety show extinct. The programmers who wish to present music shows would do well to be cautious. They should select an artist with wide demographic appeal, supply as much visual support as possible, and not let a sequence go too long; the audience may not be around for the ending.

Comedy. Comedy plays to a viewer's constant need to escape, to relieve tension, to feel better. The comedic form of choice in commercial television is the sitcom. Core characters are placed in a primary arena and respond in each program to a story stimulus. If the personalities of the principal characters are carefully constructed and sufficiently conflicting, almost any story premise has the potential to produce funny scenes. To programmers, one of the great attractions of the sitcom is its durability, its capacity to provide endless variations off the same premise.

Drama. Alan Landsburg, one of television's more successful producers in all programmatic forms over four decades, claims that there are really only three themes that count in storytelling: sex, money, and power. These are the appeals that compel people—the subjects that fascinate and motivate them. Any drama, or comedy, that explores these qualities is on a solid footing.

Sex should be understood in the larger sense of love, romance, and the eternal quest for the "right other person." This is certainly the most employed of all story appeals because of its universality. For example, in an episode of "Evening Shade," a poetic woman mechanic hopes to "get Henry's engines running with a series of erotic letters, causing a breakdown in Herman's relationship with Margaret."

Episodes about money touch all of us. The drive to get it or the fear of not having it are daily concerns. There is also the dream of sudden riches that makes us empathize with those who have come into prosperity. (In "Major Dad" Holowachuk and Elizabeth have dreams of getting rich because someone teaches them a few tricks about the commodities market.) TV game shows have been trading on the money instinct for decades. Lottery winners are news. Viewers want to see them and fantasize that they'll be next.

"Power," according to Henry Kissinger, "is the ultimate aphrodisiac."[4] People strive for it, fight for it, lie for it, and kill for it. It has been the stuff of drama from Shakespeare to "Dynasty." Pick up any weekly schedule of TV programs and read the capsule descriptions of dramatic episodes. You'll find dozens that have a power struggle as their foundation.

A host of other emotions and appeals, e.g., sympathy, revenge, nostalgia, and acquisitiveness, are valid storytelling themes but they can largely be covered by the more fundamental categories of sex, money, and power.

It should be noted that these basic themes apply to all dayparts and all forms of storytelling. Serials, whether daytime or evening, late-night adventures, and movies-of-the-week all use variations on the "big three" themes in a vast majority of their productions. So do many children's shows, albeit in more subdued form.

Qualities Tied to Success

Ever since the early days of radio, programmers have sought a "formula" for a hit. They've tried to "research" a winning combination; they've attempted to promote their way into success; they've even consulted clairvoyants and sensitives. The magic key has remained elusive, for a very good reason. There is none.

Success is a strange alchemy of timing, hunch, hard work, anticipation, professionalism, and luck. This is true for one particular show, a series, or an entire programming concept, and it holds true whether the program is format-dominant or star-dominant (see Box).

TWO BASIC FORMS OF PROGRAMMING

All programs, regardless of their objectives or their appeals, originate in one of two forms:

1. Format-dominant. The concept of the show is the key to its success; performers are selected to fulfill the requirements of the core idea.
2. Star-dominant. The star is the key ingredient; a format is designed around the skills of the lead performer.

A classic example of a high-concept, format-dominant program was "The Six Million Dollar Man," an action-adventure series in the late 1970s. The show was based on the fanciful idea that a badly injured astronaut could be rebuilt with atomic-powered electromechanical devices that would make him capable of superhuman performance. Stories were developed more to demonstrate dazzling physical feats than to provide character insights. The principal role mainly required handsome athleticism, and Lee Majors was chosen from a sizeable list of candidates.

More recent examples of format-dominant shows include "Northern Exposure," "Family Matters," "Unsolved Mysteries," and "Designing Women." Under ideal circumstances performers in these types of programs will add values that may not have been originally evident and make the series even more appealing. For example, in "Murder, She Wrote," Angela Lansbury catches just the right flavor of the Jessica Fletcher role and contributes many extra pleasures to each episode.

Sometimes a program will start out as format-dominant, but will become a star vehicle over time because a performer has shaped the show to his or her own designs, not always to the project's benefit. "Family Feud" began as a rigidly structured game show with the host serving as an affable traffic cop. But not long after the premiere, Richard Dawson changed the focus by kissing every woman contestant, reading his fan mail, displaying presents he had received, and generally favoring the audience with his view of world affairs. Many episodes required hours of editing in order to present a completed game.

The other type of show is one built around the skills and personality of a star. This is most clearly seen in talk and magazine programs such as "The Arsenio Hall Show," "Geraldo," and "The Oprah Winfrey Show." However, it also applies to fictional formats that are devised to display the performer's abilities, e.g. "The Cosby Show," "Roseanne," and all of Bob Newhart's series.

Even though a star may be exceptionally talented and enjoy a large following, the development of an appropriate format can be elusive. Such illustrious names as Mickey Rooney, Frank Sinatra, James Stewart, Jackie Mason, and Tony Curtis have stumbled in their TV series either because they weren't right for the medium or the formats weren't right for them.

FIGURE 7.2 "The Six Million Dollar Man" was the definitive example of a so-called "high concept" series. The dominant appeal of the show was the ability of the principal character to perform prodigious athletic feats because his body had been almost entirely rebuilt with bionic parts. Lee Majors, shown here, played the lead role. (Photo courtesy of MCA Television Group.)

The strength of star-dominant programs is that they automatically define their appeal. If you know the performer, you pretty much know what to expect, and sampling time is greatly compressed. On the other hand, if a star-dominant show begins to slide (or the artist is imposing impossible demands), there is no way to salvage it. You can't replace Geraldo Rivera and still call the series "Geraldo."

Although both format-dominant and star-dominant shows are capable of delivering big hits, buyers tend to favor the former. Their reasons are summarized in the experience of "You Bet Your Life," starring Bill Cosby. This syndicated series was bought by over 200 stations purely on the basis of Cosby's enormous success on the network sitcom, "The Cosby Show." "You Bet Your Life" had served as a creaky but workable format for the rapier wit of Groucho Marx in the 1950s, but when it was dusted off for Buddy Hackett in 1980, it was an instant failure. No matter. This was the 1990s and the host was Bill Cosby. How could it miss? It did, and badly. The series was canceled after one season. According to Greg Meidel, president of Twentieth Television, there was a lesson to be learned: "Format first, star second." You can have a proven big star, but you really have to have a proven format. It was a great personality in the wrong vehicle.[5]

FIGURE 7.3
Bill Cosby (shown here with Carol Burnett) was a hit in "I Spy" in the late 1960s and won an Emmy for his performance. But for reasons difficult to fathom, he didn't score big again until "The Cosby Show," which began in 1984. (Photo courtesy of the Academy of Television Arts and Sciences.)

Why did "Wheel of Fortune," after a mediocre performance as a network daytime series, become the biggest winner and money-earner in the history of syndication? Why has Dick Van Dyke, a beloved entertainer with one of the highest-rated and most admired series of the 1960s, never again been able to duplicate his success? Why did Rick Dees, a popular radio disc jockey for years, fail on TV?

Some industry leaders believe successful shows are mostly a matter of luck. In the opinion of Dick Robertson, president of Warner Brothers Domestic Television Production: "Most hits are flukes. Research can steer you more away from an obvious failure than it can steer you into a hit. If you have something that's very obviously off the mark, research can sort of tell you that. But if it's a hit, it's hard to tell. Therefore, the more times you go to bat, the more chances you have of getting a hit. We've been in the business of doing first-run shows for twelve or thirteen years now. 'The People's Court' and 'Love Connection' are two very successful shows. That's two out of a lot of programs."[6]

Despite the impossibility of predicting a hit, there are certain qualities that winning programs or concepts possess. Their presence doesn't guarantee success, but their absence almost always assures failure. Of course, it isn't as simple as going down the checklist and inserting each element into the mix. There is such a thing as artistry at work, no matter what the ingredients. A script written by James Brooks or the Charles brothers is going to be vastly superior to one written by ordinary talents. However, success embodies some combination of the following qualities.

Conflict. By far the most important ingredient in all programming is conflict. Without the collision of interests or attitudes, there is little to hold the attention of viewers. This is true even where it is less obvious, e.g., talk and comedy. The genius of successful comedy creators is the ability to design characters so sharply that any stimulus immediately places them in conflict. The laughter results from the contrast in the attitudes. In "Cheers," Sam relentlessly searches for a night of romance. Rebecca abhors casual lovemaking. Any reference to sex automatically triggers an amusing exchange between the principals. The talk shows that produce the most interest are the ones with guests who have strong opinions that conflict with other panelists or members of the audience.

The goal of programmers is to construct shows that provide collisions. Even when mounting an information show, programmers should attempt, whenever possible, to offer opposing or alternative attitudes. It is not only fair play; it is good television.

Durability. An idea that is intended to be a long-running success cannot be based on a premise that will flame out after a brief exposure. Viewers are quickly bored by gimmicks. Programmers must examine submissions carefully to be sure that the fundamental design of the show will sustain interest week after week

In the early 1970s, ABC presented "Longstreet," an hour-long weekly series that dealt with the adventures of a blind insurance company investigator. Because of his loss of sight, Longstreet had to solve all of his cases through his other heightened senses, e.g., fingering fabrics to determine a killer's clothes or hearing a distant snapping of twigs to alert companions of impending danger. Despite a fine cast and an excellent creative staff, the show survived for just one season. The concept was simply too limiting to sustain interest.

A classic nondurable form in drama is the "island" premise. Place a group of people in a distant location, surround it with water, shut them off from all communication and hope of rescue, and let their interaction provide the conflict. It doesn't work. "Swiss Family Robinson," another ABC series, was an example of this genre. It was introduced in September 1975 and disappeared the following April. There were just so many stories that could be told (the obligatory hurricane, the sudden discovery of hostile natives, the long-lost shipwrecked sailor found on the western shore) before the writers ran dry. Even the redoubtable Aaron Spelling was unable to pull off an island series. "The New People," a story about a planeload of young people who crash-landed in a remote Pacific island, slipped silently to the ocean floor after 13 weeks.

It is possible to have an island story work in cartoons or farce comedy ("Gilligan's Island") because viewers treat the premise as a springboard for fun and willingly suspend their disbelief. But drama requires consistency in plot logic and deviations destroy an audience's involvement.

One element that often helps create durability is the opportunity for many different characters to appear within particular episodes of a series. For example, in "Cheers," all types of characters can wander into the bar to interact with the regulars. In "Murder, She Wrote," Angela Lansbury can meet up each week with a new group of people who somehow manage to involve her in a murder mystery. Even if outsiders aren't used, a large regular cast can allow the main story to revolve around different characters. In "Eight Is Enough," the eight children and two adults could take turns being featured in programs. This gave a greater variety of plot possibilities than had there been only two children.

Of course, this doesn't always hold true. Large-cast shows such as "Cop Rock" and "Kay O'Brien" have vanished quickly. Conversely, well-written, small-cast series such as "Family Ties" and "Who's the Boss" have had long runs.

Even the most durable of episodic ideas eventually run their course. "M*A*S*H" and "The Mary Tyler Moore Show" both had the good grace to remove themselves from the airwaves voluntarily before their durability exhausted itself.

Information and news-based shows are, of their nature, durable. Since the events of the day and the people who make the news are ever-changing, the programs that report these activities can remain fresh indefinitely. The only threat to these series, aside from competitive challenges, is staff fatigue and boredom. Even "60 Minutes" went through one short period of lethargy and soon found itself outside the top ten ratings. Some spirited words from executive producer Don Hewitt quickly put the show back on track.

Syndication plays such an important role in the finances of network series that a show must be able to hold up for many years after its last episode has been shot. The success of "The Dick Van Dyke Show" in syndication, says the star himself, "is attributable to the wisdom of writer-director Carl Reiner, who had the foresight not to do any slang or idioms of the day or topical events." Trendy material rapidly becomes dated and hurts the long-term appeal of a program.

Whenever possible, programmers should think of specials as something more than just one-time-only productions. Many of these events have the capacity to be annual features. Some of the most highly rated specials—"People's Choice," "American Music Awards," "Circus of the Stars"—are made-for-TV events that are capable of strong performances year after year.

No discussion of durability would be complete without a bow towards the best of the breed, soap operas. "As the World Turns" has been on continuously since 1956, "General Hospital" since 1963; "Another World" premiered in 1964 and "Days of Our Lives" came on a year later. But the all-time champion is "Guiding Light," which celebrated its 40th consecutive year on the air in 1992. All of this longevity can be attributed to the remarkable abilities of such writers as Irna Phillips, Agnes Nixon, and Bill Bell, all of whom mastered the art of intertwining stories about sex, money, and power.

Likability. Viewers tune to people they like and with whom they feel comfortable. Bill Cosby, Alan Alda, Angela Lansbury, Michael J. Fox, Betty White, and Ted Koppel exemplify the personalities who generally succeed in the medium. They come across as warm, compassionate, sensitive, and friendly—the kinds of guests you wish to invite into your home.

FIGURE 7.4 "Murder, She Wrote" is a program that basically does everything right. Jessica Fletcher, as played by Angela Lansbury, is warm and likable and viewers know she will be involved in a murder mystery every week. The series is so durable that it bids fair to stay on as long as the star wishes to continue in the role. (Photo courtesy of MCA Television.)

Nowhere is the quality of friendliness more evident than with game show hosts. Some are more amusing, or better looking, or more clear-headed than others, but they all radiate cheer and good will. It is impossible to imagine a successful emcee who is downbeat, surly, and insulting to his contestants. Within a week viewers would run him out on a rail.

One might argue that Archie Bunker, J. R. Ewing, and Erica Kane ("All My Children") are hardly likable yet have developed enormous followings. However, it is important to understand they are playing the role of the burr under the saddle, the person you love to hate, in order to provide the conflict that makes the series work. And even Archie would occasionally let down his guard to allow some endearing humanity to slip through.

The amusing but insult-tossing cast of "Married . . . With Children" also appear to be unlikable. But the audience seems to understand that underneath the family's surface disrespect is a genuine affection. When the pilot program was tested, the 30 minutes of sarcasm ended with the couple walking up the staircase, the wife's hand gently patting the husband's behind. Viewer approval soared, their suspicion confirmed that deep down the spatters really loved each other. Not so with "Buffalo Bill," an arrogant, sardonic sportscaster (played by Dabney Coleman). Bill lied shamelessly and used all his friends. He gave the audience no reason to suspect any repressed decency. The series staggered through one season and disappeared.

Any programmer planning to launch a show with truly unlikable characters should be sure of two things: 1) the material is very funny; 2) a backup program is readily available.

Consistency. All viewers bring a certain level of anticipation to every program. When they tune to "Roseanne," they expect brassy humor and a touch of warmth; when they dial "Murder, She Wrote," they assume they'll see an intricately plotted murder mystery that Jessica will unravel in the last act. Deviations from these expectations disturb viewers and risk alienation. An occasional departure can be successful if the episode has a legitimate point to make. Comedies have tackled such serious themes as AIDS or the death of a parent and done them effectively. But these must be produced carefully and at great intervals. Too many laughless shows by a favorite comedy performer will in all likelihood send the audience searching for alternatives.

Diane English, creator and co-producer of the situation comedy "Murphy Brown," understands the principle of consistency very well. In the winter of 1992 the show scheduled an episode based on the televised Senate hearings of Anita Hill's allegation that Judge Clarence Thomas was not a suitable Supreme Court nominee because of his history of sexual harassment. The hearings would never have taken place if Dr. Hill's charges had not been leaked to Nina Totenberg of National Public Radio, who reported the story but refused to identify the source. Ms. Totenberg was subsequently summoned to appear before a Senate special counsel. Impressed by Ms. Totenberg's courage, Ms. English developed a story in which Murphy Brown acts similarly. Said Ms. English, "As more of our freedoms dwindle, we need to use our freedom as writers to make people know what's going on. *And we also have to make it funny.*"[7] (Italics ours)

Writers, directors, and performers must be particularly mindful of "staying in character." Cast members must always say and do things that are consistent with the roles they are playing. A writer may get easy humor from a line, but if it's at the expense of the character's nature, it could be a very expensive laugh. Sometimes broadcasters become too narrow in their interpretation of this point. There is the classic example of the executive who read a script of "My Favorite Martian," circled a line of dialogue and sent it back to the producer with the comment, "A Martian wouldn't say that."

One other element of consistency deserves mention: shows must remain true to their central intent. They cannot be all things to all people; there are limits to what programs can be and who they can reach. Broadcasters can't insert one-minute cooking tips inside Saturday morning cartoons in an attempt to attract adult women. Nor can they add hard rock acts to a Frank Sinatra/Ella Fitzgerald special to pull in kids and teenagers. Shows are what they are and any effort to broaden the base with inappropriate elements not only fails to attract the desired new viewers, but alienates the core audience as well.

Energy. Energy is that quality that infuses a sense of pace and excitement into a show. It isn't a synonym for frenzy. And it doesn't necessarily mean motion, which is oftentimes just movement without a point. Rather, it is the charging of the screen with pictures that won't let the viewer turn away, whether it is a four-way shootout, a whisper, an intense love scene, or a raised eyebrow. Actors contribute with their performances, the best ones making every scene riveting. Writers develop the dialogue and structure the acts in ways that produce rising tension and climactic endings. Directors stage the players and select the pictures that will generate the most satisfying viewer experience. If any one of these elements lets down, thereby permitting the pace

FIGURE 7.5
Alan Alda's warmth and wit have made him one of television's most likable personalities and a symbol of the sensitive, caring male. (Photo courtesy of the Academy of Television Arts and Sciences.)

FIGURE 7.6
"Murphy Brown," featuring Candice Bergen, devotes a few episodes each season to subjects of substance (unwed motherhood, the right of a journalist to withhold sources, etc.). But the show never forgets that it is a sitcom and must provide laughs while dealing with the issues. (Photo courtesy of the Academy of Television Arts and Sciences.)

to flag and attention to wander, the loss of energy will quickly result in the loss of audience.

Maintaining energy is an essential for all talk show hosts. Too often interviewers allow their guests to dictate the tempo of the program. If the host is not alert, a low-key, deliberate-speaking guest can drag down the energy level, and the two can quickly find themselves in the quicksand of boredom. The producer and director should immediately cut to a commercial and go on stage to pump up the host. Either that, or give the guest an early dismissal.

Professional Staffing. Rarely, very rarely, a show will succeed because of the novelty of its format or star even though the production is slipshod and the writing poor. But the success will be brief. The novelty will wear off, the mediocrity exposed, and the show's decline assured.

Many a promising format has been squandered because of the insufficiencies of the staff and cast. And many an ordinary idea ("Hill Street Blues," "Miami Vice," "Roseanne") has blossomed into enduring success because of the skills of the cast and the production team.

Programmers must look beyond the concept of a show and closely examine the credentials of the executors before committing to its development. Unfortunately, sometimes this means that fresh, young talent is ignored for the old tried and true. But the newcomers can work their way in gradually by attaching to (and learning from) the established.

Timing. For a program to work it must be in harmony with the times. Too far behind and the audience will dismiss it as outmoded; too far in front and viewers will rebel against it.

"Three's Company" is an example of a show that hit it just right and enjoyed a successful run from 1977 to 1984 (and then for many years later in syndication). This situation comedy dealt with a hot-blooded young man and two attractive young women who, for reasons of economy, shared an apartment. Although every episode was packed with sexual innuendo, the audience was encouraged to believe that nothing ever happened. The program was accepted because the time was right. The social upheavals that began in the 1960s and continued into the 1970s made cohabitation among unmarrieds commonplace. The network surmised that the condition was sufficiently widespread to allow the national audience to accept the premise. They guessed right. The same format, if it had been introduced ten years earlier, would have provoked a national outcry.

Similarly, it is very doubtful that "Kate & Allie," a sitcom revolving around the lives of two women who, after their divorces, decide to share a Greenwich Village apartment with their respective children, would have been acceptable in the 1960s. Single-parent family series were fine as long as the spouse had passed away. But the audience wouldn't have been ready to "glorify" a family unit (much less two) that was the product of divorce.

Trend Awareness. A capable programmer must also be aware of trends that might generate a hit or guarantee a failure. Strictly speaking, riding a trend wave is not an essential element of a successful program. But it can be a way to tap into a prevailing audience preference that will enable a show to deliver strong ratings. Similarly, an awareness of a trend that's over can help a programmer avoid a concept whose time has passed.

A programming trend occurs when producers develop shows or concepts that are similar in theme, format, or content. In network primetime television in the 1950s, there was a strong trend toward live, dramatic productions; in the 1960s the favored form was westerns; in the 1970s it was sitcoms with sharp social commentary; in the 1980s dramatic serials were prominent. TV syndication has seen similar runs. In 1982 Group W Productions introduced a first-run daily cartoon series, "He-Man and the Masters of the Universe." Never before had daily, animated programming been made expressly for the syndicated marketplace. Within two years there were *twenty-eight* first-run animated series offered for syndication.

Trends develop in two ways: 1) economic or technological necessity, or 2) the phenomenal success of a program, which suggests that variations on the form can enjoy similar rewards.

The tendency to live, dramatic fare in the 1950s was primarily the result of technological limitations. The new medium was centered in New York City; tape had not yet been invented, the Hollywood film industry by and large disdained the new industry, and film production facilities in New York were scarce. But there were excellent actors, producers, and writers available in New York theater, and they immediately gravitated to the live dramatic form.

In the 1990s there is a strong tendency towards news-oriented programs in primetime network scheduling. The reason is economics. The audience for the networks is dwindling, costs continue to rise forbiddingly, and producers are increasingly unwilling to deficit-finance because of the softness of the syndication market, particularly for action-adventure hours. In the search for affordable program forms, the networks have embraced the live or live-on-tape news presentations. Even a modest rating performance can deliver a profitable return because of the low cost of production.

Even more extreme is the trend toward infomercials—30-minute programs that masquerade as talk or interview shows but are totally devoted to extolling the virtues of some commercial product such as a diet, baldness treatment, or brand of sunglasses. Stations that might not program this material when times are good will do so in lean periods because the producers provide the programs for free and even pay the stations for the airtime.

FIGURE 7.7 On "Three's Company," starring (from left) Joyce DeWitt, John Ritter, and Suzanne Somers, bachelor Jack Tripper feigned homosexuality in order to get permission from the landlord to live (innocently) with two attractive single women. While considered somewhat spicy—but acceptable—when it premiered in 1977, it would have been soundly rejected a dozen years earlier. (Photo courtesy of DLT Entertainment Ltd.)

Often trends are started by the emergence of an enormous hit. The industry studies the soaring ratings and wonders if other versions of the theme can be developed. Frequently they can. The originator may have fed an audience appetite that isn't satisfied with just one dish. The industry will immediately offer variations and the process will continue until viewers are sated. In many instances the show which began the trend will outlast all the imitators.

"Gunsmoke" triggered the western frenzy in 1955. By 1961 there were twelve series slappin' leather each week. In 1975 only one remained—"Gunsmoke." The same pattern held true for primetime serials. The form was introduced by "Dallas" in 1978 and was an instant hit. By 1981 the number of serials had mushroomed to five. Ten years later it had receded to two, one of which was "Dallas."

Interestingly, "Dallas" was not the first soap opera to appear in primetime television. "Peyton Place," which brought Mia Farrow and Ryan O'Neal to prominence, appeared in half-hour form from 1964 to 1969, three times a week in 1965 and twice a week the other years. Although it was a huge success and dominated day-after conversation for years, it never spurred any variations.

All trends run their course. Viewers become saturated, the original idea becomes jaded and stale, ratings decline, and the search for a new appeal begins. Judging when a trend has crested is one of the programmer's more difficult decisions. One clue is rising demographics. If the younger viewers, the most volatile and easily sated segment of the audience, are drifting away, the chances are the show has seen its best days. Even if the audience size remains satisfactory, the older skew will herald a fall-off in revenue. Any new projects in the same genre are probably ill-advised.

Adequate Budget. Programmers must allocate sufficient funds for producers to make their shows. It is true that most producers will strive to obtain the largest possible budget, especially if they retain the difference between the package price and the actual cost of production. But for the most part they just want to be sure they have enough resources to mount the product they are committed to deliver.

Distributors such as networks and syndicators are paid to keep costs down. Therefore, establishing a mutually acceptable budget frequently triggers a lively dispute. The debate is well worthwhile if the result is a figure that allows the creative vision to be realized. When viewers reject a program because "it looks chintzy," everyone loses. Find the money to do it right.

Salability. In the world of commercial television, the greatest idea in the world is worthless if no one will buy it. This quality of salability must be effective at three stages: when the creator is trying to find a distributor; when the distributor is trying to find advertisers; and when both the creator and distributor are trying to find an audience. If there is a failure at any one of these stages, the program will either never reach the air or will be off after a brief stay because of a lack of revenue.

Surprisingly, the selling approach is not always the same at all three levels. Buyers of programs—networks, syndication companies, and station executives—tend to have fast-paced business days. The phone rings incessantly; associates pop in and out; and meetings are forever backed up. They also regard themselves as quick studies, their experience and natural gifts allowing them to cut through rhetoric and pierce to the heart of the show. Consequently, sellers keep their pitches short and concentrate on the "catchiest" elements.

In syndication, for example, program concepts are usually presented in tapes that proceed at a breakneck pace. If possible, they are enlivened with quick cuts of spicy or sensational material to maintain the buyer's interest and present an image of cutting-edge sharpness. In reality, the program may be successful with the audience only if it is produced in a deliberate, thorough style.

"When Group W launched 'Hour Magazine' [the hour-long women's service show] we knew from research that viewers wanted each topic dealt with fully and responsibly," said George Resing, then senior vice president with the production firm. "But we also knew that if we presented the show that way to station managers, they'd get restless, think the program was dull and not buy it. So we made a zippy bells-and-whistles 15-minute demonstrator that was paced like 'America's Funniest Videos' and used that as our main station clearance weapon. On the other hand, when we were selling the show to advertisers, we took a completely different approach: we slowed down the pace and emphasized the wholesomeness and solidity of the program to indicate that their commercials would be in a very favorable environment. This two-step strategy was very successful; the program was profitable for ten years."

One Voice of Authority. Committees rarely design anything worthwhile. The compromises required to keep peace in the conference room usually demolish the original concept. An adjustment here, a new element there, and suddenly the magic is gone. To be sure, constructive suggestions should be offered and welcomed. But it should be clear from the first meeting that one person make all creative decisions. The alternative will be chaos and camels.

The single-authority requirement is becoming more important in broadcasting. As costs rise and risks grow, there has been a developing trend towards funding by consortium, many contributors sharing in the costs of a project. However, each investor is inclined to believe he or she is entitled to creative input. It is best to get that ventilated before the papers are signed and have the authority figure identified in the contract.

In the mid-1980s a syndication program titled "America" was developed. It was an hour-long, five-a-week series that purported to present what was "hot and happening" in the country. The timing was right for the syndicated marketplace and the program cleared over 150 stations, almost all of which scheduled the show in excellent early fringe (3:00–5:00 p.m.) time periods. There was only one problem: the four corporations funding the program couldn't agree on the format and content. "I was going crazy," lamented Woody Fraser, executive producer. "No matter what I did, the next day I'd get a call from one of them telling me I was going in the wrong direction. Too much hard news, not enough hard news. Where's the comedy? Why don't we see more of so and so? We had a different fix every day. It was madness. We never did the show I originally had in mind. Big surprise: it bombed."

Innovation, Freshness. Over fifty years ago a man named Ted Bates built an advertising empire out of a single notion. Every ad prepared by his agency had to have USP, a Unique Selling Proposition. By that he meant the ad had to find one quality that could only be found in, or said about, that product. That uniqueness distinguished the product from all others and thus stimulated a buying appetite in the consumer.

The same is true for programs. If there is nothing unusual, fresh, or different about a show, why should anybody watch it? The kiss of death for any show is when the viewer says, "I get the feeling I've seen it a thousand times."

Points of differentiation do not have to be major departures from all other forms on the air. A single inventive difference is frequently sufficient. It can even be as simple as producing a conventional form in an unconventional way.

"Miami Vice," a top-ten series for a few years in the 1980s, was a standard, buddy detective show, but the imaginative music and the stylish clothes of the leads lifted it above the ordinary and commanded attention. "America's Funniest Home Videos" was a variation on "Candid Camera," but the use of average citizens' homegrown camcorder tapes was novel.

Many times the uniqueness will lie in the skills or personality of a performer. The format of "Roseanne" could not be more standard, but the singular look and attitude of Roseanne Arnold has made the series a solid hit. The same for the "The Cosby Show."

One of the major advantages of these and other innovative programs is that they promote so easily. The USP is bold and clear and permits effective promos and print advertising. Warner Brothers' Dick Robertson cited the lack of creative freshness as the primary cause of the syndication industry's disastrous 1990 season. "There were nine shows that were launched. None of them worked and the reason . . . was because none of them were very good programs. By that I mean they failed to capture the imagination of anybody. Some people were saying it was tiering [many different time periods] or [audience] fractionalization. I couldn't disagree more. I think the reason was because the shows were too derivative. There was nothing compelling about them."[8]

Two words of caution to programmers:

A. Every buyer of network programs has a file full of "breakthrough" program submissions that "can't miss because they're so different"—e.g., the one-armed midget detective who lives in a state-of-the-art tree house, or the female Siamese twins who are defense attorneys (casting was seen as a problem on this one). In the early 1990s the celebrated producer Steven Bochco attempted a weekly hour-long musical drama, "Cop Rock." It sank like one. The idea was bold and inventive, but the innovation, predictably, failed to address a need.

B. The acceptability of innovation by buyers is in direct proportion to the economic health of the industry. In hard times buyers tilt towards conservatism. Innovation means risk, and risk can mean the loss of scarce dollars. One of the reasons the 1990 syndication season was a disaster was the recession state of the business. Station owners, many of whom were highly leveraged because of the license-buying frenzy of the 1980s, were loath to cancel even marginally profitable shows for fear their replacements might do worse. Better to stay with something that isn't losing than take a chance on a big hit, and miss. Of course, as Dick Robertson pointed out earlier, the only hope of success is with a fresh, new "compelling" concept, but the decision-maker who is in a shaky economic condition is a difficult person to persuade. As George Resing observes, "They all say they're looking for something really fresh and different . . . with a track record."

What the record does show is that real success in programming can only be obtained by invention, finding new ways and new people to inform and entertain. In the 40+ years of television's history, only two animated programs had ever been successful on primetime network schedules—"The Jetsons" and "The Flintstones." And the latter left primetime in 1966. The gambling, innovative Fox Network introduced "The Simpsons," a weekly half-hour animated series, in 1990. The show quickly zoomed into top-twenty status and became a marketing phenomenon. Such success stories never happen to "safe" derivative programs. They only occur when a dreamer has an inspiration and a programmer decides to accept the risks.

Why So Many Failures?

The ingredients of programming listed above are known by virtually every producer and executive in the field. Why, then, do so many shows fail? Are the principles simply ignored or are they flouted by arrogant creators? Why, as some say, should every hit be "a fluke"?

Failure in creative endeavors is not limited to broadcasting. Consider the 1991 Broadway musical "Nick and Nora." It was based on Dashiell Hammett's "Thin Man" series, developed by the creators of such theatrical glories as "Gypsy," "West Side Story," "Applause," and "Annie," performed by three Tony award winners, designed by two more, produced with a budget of $4.3 million, and polished over a nine-week out-of-town run. It closed in one week and lost well over $3,000,000.[9]

The creative graveyards are littered with the bones of theater, publishing, and motion picture projects that couldn't miss, and did. Why?

The reason is that scores of decisions must be made at every step of development, and the outcome of those judgments cannot be predicted. The decision-maker knows perfectly well a strong production staff is necessary to make a series click. But which of the writer candidates should be picked? The one with four hits to her credit but a bomb the last time out? Or the guy who is coming off a smash but had three failures preceding it? "Of course I know my timing has to be right," a producer might say. "But you tell me, is the world ready for a mixed-race, husband-and-wife detective team, or am I a couple of years too early?" The principles are known, it's the judgments that are dicey.

Producers attempt to test the wisdom of their choices through research, out-of-town tryouts, sneak previews, and consultation with knowledgeable insiders, but in the end the public, a quirky, amorphous mass of individuals, will determine the fate of the production.

No two producers in the television industry are more savvy than Marcy Carsey and Tom Werner, developers of the incredibly successful situation comedies "The Cosby Show" and "Roseanne." The latter was a particularly inspired conception. Ms. Barr, a comedian not known for her comeliness, was a most unlikely candidate for stardom. But the audience took to her immediately. Two years later, the same producers tried the same technique, this time with a male comedian not renowned for his sex appeal. "Chicken Soup," starring Jackie Mason, was gone before the season was over. Same gifted producers, same sound strategy. Just a wrong guess.

a

b

c

d

e

f

FIGURE 7.8
When a show contains the right elements, and the cast and staff work with care and determination, the result is sometimes rewarded with television's ultimate symbol of excellence—an Emmy. (a) Sharon Gless, Outstanding Lead Actress in a Drama Series, 1986, "Cagney and Lacey"; (b) Michael J. Fox, Outstanding Lead Actor in a Comedy Series, 1986, 1987, 1988, "Family Ties"; (c) Tyne Daly, Outstanding Lead Actress in a Drama Series, 1983, " Cagney and Lacey"; (d) Robert Guillaume, Outstanding Lead Actor in a Comedy Series, 1985, "Benson"; (e) Candice Bergen, Outstanding Lead Actress in a Comedy Series, "Murphy Brown," 1989; (f) Peter Falk, Outstanding Actor in a Drama Series, 1990, "Colombo." (Photos courtesy of the Academy of Television Arts and Sciences.)

We will now move on to see how the elements of successful programming affect the other electronic media outlets.

ROLLER COASTER RIDE IN "FANTASY PARK"

Jim Cahill is an original; a colorful, iconoclastic entrepreneur who describes himself as "a song-and-dance man of the 1990s." If you want a small promotion campaign or a neat little advertising phrase, he's not your man. But, if you want the big idea, the clutch-'em-by-the-throat blockbuster, he's the guy to call.

In 1989 MTV needed a device to call attention to its faltering sister channel, VH-1. Jeff Rowe, senior vice president of programs, was given a $1 million promotion budget and a mandate to make baby boomers aware of the service. He contacted Cahill.

"The baby boom," stated Cahill, "basically began in 1953. So I looked for a way to connect that year with the boomers of today." He remembered that 1953 was the first production year of the legendary Corvette. "The 'Vette was always a hip car," Cahill reasoned. "There isn't a boomer alive today who doesn't get misty-eyed when he sees one." The big idea began to germinate. Why not give away thirty-six Corvettes to thirty-six people, one for each year of the car's life? Wait a minute, that's good, but it isn't great. A-h-h-h, got it! Give all thirty-six cars to one winner. Now that's different. That's big.

Cahill's cardinal rule is: "Always get the head honcho into the pitch meeting; otherwise they'll committee you to death." When he proposed his plan to MTV executives, the meeting included Tom Freston, president and CEO of MTV. The reaction was enthusiastic but the estimated $700,000 to buy the cars was upsetting. Freston urged that Cahill "think of a way to get some of that money back."

Cahill more than fulfilled the request. He proposed that contestants enter the competition by dialing a 900 number. "At $2.00 a call I thought we might be able to cover a few cars." For once in his life Cahill was guilty of understatement. By the end of the contest, the telephone calls had grossed over $3 million. On October 14, 1991, a computer randomly selected a telephone number from the data base, the winner drove off in one of his thirty-six vintage Corvettes, the event made the satellite news feed, highlights were seen around the country, MTV was a delighted client, and Cahill realized that he had just stumbled on to an idea for a weekly TV series.

He decided early that this idea was made for Fox. The young network was audacious and chance-taking and furiously courting a VH-1–type audience. He also decided he needed a partner, someone who could open doors and give an air of size and substance to the project. Through some work he had done in the music business, he had developed a relationship with Irving Azoff, president of Giant Records and someone able to command a meeting with anyone in the entertainment business. Azoff agreed to an association and arranged a conference with Barry Diller, chair of 20th Century Fox. (Note Cahill's head honcho law at work.)

No specific program was actually outlined to the meeting attendees. Cahill merely presented a lightning-fast, 7-minute tape that summarized the VH-1 experience. When the tape ended, Diller exclaimed, "I don't know what it is, but it's the greatest thing I've ever seen! Get the FCC to clear the lottery aspect and we'll make a pilot." "A pilot of what?" thought Cahill. "Now that I've got a pickup, maybe I ought to get a show."

He quickly made a deal with experienced TV producer Brad Lachman, and together they worked out a simple format. Every episode in the series, now called "Fantasy Park," would feature three major prizes. The fun and excitement would be derived from watching the unsuspecting winners receive their gifts from a traveling host. The heart of the show was the prizes. They had to be inventive, unusual, and, most of all, big. Cahill and Lachman thought up such mind-bogglers as:

- Two Hawaiian condos for the price of none
- A gold Amex card, a $200,000 line of credit, and you never see a bill

A Ticketmaster terminal in your own living room. Want two tickets to the Rams' game? Just dial 'em up.

In the view of the producers, it was essential that the entire show be shot on location to maintain a fresh, distinctive look. Advance crews would be sent to every winner's town to secretly collect background footage and information. This material would be edited and then aired right before the host arrived to bestow the prize. In this way viewers would know the winners-to-be and empathize with their excitement. The host would be seen flying to each destination and then "ambushing" the winners at home, or in the office, or at some spot surreptitiously arranged through friends and relatives. After the tears and jubilation, the host would climb into the chopper and head off to the next surprised winner.

Obviously, a program of this magnitude would be costly: "around $400,000 a week" estimated Lachman. But the producers were confident that the expenses could be covered, and then some, by the 900 number. Of the gross dollars generated by the calls, 23 percent was to be deducted for the telephone company's fee. The next $400,000 would be used to pay for the prizes, and the remainder would be divided 30 percent for Fox, 30 percent for the Fox affiliates, 30 percent to the production company, and 10 percent to charity (to help soften the lottery tone of the show).

If each episode attracted 500,000 calls ($1,000,000), the telephone company would pocket $230,000; the $400,000 prizes would be paid for; a charity would receive $37,000; and Fox, its affiliates, and the producers would each walk away with $111,000. Over the course of a season it amounted to some handsome numbers for everyone.

Fox loved the concept and the arithmetic, but Peter Chernin, President of Fox Entertainment, expressed some misgivings about the format. He wondered if it wouldn't be "a warmer, more personal show" if most of the winners were brought to a studio. Other Fox executives shared that view and were also concerned that the absence of a home base might cause the show to lose identity. Perhaps, one suggested, the show should take over an airport hangar, and each week winners would be seen taxiing into this novel "studio."

At an early point in the discussions Fox talked about possibly upping the ante from a single pilot to a six-episode on-the-air commitment. Although this was heady wine for Lachman and Cahill, they continued to resist the home-base proposal.

As the days flowed on, so did the format recommendations. Lachman had the impression that "Fox wasn't confident about the simplicity of the concept. They kept suggesting more and more elements to give the show additional appeal." Among the suggestions were: present four prizes instead of three; add a live giveaway; find a game show element.

Cahill and Lachman were exhorted to price out all of these proposals in order to arrive at a budget on which a contract could be based. This was an arduous, time-consuming exercise for the producers, made even more burdensome because of their disagreement with many of the suggestions.

While all of these computations were being processed, Chernin announced that the premiere would take place on October 28, 1990 (six weeks away), and that the show would appear for six consecutive weeks. Before Lachman could acquaint him with the impossible logistics of such a short preproduction period, the date was scrubbed. The Fox Promotion Department noted that October 28 was in a sweeps month. It was imperative that Fox' new programs do well in this key rating period. But the Fox stations, who were profit participants in the "Fantasy Park" 900-number financial pool, were sure to divert promos from other shows to increase their take on "Fantasy Park." The diminished promotional support for the rest of the lineup could be devastating.

The premiere was rescheduled for January 12, 1991, which gave rise to a new problem. Lachman was concerned that starting this highly produced new venture in the dead of winter might jeopardize the stunts and surprises. Snow storms and icy runways could cause risks and delays that might make it impossible to deliver shows by air dates.

Once again the premiere was postponed, partly because of Lachman's concerns and partly because the matter of a studio origination had not yet been resolved. Cahill and Lachman were still battling for a 100 percent on-location show, and the Fox executives had become even more determined to work from a home base.

By now, December 1990, everyone was stretched thin. Chernin was angry over the delay of air dates and the uncertainty of the format, other Fox executives were miffed at the producers' reluctance to accept new program elements, and Lachman and Cahill were dismayed by the demands to depart from their original concept. The ever-expressive Cahill muttered: "I was beginning to envision the postnuclear rubble of a dying dream."

Around Christmas everyone, perhaps warmed by the spirit of the season, became more conciliatory. The FCC added to the tone of detente by officially approving the lottery device with only a few easily-handled modifications. A new air date, February 9, was locked in.

In early January a new Fox executive, Sandy Grushaw, was assigned to supervise the show. A self-proclaimed "old promotion man," Grushaw declared the show lacked "life-style interest." As Cahill remembers, "The new guy said it wasn't enough just to present the two Hawaiian condos and some background on the winner. The prize had to be shown in its local environment. By that he meant luaus, leis, hulas, and those funny boats with the side pontoon. And to pay for all of this extra business we could insert some plugs for a bunch of small appliances."

Thoroughly drained by the daily wrangling over the show's structure, Cahill and Lachman capitulated on the home-base dispute and a number of other additions requested by the Fox staff. They placed their hopes in the telephone response; a huge turnout of calls would help soothe the bruises. Instead, what they got, on January 16, 1991, was the Persian Gulf conflict, and another postponement.

At the conclusion of the brief war, Fox executives examined the schedule and selected Saturday, April 13, as the final, irrevocable, set-in-cement air date. By now all thoughts of a short flight of six episodes had been abandoned. Prizes for all these episodes would have involved a commitment in advance of $2,400,000, and the network was disinclined to take on such an obligation without some indication of the show's popularity. Better, they reasoned, to do one program, see what they had and then proceed or abort depending on the results. The downside of this caution was the loss of any momentum the premiere might develop and the need to start the promotional process all over again if additional shows were committed in the future. And it was that selfsame subject, promotion, that was causing the latest rupture in the Cahill-Fox relationship.

For the program to succeed, it was essential that the audience understand quickly and clearly the unique concept and the requirement to call a particular 900 number. The most effective way to get the message across was on-the-air promotion. The spots had to be numerous and persuasive. Cahill pleaded to be allowed to produce them. The straight-ahead approach that he had developed for the VH-1 Corvette campaign had been simple, crystal clear, and wonderfully productive. "No," he was told, "promos are a network function; we're the pros, we'll do them."

In a further stroke of bad luck for Cahill, Diller himself took a personal interest in the promotion. The chair wanted the spots to have some flair, an artistic quality that would command attention through style. He was influenced perhaps by the launch campaign for the Infiniti automobile, which consisted of a series of nature scenes accompanied by a sonorous voice explaining the glories of the universe. The product itself was never shown, which prompted comedian Jay Leno to remark, "Nobody's buying the car, but rocks and trees are sold out."

The announcements were introduced the weekend prior to the premiere date. They drew 15,000 telephone calls. Cahill was suicidal. He bellowed and begged until he finally found a Fox executive who was willing to risk the chair's wrath and authorize Cahill to design new spots. In four days these promos attracted 500,000 calls.

The program that aired was a distant relative to the format originally blueprinted by Cahill and Lachman. It originated from a small stage at KTTV, Fox's owned station in Los Angeles, California. Four hosts presided, one at home base and the other three on the road. Four major prizes were given away. Two winners received their rewards on location, but not with any of the personal background footage that Lachman considered essential. Another viewer was flown to L.A., assured that he had won a prize but not told which one. The actual gift was revealed on the show. The final award also contained a surprise element. Three people whom the computer had selected for a cash prize were invited to the studio for the presentation of the checks. But during the ceremony they were told that the first one who completed a telephone call to a particular number in Hawaii would win two condos. As a special feature for the home audience, the home-base host gave away $10,000 in cash to a registered viewer picked by the computer while the show was on the air. All the principal gifts were given "life-style" production values as desired by the network.

When the returns were in, "Fantasy Park" was neither a smash nor a disaster. Its rating was just about the same as the show it replaced. There was no particular sizzle to the postshow word of mouth, no exceptional mail or telephone response, no unusual press reaction; in fact, there wasn't much of anything. It most certainly had not been a big event.

The part the Fox people liked the most was the contest among the three studio competitors to see who would be the first to connect with the Hawaiian telephone number and win the two condos. When one of the executives began to talk about "making it a game show," Cahill and Lachman decided "Fantasy Park" had run its course with Fox and maybe they should try their luck elsewhere.

In a postmortem that was more mournful than angry, Cahill remarked: "They never got what they bought. They killed it with encouragement. In the end it simply had no soul."

Postscript: The "Fantasy Park" story, told completely from the perspective of the creators, Cahill and Lachman, could be interpreted as an indictment of buyer denseness and insensitivity. That is not the intent. It is by no means certain that the original format of "Fantasy Park" would have produced a hit. Perhaps some, or most, of the network's suggestions were necessary to infuse some variation into an otherwise static design. No one will ever know until the program is aired in its original form—which means it will probably never be known. What is known is that a grab-bag version packed with the bits, pieces, and oddments of many contributors didn't work.

The real lesson for programmers is: Don't lose the original idea that got you interested in the first place. Refine it, sand it, buff it, but don't lose it. As we observed earlier in this chapter, shows cannot be all things to all viewers. Each program has an essence, and any attempt to expand it will probably destroy it. After a few weeks of development, take a step back and objectively examine the format. If it differs substantially from the original concept, chances are you're heading off course. Call a meeting immediately and announce those "improvements" that are about to be abandoned.

NOTES

1. "Marketing Programming in the 1990s," *College Broadcaster* (November/December 1990): 5.
2. "The Goldenson Year," *Los Angeles Times Calendar* (February 3, 1991): 78.
3. *The Broadcasting Yearbook* (Washington, DC: Broadcasting Publications, Inc., 1991): F-26–F-30.
4. *Bartlett's Familiar Quotations*, 15th ed. (Boston: Little, Brown and Company, 1980): 902.
5. "TV Convention Draws Record Crowd," *Los Angeles Times* (January 28, 1993): F-4.
6. "At Large: Dick Robertson: Surveying the Supply Side," *Broadcasting* (January 14, 1990): 67.
7. "NPR's Totenberg: Pushed into the Limelight," *Broadcasting* (November 4, 1991): 58–60.
8. "At Large: Dick Robertson," 66.
9. "B'way Case Closed for 'Nick & Nora,' " *The Hollywood Reporter* (December 11, 1991): 4.

8 Elements of Successful Programming: Part II

CABLE TELEVISION

Cable TV, like its commercial kin, deals with a variety of programming objectives, fundamental appeals to an audience, and qualities tied to successful programming. Many of these traits are identical to those of commercial broadcasting, but with a slightly different flavor because of the heterogeneity of the cable services.

Programming Objectives

Cable differs from commercial TV in that most of the networks do not attempt to gain the widest possible audience. Only the broad-based services, such as USA Network and the superstations, aim to attract a large general audience.

A Specific Target Audience. Because of the high degree of narrowcasting within the cable services, particular networks are much more likely to aim for very specific demographic audiences. MTV's target audience is quite different from that of Nostalgia Television in terms of age and general life-style. CNN's stated target audience consists of intelligent, well-informed people with one or two college degrees who read two newspapers a day. Nickelodeon caters to the very young, Black Entertainment Television targets African Americans, Galavision's programming is meant for the Spanish-speaking population, The Nashville Network is for country fans, Lifetime is aimed at women. Each of these services must consider its target audience as it assesses program ideas and its advertising possibilities.

For cable services that do not deal with advertisers, the story is a little different. Pay movie channels, such as HBO and Showtime, receive their money from cable subscribers who decide to pay for their programming. Therefore, their specific audience is the people who have subscribed. The services must program in such a way that they keep these people happy so they do not become disenchanted and disconnect. One way they accomplish this is by providing programming that the "free" television networks can't match, such as feature films fresh from their theatrical release, made-for-TV movies, high-profile sports events, and series and miniseries with adult themes and language. Pay-per-view movies or specials are successful if enough people sign up to bring a profit.

Another type of specific audience that some cable systems try to attract is an "in-and-out" audience. The Weather Channel, for instance, does not presume that people will sit and watch it for hours, although they will happily accept someone who does. Cable News Network and the various shopping channels are likely to capture audience members for a limited amount of time. What they are aiming for is a large total number of viewers, even though any one person might not stay with the service for a long period.

Prestige. A few of the cable systems, such as Arts and Entertainment, Bravo, and C-SPAN, were founded with the idea that their programming would be prestigious. Peter

a

b

c

d

FIGURE 8.1
Many cable networks target specific audiences: Nickelodeon, with cartoon shows such as "Rugrats" (a), is meant for the very young; Lifetime, featuring programs such as "Your Baby and Child with Penelope Leach" (b), targets women; BET is directed toward African Americans and includes such programs as "Our Voices" (c), a talk show hosted by Bev Smith (seen here with guest the Rev. Jesse Jackson); and CNN, with hosts such as Larry King (d), feels it offers material to appeal to intelligent, well-informed, college-educated people. Photos courtesy of MTV Networks (a); Lifetime (b); Black Entertainment Television/Roberto Brooks (c); © Cable News Network (d). All rights reserved.

Hansen, vice president of programming for A&E, notes that his service "must maintain its cultural image because cable systems use A&E extensively when they are applying for franchise renewal. The fact that many educators tape A&E programs off the air and used them in classrooms is of positive prosocial value when systems are illustrating their community service."

FIGURE 8.2
A&E, with its dramas such as "The Last Act Is a Solo," has prestige as one of its major objectives. (Photo courtesy of Arts and Entertainment/Linda Alaniz.)

Awards. Awards are important within the cable world, just as they are within commercial broadcasting. In 1977, shortly after the very first wave of cable networks was established, the National Cable TV Association began bestowing ACE (Awards for Cablecasting Excellence) awards on programs. The National Federation of Local Cable Programmers also established an annual awards contest for programs aired over local cable system channels. The cable people lobbied long and hard to be allowed to submit their programming for Emmy awards and were eventually granted entry in 1987.[1] Their persistence paid off. In 1993 cable programs garnered 55 nominations for Emmy awards.

Particular Local or National Purpose. Local cable systems, through their local origination and public access channels, program primarily material intended for specific local purposes. Often the cable companies themselves do not set the local objectives. They are set by the individuals or community groups that produce the programming. If the city council decides that cablecasting its monthly meetings serves the local community, it arranges for the meetings to be shown on a government access or public access channel. In other instances, the cable system management sets objectives for local origination channels. It might, for example, determine that a series about crime watch would benefit the community. Or local management, without prodding from the city government, might decide to cablecast the city council meetings as part of a service to the community. In many instances, cable systems are required to engage in some sort of local programming in order to keep their franchises, a situation very similar to that of broadcast stations needing to serve the community in order for their FCC license to be renewed. Cable system programmers, like station programmers, should give thought to what really serves the community and not opt for the cheapest and most convenient form of programming they can find.

The cable service that is most likely to fulfill particular national needs is CNN. In fact, there are those who claim that CNN is really an international service. As a *Time* article honoring CNN's founder, Ted Turner, stated, "CNN demonstrated that politics can be planetary, that ordinary people can take a deep interest in events remote from them in every way . . . and can respond to reportage in global rather than purely nationalistic terms."[2]

In addition to bringing the nation and the world important news, CNN has at times become embroiled in the events it covers. For example, when the Soviet Union wished to protest the U. S. invasion of Panama in 1989, it did not complain to the embassy but rather to the CNN bureau in Moscow. CNN was the only news organization allowed to remain in Iraq during the 1991 Persian Gulf conflict, partly because Iraq's leader, Saddam Hussein, was a CNN fan.

Because cable TV has so many more channels than broadcast TV, its objectives are more diverse. Nevertheless, as with broadcasting, meeting the objectives is a primary sign of success.

Fundamental Appeals to an Audience

For the most part, the various cable services have adopted the same basic programming appeals as broadcasting—hard news (CNN), soft news (The Discovery Channel), music (Country Music TV), comedy (Comedy Central), drama (The Movie Channel). The channels that are broad-based (USA, WTBS, WOR) tend to feature programs in each category, just as broadcast networks do. Religious programming is given more emphasis in the overall scheme by specific cable networks, so inspirational programming could be added to the list of appeals used by cable. And, overall, cable presents more stand-up comics. But for the most part, appeals of cable programming are identical to those of commercial TV.

However, although the fundamental forms (such as comedy and drama) may be identical, the treatment, especially on the pay networks, often is not. Viewers will not pay monthly fees for more of the same. Thus, a sitcom on Showtime is not the same as a sitcom on CBS (see Box). Cable network operators must be ever mindful of the need to be unique or they won't be operating for long.

"BROTHERS": TOO HOT FOR ABC, JUST RIGHT FOR SHOWTIME

No program more clearly demonstrates the difference between the pay cable channels and the traditional networks than the comedy series "Brothers."

At the end of the 1980–81 season the most applauded and awarded comedy on television was "Taxi," on ABC. It not only won an Emmy as the Outstanding Comedy Series, it also brought statuettes to lead actor Judd Hirsch, director James Burrows, and writer Michael Leeson. It was, in short, the hottest comedy on the air.

Marcy Carsey, ABC's head of comedy development, was delighted with the show's success. She became even more excited a few months later when Gary Nardino, president of Paramount Television, "Taxi's" production company, called to say that Ed Weinberger, the show's producer, had a new program idea for her. He sweetened the news by adding that Weinberger's partner in the venture was David Lloyd, generally accepted as the best comedy writer in the business. Lloyd won two Emmys for his work on "The Mary Tyler Moore Show," one of which was for the now legendary episode, "Chuckles Bites the Dust."

As Nardino described it, the new program, titled "Brothers," featured three male siblings: a former football player now running a restaurant; an ultraconservative construction worker; and the youngest of the three, a likable young man who had just announced his homosexuality. Carsey knew that the going would be uphill for such a show to make the schedule, but she liked the concept, admired the creators, and decided to commission a script. Predictably, it was brilliant, funny, sensitive, provocative—and unacceptable.

Despite Carsey's eloquent and exuberant support, network officials would have none of it. They remembered all too vividly the furor raised a few years earlier by "Soap," a slightly off-the-wall comedy that included a homosexual, played by Billy Crystal. Even though the Crystal character was somewhat marginal to the series and was clearly presented as a comic send-up, the role produced a season-long bombardment of criticism, threats, and boycotts. To make matters worse for network executives, "Brothers" was intended as an honest presentation of gay life, and the homosexual character was designed to be sympathetic and believable. Veterans of the "Soap" campaign had little stomach for another pitched battle, and "Brothers" was rejected quietly but firmly.

For a couple of years the script moldered in Paramount's dead-storage bin. It had been all but forgotten until Peter Chernin, an executive at Showtime, called Paramount officers to see if the company had available any unusual, different kind of comedies. The pay channel was eager to start scheduling series programming but obviously couldn't present conventional sitcoms because subscribers wouldn't feel justified shelling out money for what they could see for free. Paramount programmers remembered "Brothers" and gave Chernin a brief description. He loved it. Nardino, now an independent producer attached to Paramount, was asked if he could put all the pieces back together again. He could, and did, and was rewarded with a commitment for fifteen episodes.

The show premiered on July 13, 1984, and immediately began to accomplish Showtime's goal: it created excitement, it got talked about, and it increased subscriptions. In fact, it did so well that over the next four years Showtime ordered an additional 100 episodes. Its run ended in July 1989.

In all that time, according to Nardino, the Showtime executives never demanded a single script change or tried to alter program content in any way. The series tackled such tinderbox themes as AIDS, gay-bashing, and emerging from the closet, and received scores of plaques, awards, and kudos for courage and sensitivity.

Would there have been a similarly positive response if the show had appeared on ABC? Probably not. Networks are subject to so many controls and influences (see Chapter 9) that these many pressures would surely have caused substantial script modifications and greatly lessened the show's boldness and effectiveness.

Although "Brothers" was rewarding in every way to Nardino, he did experience one major disappointment. The program was a dismal failure in syndication. After the run on Showtime, he attempted to sell the package of 115 episodes to local stations. Only a dozen committed. It is Nardino's belief that the much-publicized spread of AIDS "created an atmosphere of homophobia and killed any chance we had. Stations were too scared to go near it. But the five seasons on Showtime were great. We made some money and did some good things. Can't ask for more than that."

FIGURE 8.3
The "Brothers" premise that a gay lead character could be presented realistically and sympathetically was too problematic for its original developer, ABC. However, it neatly fulfilled Showtime's need to be different and unusual. The large cast included (front, from left) Mary Ann Pascal, Robert Walden, Hallie Todd, and (rear) Philip Charles MacKenzie, Brandon Maggart, and Paul Regina. (Photo Courtesy of Gary Nardino Productions.)

BIGGEST LAUGH IN "BROTHERS" FIRST EPISODE

Clifford, the youngest of the brothers, has recently revealed that he is a homosexual. The subject is being discussed by Joe, the oldest brother, and his daughter, Penny:

JOE: Your Uncle Lou is on the way over to discuss the situation. He doesn't know anything about it.

PENNY: Well, it's not a disgrace, you know. It's not something to be ashamed of.

JOE: Who said I was ashamed?

PENNY: So, he's gay.

JOE: Shhhh!

PENNY: Daddy! Some of the greatest people in history have been gay. Alexander the Great, . . . uh, Michelangelo . . . Plato.

JOE: Plato? Plato?

PENNY: Yeah.

JOE: Mickey Mouse's dog is gay?!

Qualities Tied to Success

Because the appeals of programming in cable are similar to those in broadcasting, many of the qualities are the same also. Conflict within individual drama and comedy programs is needed in order to hold an audience. Personalities should be likable and have energy. Innovation, professional staffing, and trend awareness are all necessary. Consistency should be maintained. (In the mid-1980s, ESPN tried programming financial news for part of the day, but this just didn't work with its all-sports image and was rather quickly abandoned.) Cable programmers are becoming more and more dependent on advertisers and sometimes place their original programming in the syndication market, so salability is an increasingly important factor.

But there are a few of the qualities of success that have unusual or particularly important meaning for cable.

Durability. For example, the concept of durability has some interesting peculiarities within the cable world. Because much of the programming is so narrow in terms of concept and intended audience, many have doubted that cable programming would have durability. And in some cases, they were right. A channel devoted to nothing but health information, Cable Health Network, was not successful and eventually merged with Daytime, a network featuring women's programming, to form Lifetime. The addition of health concepts to the women's programming has made for one viable network instead of two weak ones. Linda Fishman, vice president of original programming for Lifetime, notes, "We still find it challenging to define women's programming, but we have emerged as a very successful network."[3]

FIGURE 8.4 When MTV was started in 1981, many critics scoffed that the concept of music videos introduced by video jockeys—such as Steve Isaacs, shown here—would never last; by 1991 the critics had changed their minds. MTV had proved to have great durability. (Photo courtesy of MTV Networks.)

But those who were skeptical of the concept of 24 hours a day devoted totally to news or to videos of popular music were proved wrong. Cable has also shown that narrowcast ideas such as children's programming, all-movies, sports, religion, and weather can all have durability.

Timing. The idea of a network devoted to documentaries was discussed for many years but never received enough support to be executed until John Hendrick, a history professor, founded The Discovery Channel in 1985 as a vehicle to cover people, places, travel, and nature. Home shopping took hold because of the deregulatory relaxation regarding commercials; when there were no longer regulations regarding the amount of time that could be devoted to commercials, all-commercial channels were free to develop.

Adequate Budget. Qualities tied to the production process have undergone evolution in cable. In the early days, cable programs rarely had adequate budgets or professional staffing. It was an area where new people could break in because the established people were not willing to work for the fees cable could provide. However, over time, cable has improved its economics and now is often on a par with network program budgets. The same people who pitch to networks also present their ideas to the appropriate cable networks.

One Voice of Authority. Because of the volatile nature of cable, the idea of having one voice of authority has often been violated. Original concepts for programs and networks were often the idea of one person or a small group of like-thinking people. But because the concepts needed capital in order to become a reality, other companies or individuals were brought in, each eager to contribute creatively. Playboy started in the cable business by joining forces with an established adult service, Escapade. The two companies soon found, however, that they had differing ideas as to what constituted programming. Playboy had a well-established reputation based on its magazine and felt the cable programming should closely mirror the magazine content. Escapade had already been programming material that was racier. The two philosophies didn't mesh, and Playboy decided to go its own way, a divorce that cost the company $3 million.[4] Quite a price to pay in order to obtain one voice of authority.

COMMERCIAL RADIO

Commercial radio is quite different from television in its program objectives. Yet, despite this, the qualities that make for successful programming are quite similar.

Programming Objectives

Rarely is a radio audience a large one. Because people have such a wide divergence of musical tastes, a radio station that tries to program to reach everyone will probably attract no one. Radio stations target a specific audience—teenage males who like hard rock, older women who enjoy listening to talk, people who want 10 minutes of capsulized news. Because radio is relatively inexpensive, stations can make a profit even though they do not have a large audience.

Prestige and awards are not major radio objectives, although stations certainly tout awards when they win them. The few remaining commercial stations that program classical music do consider prestige. These stations do not expect large audiences or large amounts of advertising, but the owners are willing to continue the format because the prestigious nature of the audience attracts enough specialized advertisers to generate a profit.

Two programming objectives deserve special consideration. One, fulfilling a particular local purpose, is more dominant in radio than in TV. The other, creating a particular mood, is peculiar to radio.

Particular Local Purpose. Fulfilling local purposes is much more predominant in radio than in TV. Interacting with the community helps with license renewal, but it is also an essential element of station operation because radio is primarily a local medium. Radio stations often engage in activities that are part promotion and part local public service.

For example, KFI in Los Angeles sponsored a bike ride to raise money to clean up the ocean. People paid $10 to enter and all the money went toward the cause of making the water more pure. On a particular day, everyone who had made a contribution met at a predetermined place along the beach and took a bike ride with the KFI radio personalities and staff employees.

Many stations program short segments that describe upcoming weekend happenings. Mike O'Reilly, program director of WPDC in Elizabethtown, Pennsylvania, places these on his station four to six times a day Wednesday through Saturday. "We find these announcements keep us in touch with the community and the community in touch with us," he says.

Additionally, call-in shows concerned with local issues are often top-rated on stations with all-talk formats or even stations that do only occasional talk shows. Talk show hosts and disc jockeys appear at many charitable functions in the community, in part to add allure to the community event and in part to build awareness (and ratings) for the station. Even the numerous contests where listeners can win money, tickets, or CDs often have local tie-ins. For example, the tickets may be for a concert by a local music group. Hometown involvement makes good, sound business sense for most commercial radio stations.

A Particular Mood. One of the main programming objectives of radio, which is not really apropos to television, is that it strives to create a particular mood. Although individual TV programs may make a person sad or angry, the moods do not usually affect what a person is doing other than watching TV. In order for a TV program to put you in a certain mood, you must give it your attention. But such is not the case with radio, which actually encourages inattention. Music is background-oriented, and can influence how people feel about other activities. Radio programmers try hard to select music that will fit the predominant moods of their target audience.

Rhythm, pitch, loudness, instrumentation, melody—all these aspects of music can affect mood. For example, high-pitched music is more pleasant and playful than low-pitched music, which tends to be serious or sorrowful. Saxophones are more romance-inducing than trumpets. Loud music is more capable of providing isolation—a sound wall to keep out other people. Programmers must think of the music they select in terms of its general appropriateness for the target group and the specific activities that the audience might be engaged in at various times of the day. For example, young men like fast, hard music and older women prefer something slower and softer. But early in the morning, when everyone must move rather rapidly, the music played on a station appealing to older women should be more sprightly than it is late at night, when audience members are more likely to be relaxing.

Overall, the primary objectives that radio programmers deal with are those of attracting a demographically specific audience, fulfilling local needs, and establishing a mood.

Fundamental Appeals to an Audience

Music, hard news, and soft news are the primary appeals related to radio. Comedy and drama hardly ever appear on commercial radio these days.

Music. There is no question about the audience for radio music. It is just a matter of selecting the right sound for the right group. This is easier for stations than for networks or syndicators because stations can take into account the mores of the local community. Networks and syndicators that program rock music, in particular, have trouble because stations in certain parts of the country do not want to broadcast profane or sexually oriented lyrics. But other formats can run into this problem, too. Doug Peck, president of House of Music, found that several stations in the Bible Belt became upset with his choice of easy listening music when it contained Frank Sinatra's "Strangers in the Night," a song about two people meeting and going home together. They were equally disturbed by John Lennon's line, "Imagine there's no heaven."

Hard News. Radio is a vital factor in many people's lives for obtaining information. Waking up to radio news is quite common. People want to learn what has happened during the night—or sometimes they just want assurance that a major disaster has *not* happened. Once they are convinced that life is going on as usual, they are willing to switch to another appeal, such as music. Often the main elements of news broadcasts that attract audience members are the ones that affect them most personally—weather and/or traffic. All-news stations provide hard news throughout the day so that people can know about the latest happenings at any time.

When radio was deregulated in the 1980s, stations were no longer required to broadcast news. Although some have dropped all newscasts, many others have continued to bring listeners news at least once an hour because people want to know what is going on in the world.

Soft News. Talk shows are the most profuse supplier of soft news. Call-in shows where people discuss their problems, their questions regarding sports, their opinions on current events—all of these cater to curiosity. Soft news is also programmed from features supplied by syndicators and networks such as commentaries on controversial subjects, background information on stories on the news, and interviews with celebrities.

Qualities Tied to Success

Although radio is an aural medium and TV is basically visual, the qualities that make for successful programming are very much the same. For example, conflict is often an important element in talk radio. These programs really come alive when the caller takes deep issue with the host or other callers. Likewise, radio stations need adequate budget and professional staffing. The latter often translates into the need to have people who really understand the type of music the station plays. A very professional, high-minded person who really knows jazz will be of little use to an easy listening station.

Formats need both durability and consistency. If a format is too narrow in scope, it will not endure. For example, one radio station decided to try an all-Elvis format—it didn't last long. Disco music was short-lived also because the music wasn't varied enough—all of it was sweet and glittery. If a particular station sound is not consistent, listeners will change the dial. Imagine the result if a classical music disc jockey, bored with Beethoven, decided to send out to the listeners Madonna's latest hit. Or vice versa. Similarly people who have set a rock station on their car radio buttons will tune out very quickly if they hear a cooking show when they push the button.

Timing and trends can help bring a radio station success. A station that switches to an all-news format just before a major international or national crisis has the fortune of good timing. One that latches on to a new sound in music that becomes big has cashed in on a trend.

But of all the qualities tied to success, the two that are probably most important for radio are likability and innovation.

Likability. Likability is very important for disc jockeys. They are the single most distinctive element in radio programming and are largely responsible for attracting the audience, especially if several stations in one market program the same type of music. One of the major elements that makes for likability of a disc jockey is energy. This is especially true for radio formats that feature music with high energy.

A number of program directors were asked what they looked for in a disc jockey. Most of the answers related to likability: "upbeat people who don't feel they have to talk a lot," "someone who sounds natural and not like a radio personality," "the ability to relate locally and know the town," "a sense of humor," "energy and a knowledge of music."

Talk show hosts must either be likable or be people you love to hate. Caustic talk show hosts who insult callers certainly do exist on radio and are instrumental in creating conflict. They often engender sympathy for the callers. But the talk show hosts with the greatest longevity are the ones who are the type of people you would like to invite into your home—or automobile.

Innovation. Innovation is supplied primarily by changes in musical tastes. Each generation tends to develop its own style of music. A radio programmer must be attuned to the changes and be aware of when a music format is reaching a midlife crisis and needs a change. The world moves on and what works in radio today may be totally out of date tomorrow.

Usually a programmer has more free rein to be innovative in good economic times than in bad. If enough money is available to cover a bad guess concerning the fickleness of the public, a programmer is more likely to be given the go-ahead to try something new. When economic times are rough, radio, like many other endeavors, tends to be conservative. As Joe Garner, senior national affiliate relations manager of Westwood One network, said during one of the down periods: "Radio is now the playground for the proven rather than the playground for innovation. Programs don't have a chance to grow and find an audience; programmers go with the proven rather than experiment."[5] This should not become the standard operating procedure for radio. Innovation is needed to keep the medium healthy—and wealthy.

PUBLIC BROADCASTING

The noncommercial nature of public broadcasting makes it unique because, although it must have enough money to cover costs, it does not have to concern itself with profit or advertising.

Programming Objectives

Often public broadcasting success criteria vary for different shows. Peter Pennycamp, vice president for cultural programming and program services, explains the different program goals on NPR. "The goal for 'Car Talk' is a large audience because it is a comedy. It's not underwritten because the hosts attack everyone. It doesn't generate money for the stations, but it is very popular with the audience. It's the nearest thing we have to a commercial show. The goal for 'Fresh Air' is that it be picked up in major markets because, with its multitude of segments, it is expensive to produce. The larger markets pay us more. Our goal for 'Wade in the Water' is that 100 stations take it, but it doesn't have to be 100 large stations. 'Bob and Bill' are a comedy team who produce a classical music program in Pullman, Washington. Their program makes classical music accessible, so we look at it as a service. We're happy if 70 stations air it; we have no expectations beyond that." Because goals for particular shows are different, public broadcasters have a wide range of programming objectives.

However, in general, public broadcasters do not seek (or attain) a large audience. The broadcasters need to prove their worth to their funders, but they do not need to do so by delivering an overall enormous audience. They want to attract a large enough audience to make the costs worthwhile and to have a base of people who will donate to a station, but the nature of the audience is often more important than its size.

For this reason, the most important of the objectives to public broadcasters involve a specific target audience, prestige, awards, a particular local or national purpose, and a unique objective that public broadcasters refer to as "units of good."

PUBLIC TV AUDIENCES

Public TV audiences are somewhat different from the population as a whole in terms of education, occupation, and income.[7]

- Education—4+ years of college:
  - General U.S. population—21%
  - Overall public TV audience—24%
  - Public TV primetime audience—27%
- Occupation—Professionals, Owners, and Managers
  - General U.S. population—26%
  - Overall public TV audience—28%
  - Public TV primetime audience—31%
- Annual Family Income—$40,000+
  - General U.S. population—24%
  - Overall public TV audience—28%
  - Public TV primetime audience—31%

A Specific Target Audience. Public broadcasting audiences are intellectually elite and politically influential. Radio listeners are 40 percent college educated compared to 19 percent of the general population. About one-sixth have a graduate degree, and 35 percent earn at least $50,000.[6] PBS publishes a brochure *Who Watches Public TV* that highlights the upscale nature of its audience.

Sometimes public broadcasters have found themselves between a rock and hard place regarding their upscale audience. On the one hand, they are supposed to be an alternative to the commercial networks. They are the closest thing we have in this country to a media entity charged with "giving the people what they need and not necessarily what they want." Public broadcasting's mandate is to be uplifting and to instill in the people a desire to reach a higher cultural plane. But by programming in this manner, it attracts only a small core (usually 2 or 3 percent) of the audience—generally those who are already interested in reaching a higher cultural plane. After all, no one is forced to watch public TV or listen to public radio, and there certainly are many less demanding alternatives available.

But public broadcasting receives most of its funding from the common till—taxes. "Why," ask the members of Congress who must approve its allocations, "should everyone pay for public broadcasting when only a small percentage of the people use it?" Any number of times various groups have pounded on public broadcasting to lower its standards and produce programming that is more accessible to the masses. But when it does so (with such programming as tennis matches, old movies, or sitcoms), it gets hit from the other side. "Tennis matches replacing British drama?" ask the raised eyebrows. "What is public TV coming to?" Unfortunately public broadcasters do not have a middle ground they can retreat to, so they take the heat and generally wind up programming more for the elite than the masses.

Prestige. Because of the type of material it programs, public broadcasting, more than any other media form, is cloaked in prestige. Often companies that underwrite programs on public broadcasting do so because of the prestige involved. They receive, for their money, air credit that many would say very closely approximates a commercial. But because it is on public broadcasting it does not suffer from a hard-sell stigma. As Bill Buzenberg, NPR's vice president for news and information, puts it, "Instead of delivering a mass audience of consumers to advertisers, our purpose is to speak to citizens of a democracy, providing information that a free and educated society cannot live without."

Awards. Public broadcasting's type of programming does win awards, which the broadcasters cherish. Chris Fennimore, program director of WQED, says: "Awards are great for the morale of the people who worked on them. They also bring in the audience." Favorable articles about particular shows, or public broadcasting in general, are also considered a type of award.

Particular Local or National Purpose. The specific local and national programming that public broadcasting is most likely to undertake involves education. Programs are aired to be shown to schoolchildren as part of their classroom studies and to be part of formal college education. Some of these programs are selected and aired by local stations and some come from a special PBS service called Adult Learning Service (ALS). ALS is a cooperative project between public TV and over 1200 colleges. PBS researches, acquires, and transmits (by satellite) college credit courses and helps local stations and colleges schedule and promote the courses so that they can be taken for college credit by citizens. About 250,000 students a year enroll in these courses.[8]

"Units of Good." A buzz phrase within public broadcasting is "units of good." This is acknowledged to be a rather fuzzy term, but it underscores public broadcasting's mission of being an uplifting entity. Dan Everett, director of broadcasting at WGBH, explains, "In order to determine success, we often ask each other 'How many units of good does this program do?'" John Grant, of PBS, echoes, "A program might have a 1 rating, but if we think it has many units of good, we'll keep it on."

One type of program that is considered high in units of good is one designed to make classical music accessible. Dan Jensen, of APR, notes: "Children don't listen to classical music in schools anymore, so they are not familiar with it. They need to be educated in a subtle manner. That's why we are airing 'Schickele Mix' with Peter Schickele. He plays all types of music on his program, comparing classical to rock and roll, Broadway music, and many other forms." NPR's "Bob and Bill," classical music with a comedy twist, is high in "units of good" for the same reason.

Fundamental Appeals to an Audience

Hard news, soft news, music, comedy, and drama are also the basic forms of programming on public TV, but the emphasis and specific content are different.

Hard News. The hard news is more analytical and in-depth than a commercial network newscast. An NBC newscast might devote one minute to the story of a hurricane, emphasizing the death and destruction it has caused. A report on PBS's "The MacNeil/Lehrer Newshour" or NPR's "All Things Considered" might last for 15 or 20 minutes and cover the nature of hurricanes, interviews concerning the last several hurricanes that have occurred, information about how hurricanes are predicted, and/or tips on how to avoid damage from hurricanes.

Soft News. Many more soft news programs are presented on public radio and TV than on commercial TV. In fact, informational programs such as "The Frugal Gourmet," "The World of Ideas with Bill Moyers," "Nova," and "Wall Street Week with Louis Rukeyser" dominate the PBS schedule. Typical public radio informational program titles are "Marketplace," and "Ideas in the Air."

Music. Music presented on both public radio and TV tends to be classical or offbeat—rarely anything that is an up-to-date hit. This, of course, is in line with the fact that public broadcasting is supposed to be an alternative to commercial broadcasting.

FIGURE 8.5
Peter Schickele, host of "Schickele Mix," plays music from various eras to illustrate different concepts. For example, he shows how musicians have skipped up an octave at an unexpected time by using examples from the Beatles, Mozart, and Elton John. (Photo courtesy of American Public Radio.)

Comedy. Comedy is less predominant on public TV than on commercial television—especially that of the sitcom variety. The comedy that is presented tends to be primarily British and hence more subtle in nature than typical American sitcom comedy. Public radio stations often program radio comedies from the 1930s and 1940s—Jack Benny, Burns and Allen, Fibber McGee and Molly.

Drama. The dramas are less action-packed and more relationship-oriented than most of those that appear on commercial broadcasting. They generally take their time to explore facets of character thoroughly. Public radio is about the only place where drama can be heard on radio—some of it from the American past, some of it from the BBC, and occasionally drama programming that is produced by a modern public radio station.

Qualities Tied to Success

The qualities of successful programming that are applicable for commercial and cable TV are equally appropriate for public broadcasting. The upscale, prestigious nature of the programming means some of the qualities are softened a bit—e.g., conflict is a little more slow-moving and psychological, durability has an intellectual element, energy is more cerebral than physical. The area related to qualities of successful programs where public broadcasting has had the most difficulty revolves around one voice of authority.

One Voice of Authority. Public broadcasting (and other educationally oriented institutions such as universities) thrive on committees and advisers—and sometimes advisers to committees. Although some public broadcasting concepts are conceived and executed primarily by one person (e.g., Ken Burns' "The Civil War"), many others are the work of committees. Content experts, educational evaluators, community leaders—all are brought in to give opinions. And usually they are not window dressing. They are leaders in their respective fields who are used to

FIGURE 8.6
The PBS series "The Civil War," which featured photographs such as this one of Camp Griffin in Virginia at the beginning of the war, had one basic voice of authority—that of Ken Burns. It was also high in prestige, awards, and "units of good." (Photo courtesy of Florentine Films.)

having their ideas taken seriously. If ignored, they can cause trouble.

One college-credit science course was undermined by a university professor who felt his ideas had been neglected. This professor, whom we will call Dr. X for anonymity, was one of a committee of professors called in to advise on content. He tried to dominate all the meetings to the extent that the other professors asked the producer to remove him from the committee. Upon hearing rumors of this, Dr. X listed all the problems he saw with the design of the series and then resigned from the committee. When the series aired, he nitpicked the content and presentation, called the press, and "proved" that the course was basically inaccurate and ill-conceived. Although some of his accusations were totally false, the publicity that he stirred up kept the course from ever airing again.

But despite negative incidents of this type, public broadcasting does not seem to want to drift toward one voice of authority. In fact, some of the agencies that fund public broadcasting require in their applications the names of people who will be consulting on program concepts. The longer the list (especially if it is prestigious), the better the chances of being funded.

One of the reasons American Public Radio was formed was that public radio stations thought NPR was too much of a voice of authority. They wanted to have more of a say in their own programming. PBS has recently switched from a programming form with a great deal of input from local stations to one with centralized Washington control. It is too early to say whether this will succeed, but there are indications that the troops are already somewhat restless.

Although the result of the loss of one voice of authority is often chaos, chaos seems to be much more acceptable within public broadcasting than within the commercial broadcasting realm. There are even those who profess that chaos leads to art.

Programmers labor mightily to put their shows together. Juggling the creative elements is a time-consuming and at times exhausting task. But the job is even further complicated by the need to accommodate a host of other considerations that are enumerated in Chapters 9 and 10.

NOTES

1. Interview with John Leverence, Awards Director, Academy of Television Arts and Sciences, January 14, 1992.
2. "Man of the Year," *Time* (January 6, 1992): 23.
3. Speech by Linda Fishman of Lifetime at the Academy of Television Arts and Sciences' "1991–92 Network and Cable Programming Preview," July 13, 1991.
4. "Can Playboy Save Its Skin?" *Channels* (November 1986): 37–40.
5. "Marketing Programming in the 1990s," *College Broadcaster* (November/December 1990): 5.
6. "NPR: The Kids Turn Pro," *Washington Journalism Review* (September 1991): 32.
7. *Who Watches Public Television* (Alexandria, VA: Public Broadcasting Service, 1986): 13–15.
8. *Facts About PBS* (Alexandria, VA: Public Broadcasting Service, 1990): 16.

9 Influences on Programming: Part I

The shaping of a radio or television program involves many hands, some of which are concealed or merely implied. The principal molders, of course, are the creators and their staff people and the programmers at networks, stations, and cable systems. But beyond them, there are a number of others who exercise some influence on the eventual product.

Sales staffs want programming that will attract advertisers; finance departments favor programs that are inexpensive to produce or buy; most sponsors want to avoid controversy; pressure groups object to various aspects of program content; the government has created laws and regulations that affect programming; and on and on.

The purpose of this chapter is to examine the primary groups that influence programming. Generally, it is the person in charge of production who must examine all the contributions and evaluate them. Often the ideas presented are valid, and changes should be made in content or scheduling to accommodate them. At other times, the programmer may feel that the proposed changes are not essential and could jeopardize the project. A programmer who believes that a show feels right the way it is should fight to keep it that way.

COMMERCIAL TELEVISION

First, a preliminary comment about the manner in which producers and distributors differ in their response to outside influences. People connected with the making of a show like to function unfettered. Any interference with their creativity is viewed as a suppression of artistic freedom. Words like "censorship" and "cowardice" are routinely bandied about, and producers find a supportive press ready to publish the charges against the lords of the airwaves. Why not? It's good copy.

However, it is important to realize that producers have little to lose in this struggle against "censorship." The penalties are all borne by the distributors. They are the ones who run the risk of loss of license if the content violates federal rulings, or financial disaster if the program fails to attract advertisers. Conversely, the production company receives its full license fee even if the program doesn't contain a single commercial. To be sure, networks, syndicators, and other large distributors have been conservative to a fault over the years. Profits were huge whether or not the programming broke new ground, so why invite problems? It was only through the determination and feistiness of producers such as Norman Lear, Herbert Brodkin, and Steven Bochco that the medium was lifted into more mature and satisfying productions. However, those advances could not have been made without some people at the network giving a green light with the full knowledge that if something went wrong the downside was mostly theirs.

Station Influence

The symbiotic relationship between networks and stations exercises a certain control on program content. The network may produce the program, but the stations command the distribution base. Both must be satisfied with the product for the system to be successful. If a large number of stations refuse to present a network program, the audience will be dramatically reduced and a huge financial payback to the advertisers will be required. For that reason networks are sensitive to the program evaluations of station programmers and managers. Making decisions in this area is a difficult juggling act because of the enormous range of tastes across the country. A program that is perfectly acceptable in San Francisco may be boycotted in Little Rock. When a network is developing a potentially controversial movie, series episode, or documentary, it will frequently contact a committee of key affiliate managers to take a temperature reading. Many times this advance notice will produce some script adjustments that will avoid a wholesale bailout and an unseemly dispute.

The relationship between syndicators and stations is not nearly as close. Networks and affiliates are joined at the hip throughout the broadcast day, every day, week in and week out. Because stations and syndicators come together just for particular shows, their association is a lot less familial. Also, there is no mechanism set up to notify stations of controversial material, nor are there subcommittees among the stations to monitor and protest content. However, if a production company consistently supplies programs that cause trouble in local markets, the station managers are not shy about voicing their displeasure; and, of course, they hold the ultimate weapon, a cancellation.

The Sales Force

The input from the sales department is often valid—but treacherous. In commercial television, a program must be salable or it won't get on the air. Or, if it does find a place on the schedule, it will quickly disappear if it fails to generate advertiser interest. For this reason, the sales department is in a position to exercise a strong influence on the selection of programs.

In syndication two sales voices must be heard, the one in charge of clearances, the other in charge of barter. If either states that the program is unsalable, the chances are the project will be abandoned. But the ultimate decision-maker, presumably the president of the production company (or distributor), must be very alert. Many salespeople tend to evaluate programs on the anticipated degree of difficulty in selling the show. A program that will sell easily "on the come" will be supported enthusiastically while one that looms as a tough sell will be received coolly. None of this may have any relationship to the inherent strength of the show.

An example of this phenomenon is "USA Today," a syndicated series launched in 1988. When the announcement was made that Gannett was going to produce a daily series based on the style and content of its newspaper, *USA Today*, the fever among distributors was white hot. The publication's flashy graphics and sound-bite features made it appear a natural for television. Since Gannett had no station clearance arm, the distribution of the show would have to be awarded to an outside company. The sales departments of scores of distributors pleaded with their managements to guarantee millions of dollars to the publisher to obtain the account. The buzz in the trade was so positive about the show that salespeople salivated at the thought of representing it.

Program-oriented executives in the industry were less fervid. They remembered that the newspaper was developed as a print version of TV news programs. In effect the TV "USA Today" was a television show based on a newspaper that was based on television shows. Where was the new appeal to the viewer? Furthermore, Gannett insisted that the program appear in primetime access time, which meant it would be seen immediately following local and network newscasts. It was hard to see what content the new series would bring to the audience that hadn't already been presented just minutes ago.

Eventually, Gannett decided to form its own distribution staff, and the program was cleared in virtually every market on excellent stations and in attractive time periods. But it never caught on. Amidst endless program adjustments, it staggered along for sixteen months then finally collapsed at the loss of many millions of dollars.

The experience proved the old adage: there's only one thing worse than making a program you can't sell, and that's selling a program you can't make.

The Finance Department

Members of the finance department used to be dismissed as "those bean counters" and were given very little credence or respect by those in the "glamorous" programming area. But as the economy has tightened and more and more company presidents and station managers have come up through the finance ranks, these cost-conscious executives are now in a position to affect programming decisions.

As with the sales force, their ideas are often valid, but the programmer must be wary that, in an effort to improve the bottom line, the financial people are not allowed to jeopardize the future health of the entire programming schedule. ABC's news producer, Christopher Isham, worried about this in 1986, shortly after Cap Cities, known for its penchant for running a lean shop, took over the network. "When I first got to ABC, if you had a story, you were told to do it. Now for an investigative story you have to go through enormous hoops with the accountants. I don't mind working hard. What I mind is being nickeled and dimed. The accountants have more influence on how we cover the news. They ask whether it's cost effective. There are whole parts of the world we don't cover because it's not cost effective. We have not covered Haiti in the last four months. That was a financial decision."[1]

Controllers want to keep costs down—that's their job. Sometimes programmers can use the finance department as the scapegoat for refusing, or at least negotiating, high actor or anchorperson salaries, license fees, or syndication program expenses. But when costs are cut arbitrarily and quality begins to suffer, programmers should challenge financial dictates. Low quality programs lead to lower audiences, which lead to reduced sales, which lead to an even more devastating bottom line. There is no such thing as a cheap disaster.

Broadcast Standards and Practices

Network programs are subject to certain legal requirements. For example, if an airline has helped to defray the cost of a production, that fact must be disclosed on the air. In addition, network programs must conform to the company's own self-imposed codes of taste and sensitivity. The broadcast standards and practices department (sometimes referred to as "the censor") is responsible for these two functions.

By far the more celebrated of the department's activities is the supervision of program content. Each network has drawn up a booklet that defines the guidelines for the acceptability of material. While some of these policies are hard and fast (no frontal nudity), others are flexible and subject to the context of the production. Over the years interpretations have become looser or more stringent depending on shifts in national attitudes. According to Ted Cordes, NBC's vice president of broadcast standards and practices, "Networks are far more liberal today on language than they were fifteen years ago, and far tougher on matters involving drugs and alcohol."

Because judgments about program substance are largely subjective, programmers and producers frequently view BS&P representatives as the enemy and accuse them of being arbitrary and destructive. "Not so," says Cordes. "We always try to work with the program people to find another way to make the scene effective." Nevertheless, many a late-night edit session has been enlivened by a debate over the exact point at which the director must cut away from a curvy lady's tumbling teddy.

To operate profitably, a network must avoid offending any of its many constituencies: government, advertisers, affiliated stations, and the scores of groups that make up the general public. The wrath of any one of these can inflict heavy losses on the broadcaster. It is the job of BS&P to catch an offensive word or scene before it causes trouble.

In this regard, the experience of NBC is instructive. In 1988 the new management of the company decided that an extensive standards and practices department was unnecessary, and the unit was drastically downsized. The theory was that the executives of the program department, working cooperatively with the producer, would exercise the required taste and judgment to air suitable material. It was an egregious miscalculation. It ignored the fact that both the programmer and the producer were primarily rewarded for attracting large audiences. Whenever they had to choose between a juicy scene with a high rating potential and a softer scene with the probability of lower ratings, they tilted towards the former. During the first twelve months of this new joint "supervision," the network was forced to forfeit $12,000,000 in lost revenue due to sponsor and/or affiliate defections. The BS&P department was substantially increased the following year.

Although programmers may occasionally bristle at the elimination of pet scenes or phrases, they should be happy that someone is there to protect them against themselves.

Top Management

Programmers have bosses. They usually report to station managers, general managers, presidents, or chief executive officers. Often the person the programming head reports to has to go further up the organization chart. For example, the network president may report to the CEO (chief executive officer) of a parent company (NBC to General Electric, ABC to Cap Cities). At the station level, the general manager may be responsible to someone at the group owner (Westinghouse, Gannett, Tribune). These top executives have every right to overrule the decisions made by the program department. Sometimes they do. In fact, occasionally group owners will dictate what syndicated programming a station must play because they have purchased the programs, at a discounted rate, for all their stations.

Influence from top management is not necessarily bad. Programmers can get so bogged down in stress and minutiae that they lose perspective. A good top manager can point out new directions or keep the programmer's eye on the target.

But members of top management who micromanage the programming department or constantly Monday-morning-quarterback the head of programming can be destructive. So can top managers who are timid and unwilling to take risks with new programming concepts.

It should be pointed out, however, that executives (top management, finance heads, sales managers, standards and practices people, and programmers) have shown a good deal of courage over the years in going forward with breakthrough material. "Soap," a 1970s sitcom with a recurring homosexual character; "Maude," a series featuring Bea Arthur in which the lead character had an abortion; and *Something About Amanda,* a movie-of-the-week that dramatized a case of incest, are just a few of the controversial themes that made the air because programming people, supported by top management, decided the topics were important enough to be seen despite threats and acrimony.

Sponsors

One of the strongest influences on programming is the advertiser. Shows that are unable to attract sponsors, or hold the ones they have, will lead short, unhappy lives. Advertisers are extremely sensitive to program content that may alienate or infuriate potential customers. Their goal is to expand gross sales, not extend artistic boundaries. Although there are some who are genuinely concerned about the issue of free expression, their primary obligation is to stockholders, and this imposes severe limits to their programming boldness.

ABC reported that it lost more than $1 million in revenue in 1990 when advertisers pulled out of an episode of "thirtysomething" that featured two guest-starring gay men talking in bed, apparently after having sex. To avoid further financial setbacks, the network decided not to rerun the episode. In the following season the network said it lost $500,000 in another "thirtysomething" episode that depicted the two men accidentally meeting for the first time since the one-night stand.

The dilemma of the broadcaster who must attempt to reflect the realities of contemporary society within the confines of advertiser acceptability was eloquently addressed by Robert Iger, president of ABC Entertainment: "The danger of having to impose some form of content standards because of advertiser pressure is one of the more disturbing parts of my job. I am running a division that has a fiscal as well as a creative and social responsibility, and to maintain a balance between them is sometimes very difficult." His counterpart at NBC, Warren Littlefield, urged advertisers to "treat network programming the way you treat other outlets where you place your message. I know some themes are forbidden, but you have created bold, distinctive advertising messages for the public; we're trying to create a program environment that does the same. All that we're asking is that you take a close look and not run away from themes the audience has a tremendous desire to explore."[2]

Not every wish of an advertiser is honored. When the sponsor's requirements conflict with the integrity of a production, the demands are frequently rejected. A classic example occurred with "Missiles of October," ABC's superb three-hour special that recreated the events of the Cuban Missile Crisis. The incident began when the United States discovered that the Soviet Union was installing offensive nuclear weapons 90 miles off Florida's shore. President John F. Kennedy demanded that Soviet president Nikita Khrushchev remove the launch sites without delay. Khrushchev refused and for thirteen days the world stood poised for a nuclear holocaust.

A Japanese automaker was impressed by Stanley Greenberg's teleplay and offered to sponsor half the program, a purchase worth slightly over $1 million. However, the deal was contingent upon the deletion of two lines in the script. Both were references to Japan's participation in World War II, the most important of which was a statement by Robert F. Kennedy that the president insisted that the air force immediately disperse American bombers and fighter planes. "My brother doesn't want them lined up wing tip to wing tip the way they were at Pearl Harbor when the Japanese nearly wiped out our air force."[3] Greenberg and producer Herbert Brodkin objected to the change, pointing out that it was a major concern of the president and that its omission would make Kennedy look naive and inattentive to the lessons of history. The network agreed, the lines stayed in, the automaker withdrew its sponsorship. The sales department sold the time to a scattering of clients, and the network lost in excess of $500,000 in billings. But to this day the show can be viewed exactly the way the author intended.

Pressure Groups

There are a number of organizations specifically and exclusively set up to monitor television programming. Their mission is to encourage programming that conforms to their standards of taste, political correctness, and morality, and to eliminate programming that they find offensive.

Action for Children's Television (ACT) was one of the most durable of such groups. It was formed in Boston in the 1960s by a group of parents who wanted networks and stations to improve the quality of children's programming, which at the time consisted mainly of cartoons, many of them violent. The organization, led by Peggy Charren, succeeded in convincing the FCC that it should develop guidelines for children's programs that broadcasters were to adhere to concerning children's programming.

These guidelines, issued in 1974, stated among other things that stations would be expected to present a reasonable number of programs designed specifically for children and that they were to use imaginative and exciting methods to further children's understanding of such areas as literature, fine arts, history, science, the environment, human relations, reading, and math. For a number of years these strictures were observed and both stations and networks developed programs such as after-school dramas, science shows, and news programs for young people. Although these efforts were meritorious, they attracted smaller audiences than did traditional cartoon fare and the broadcasters' bottom line was adversely affected. When deregulation came into vogue in the 1980s, the FCC stopped enforcing the 1974 regulations, and fast-paced adventure cartoons returned to dominate the schedule as before. ACT sprung into action again. Through its lobbying efforts, it succeeded in convincing Congress to pass a 1991 law stating that when stations come up for license renewal they will have to prove that they are serving the needs of children. After this success, Ms. Charren declared her mission achieved and disbanded the organization.

Another pressure group is Media Research Center, whose principal concern is the invasion of leftist doctrine in TV programs. Each year it publishes its list of the ten "most biased" shows, programs that "contain liberal messages on issues ranging from the environment and gun control to business, civil rights, and abortion." Among the shows cited in 1991 as "revolving around the left's agenda" were: "A Different World" ("often takes a militant leftist look at racial matters"); "MacGyver" ("everything he does revolves around the left's agenda on the environment, gun control, and foreign affairs"); and "The Trials of Rosie O'Neill" ("almost every episode has some liberal statement to make on feminism").[4]

Probably the most prominent conservative activist is the Rev. Donald Wildmon, founder of the American Family Association. A Methodist minister from Tupelo, Mississippi, the Rev. Wildmon has attempted to influence program decisions primarily through pressure on advertisers. He has established an organization, Christian Leaders for Responsible Television (CLeaR-TV), to monitor programs and administer advertising boycotts. Over the years he has targeted such shows as "L.A. Law" ("perverse and illicit sex"), "Growing Pains" ("perverted family entertainment"), "48 Hours" ("subtle swipes at Christianity"), and "thirtysomething" ("homosexual scene"). He numbers among his boycott successes NBC's *Roe v. Wade,* a TV movie from which advertisers withdrew over $1 million in advertising when it did not support his views against abortion. He also persuaded Pepsico to abandon a $5 million advertising campaign featuring Madonna because of the entertainer's alleged "anti-Christian symbolism in her songs and videos."[5]

Resisting the efforts of these conservative organizations is People for the American Way, a liberal group that encourages the media's freedom of expression. Founded by TV producer Norman Lear, the group carries on a relentless struggle against the "religious right" in law courts, Washington, and the nation's press. The conflict between Lear and Wildmon took an ironic twist in 1991 when the producer introduced a sitcom, "Sunday Dinner," in which one of the lead characters took a moment out in each episode to have a talk with God, thereby establishing her belief in a divine existence. The Reverend was wary and advised his constituents to be alert for "subtle secular humanist sentiments" in Lear's "spiritual odyssey." The matter was rendered moot when CBS canceled the series after six episodes.

Added to all these formal organizations are the ad hoc groups or individuals who protest only when a program offends their particular interest. For example, Terry Rakolta, a Michigan homemaker, didn't like a "Married . . . With Children" episode in which Peg Bundy shopped for a bra. She tried to incite a boycott of the sponsor's products. The movement backfired when many new viewers tuned in to see what all the hollering was about. Undeterred, Ms. Rakolta formed Americans for Responsible Television, a group dedicated to the restoration of family values in television programs.

Any show that touches on a sensitive theme will draw fire. One day it may be the NAACP, the next day the Veterans of Foreign Wars, followed by the National Organization for Women, or the Anti-Defamation League, or the Knights of Columbus. What's a poor programmer to do? "Show some courage," says Steve Mills, for many years CBS's vice president of movies made for television. "Every program will step on somebody's toes. All you can do is make it honest, make it fair, and make it good. And when the heat comes, tell the complainers why you think it's a picture that should be seen, and hold firm to your convictions."

TV Critics

Although frequently derided by executives, TV critics serve as one of the consciences of the industry and sometimes exert an influence on program decisions. The complaint of the producers and executives is that the critics are too lofty in their expectations and fail to understand the realities of public taste. Viewers don't want *Macbeth*; they want "Married . . . With Children," say program makers.

Critics counter by pointing out the audience appetite for such well-crafted shows as "Hill Street Blues," "Cheers," "L.A. Law," and "M*A*S*H." As long as the objective of commercial television is maximum return on investment, the common denominator of programming will always be lower than desired by media critics who tend to be better educated and better read than the average viewer.

However, executives do not like to see their work and their product constantly pilloried in the press. A string of negative reviews can make them more receptive to a quality submission, just to relieve the heat. The net gain might be the viewer's.

Academic and Nonprofit Studies

Universities and foundations frequently examine aspects of television and publish their findings. Although programmers rarely make decisions based on these reports, the studies can create an environment that will eventually influence program content. The National Coalition of Television Violence, a nonprofit organization made up of mental health specialists and media researchers, periodically advises the nation of trends in television violence. For three decades Dr. George Gerbner, of the University of Pennsylvania, has had a staff counting incidents of violence on television programs and conducting various research studies about violence. His studies have helped to create an awareness of the daily mayhem, but in all candor it is hard to see that he has caused any reduction.

One group of studies that probably had an effect on programming were those solicited in 1969 by the U.S. surgeon general to explore the effects of violence on children.[6] These studies, supervised by twelve prestigious researchers, concluded that a modest relationship exists between viewing violence on TV and aggressive tendencies in children. These studies were undertaken at the same time that ACT was promoting reforms in children's television. The reforms that came were the result of various factors, but these research studies were among them.

Sometimes studies are conducted somewhat tongue in cheek. The National Citizens' Committee for Broadcasting undertook a study to determine how many years each network would have to spend in jail if convicted of all the crimes it portrayed in one week. The range: 1063 to 1485 years.[7]

Government

Commercial television transmits its product via signals that pass through the nation's airwaves. These airwaves belong to the public and are therefore subject to the supervision of government. In 1934 Congress passed the Communications Act, which delineated the rights and limitations of broadcasters and established the Federal Communications Commission to carry out the regulations prescribed in the act. The FCC is still the broadcasters' main interface with government.

The Federal Communications Commission. This body is an independent executive agency comprised of five commissioners, no more than four of whom may be members of the same political party, appointed by the president with the consent of the Senate for five-year terms. Among its powers is the ability to grant, renew, revoke or modify broadcast station licenses.

Although networks as such are not licensed, all stations are. Since all of the networks own and operate a number of stations, the government can influence their performance by threatening them with the loss of one or more of their owned and operated stations. In addition, networks do not want to cause problems for their affiliated stations by sending them programming that might give them difficulties at license renewal time. Syndicators, too, are aware that the stations would not be happy with syndicated material that placed their licenses in jeopardy.

License renewal is dependent on many factors such as fulfilling equal employment obligations and broadcasting on the right frequency with the right power. But the quality of a station's programming is also a factor in licensing decisions. The FCC does not proscribe any programming ahead of time; a clause in the Communications Act of 1934 prevents such prior censorship. But another clause in the Communications Act states, "The Commission, if public convenience, interest, or necessity will be served thereby, subject to the limitations of this Act, shall grant to any applicant therefore a station license provided for by this Act."[8] This "public convenience, interest, or necessity" clause has become the keystone for license renewal. Its definition is sufficiently broad for any administration to make life uncomfortable for any station that, in the FCC's judgment, is misbehaving.

TV stations are up for renewal every five years. This makes broadcasters ever mindful of Washington's reaction before approving boundary-stretching concepts. Very few TV stations have actually lost their licenses because of their programming, but it has happened. For example, in 1990 a Chicago station was denied license renewal, in part because the station at one time aired pornographic movies. In 1964 the FCC issued a short-term license to WLBT in Jackson, Mississippi, because of the manner in which it presented racial issues. This station's license was later revoked due to a court ruling.

If the FCC feels a station is doing something improper, whether it is at license renewal time or at some other time, it can take actions other than revoking a license. It can fine a station or it can issue a cease and desist order that puts the station on notice that it is to stop a certain action or it may receive further punishment. The FCC also has the power to issue licenses for less than the full five-year term in order to "encourage" the station to mend its ways so that it can receive a full-term license. All of this should influence programmers who are charged with serving the public convenience, interest, and necessity.

Although the FCC is the main government body that broadcasters deal with, various facets of the executive, legislative, and judicial branches of government can have an impact on programming policies.

Congress. Congress, as already mentioned, approves FCC commissioners and passes laws (such as the Communications Act) that govern or affect broadcasting. In addition, various congressional committees occasionally investigate various aspects of broadcast programming.

One subject that has been investigated at least once a decade is violence on TV. As far back as 1950, Senator Estes Kefauver mused on the Senate floor that perhaps there was too much violence on TV. An investigation into violence that occurred in the 1961–62 season was mainly the result of ABC's attempt to reach its target audience of 18- to 34-year-olds by cranking up sex and violence in action-adventure shows. As the mayhem and heavy breathing increased, so did the audience. In response the other networks went up a few decibels themselves and the nightly din was deafening. Viewers contacted their representatives in Congress, and network leaders were summoned to Washington to be grilled before Senator Thomas J. Dodd and his committee members. The ABC president, Oliver Treyz, was a particular target. His series, "Bus Stop," had presented an episode in which rock singer Fabian had portrayed a psychopath who lived to kill and torture. When a senator asked the beleaguered executive if he allowed his own children to see such a show, the answer was a stammered "no." Shortly thereafter, "Bus Stop" was canceled and so was Mr. Treyz.

In 1993 intense congressional heat was once again being applied to the industry, this time by Senator Paul Simon, D-Ill, author of the 1990 Television Violence Act. The agreement by the four networks earlier in the year to affix viewer advisories to self-evaluated violent programs apparently was not satisfactory to the legislator. "If there is not some indication that [the TV industry] is moving in the right direction, then my colleagues are going to be pushing, and pushing hard," stated the senator to a packed convention of broadcast executives.

Congressional hearings have covered many other programming-related subjects. During a 1950s House subcommittee hearing, Charles Van Doren, the most famous of the "Twenty-One" quiz show winners, confessed that he had been helped with answers in order to defeat contestants who were less appealing to the public. At the time there was no law that prohibited this practice, but it was certainly unfair to other contestants and unfair to the audience, which believed the competitions were genuine. As a result of its hearings, Congress amended the Communications Act, making it unlawful to give such help to a contestant. One of the most recent hearings involved the effects that the network budget cuts of the late 1980s and early 1990s might have on the ability to cover news effectively.

The President. The president and other members of the executive branch do not have as much formal power to influence programming as does Congress. Presidents can veto bills related to broadcasting, just as they can any other bills, and they appoint the members of the FCC. However, the real power wielded over programming by the executive branch tends to be informal.

For example, in the late 1960s and early 1970s, when anti–Vietnam War activities and civil rights demonstrations dominated the evening news, the Richard Nixon administration applied considerable pressure to the networks and stations to present "a more balanced" perspective of the national mood. In the administration's judgment, the "silent majority" of Americans supported the president's policies, but the dissidents were receiving a disproportionate amount of attention because their demonstrations supplied the media with interesting pictures. Although the government never initiated any formal action against the networks, the capacity (some said "implied threat") to do so caused industry executives to monitor very carefully all footage with political implications in order to avoid accusations of bias.

The Courts. The judicial system becomes involved with broadcasting when cases are brought to it. These cases can be brought by individuals or companies that feel they have been wronged by some aspect of commercial broadcasting, or they can be brought by one media organization against another. In addition, if a broadcaster does not like an FCC decision, that broadcaster can appeal through the U. S. Court of Appeals.

When cases related to programming are tried in court, they are handled by lawyers rather than programmers. But programmers are sometimes called as witnesses or they are asked to give background information to the lawyers.

Often the various branches of government interrelate. Congress may pass a general law that is then interpreted by the FCC in a manner unfavorable to broadcasters. The broadcasters can take the FCC ruling through the courts and, if the ruling is still unfavorable, they can go back to Congress to attempt to have the basic law changed. The interrelationship, involvement, and influence of the various branches of government can be seen by looking at several programming issues that are dealt with frequently in the commercial broadcasting world.

Equal Time. During periods of political campaigns, broadcasters are required to make time available in their programming schedules to all candidates running for federal office. They are also strongly encouraged to allow time for state and local candidates. Whenever time is made available to one candidate for a particular office, equal time must be allowed for all other candidates running for the same office. This applies to political commercials as well as programs. If one candidate buys $100,000 worth of time to run political ads, all other candidates must have the opportunity to buy an equivalent amount of time. The "equal time" provision comes from Section 315 of the Communications Act

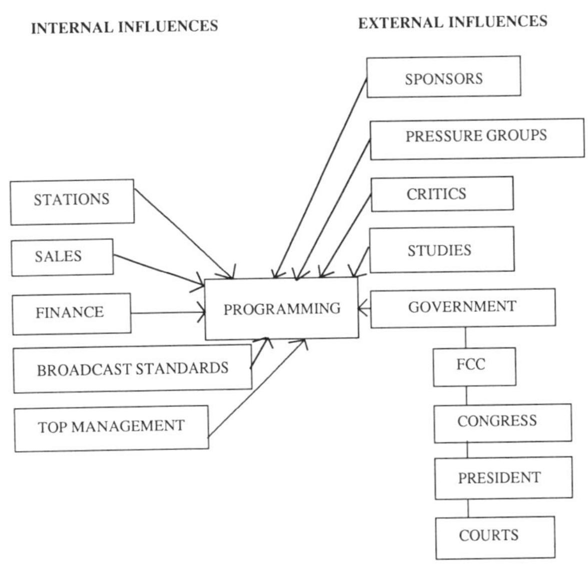

FIGURE 9.1 Various internal and external forces influence programming.

passed way back in 1934. Section 315 actually states that broadcasters "shall afford *equal opportunities* to all other such candidates for that office,"[9] but over the years this section has been commonly referred to as the "equal time" provision.

Interpretations of Section 315 have kept an army of lawyers employed. Various organizations have printed volumes of guidance material for stations to follow, but nevertheless each election campaign usually turns up at least one new wrinkle that must be decided by the FCC and/or in the courts.

The ramifications of Section 315 on news programming have been particularly complicated. If a candidate is legitimately in a newscast for some reason other than his or her candidacy, does a station need to give equal time to all other candidates? At one time, the answer to this was yes. But Lar Daly changed that. Mr. Daly was an eccentric soul who dressed in an Uncle Sam costume and ran (unsuccessfully) for just about every office imaginable. At one point when he was running against the incumbent mayor of Chicago, he demanded equal time because the mayor was seen on a news program. To keep peace, he was given his time, but Congress then amended Section 315 to exempt bona fide newscasts and other news-related programs such as documentaries and news interviews.

The political debates seen on network TV between the Republican and Democratic presidential candidates have also had to undergo exemptions from Section 315 so that minor presidential candidates do not need to appear. In the first year of such debates, 1960, Congress simply suspended Section 315. It refused to do this again in 1964, but the FCC declared that the debates could be arranged by the League of Women Voters and then covered by the networks as a bona fide news event, exempt from Section 315 because of the aftermath of Lar Daly's antics. Eventually the FCC ruled that the broadcasters themselves could stage the debates. The League of Women Voters appealed this ruling to the courts; they had come to like the prestige associated with staging the debates. But the courts sided with the FCC. Now networks or stations can produce debates on their own.

Another thorny problem arises if the local weathercaster decides to run for city council. So far, people with on-air broadcast employment are not exempt from Section 315. Unless the weathercaster can obtain a written waiver from the competing candidates, each time he or she delivers the weather forecast, the station must give equal time to all opposing candidates.

Now that a number of actors have wended their way into politics, a similar problem arises with entertainment programming. When Ronald Reagan ran for president, stations were legally obligated to give equal time to opponent Jimmy Carter anytime they ran an old movie with Reagan in it. Most stations avoided the problem by making sure they did not run any old Reagan movies for the duration of the campaign, but Carter also made light of the situation, joking that Reagan would probably lose votes if his old movies were shown.

These are but a few of the problems that have arisen regarding Section 315. Programmers should be well versed in the provisions, exclusions, and aberrations of this section of the law—and they should keep legal counsel handy during election times.

The First Amendment. Obviously, the First Amendment was passed by Congress long before anyone envisioned commercial broadcasting. However, the provision for freedom of the press, intended in those days for newspapers, has been extended to radio and television.

One of the main issues that people involved with news programming have to address rather frequently is the placement of the line between censorship and clear and present danger. When newspeople feel the government is censoring the information given to them, they cry that their freedom of press is being violated. The government often retorts that it is withholding the information because releasing it would place the country in a dangerous position. This issue was debated frequently during the Persian Gulf conflict of 1990. The government refused to give out certain information about location of troops and numbers of casualties. Reporters, feeling the public had a right to know, tried to gather the information themselves—often endangering their own lives in doing so.

Reporters, in the course of their duties, frequently gain access to information that is considered confidential. News editors, producers, and top managers must decide when to release such information publicly and when to hold it in confidence. This whole confidentiality issue is further complicated by the fact that sometimes reporters are asked by the courts to divulge information about facts and sources they promised to keep secret. Most states have passed "shield laws" to protect reporters from having to divulge material on the whim of the courts. But most of these laws state that broadcasters must divulge the information if it would prove to be crucial evidence in a case and if it cannot be obtained any other way. What this boils down to is that reporters must be very selective about offering confidentiality.

The First Amendment has been used in broadcasting for many issues other than the release of confidential information. Journalists cited the First Amendment in order to retain the right to interview people as they were leaving polling places. These so-called exit polls had been opposed by politicians because they felt it affected the votes of those in the western time zones. One of their points was that if people in the West felt that the major candidates had already won or lost, they would not even bother to vote. But broadcasters argued successfully that under the First Amendment, they had the right to conduct exit polls.

In the area of entertainment programming, the First Amendment was used successfully to abolish "the family hour." For a little over a year in the mid-1970s, the networks, in a self-regulatory effort to quell criticism, required that primetime programs aired before 9:00 p.m. contain little or no sex or violence because a large number of children constituted the audience. Writers, producers, and directors opposed this family hour concept and said that it had come about through subtle pressure from the FCC. The writers and others successfully used the First Amendment to claim that the FCC, National Association of Broadcasters, and networks had conspired to censor primetime programming by requiring content that was suitable for children.

Obscenity and Indecency. Broadcasters are often confronted with situations in which obscenity (outlawed by the U.S. Criminal Code) and the First Amendment clash. The difficulty usually revolves around the definition of obscenity. Its present definition stems from a 1973 court case, *Miller* v. *California*. According to this decision, in order for a program to be obscene it must contain the depiction of sexual acts in an offensive manner, must appeal to prurient interests of the average person, and must lack serious artistic, literary, political, or scientific value. The difficulty is that material is perceived differently by various individuals. One person's art is another's obscenity. Concepts also change over time. What was offensive in one decade may be perfectly acceptable in the next.

Another problem is that obscenity and indecency are often confused. Indecency is not as serious as obscenity and is generally considered to occur when something is broadcast that is offensive in relation to the standards of a particular community. This is why station programmers should preview network material and, in some instances, legitimately preempt it for something more in line with community acceptance. Both obscenity and indecency can lead to the wrath of the FCC. Its rulings totally bar the broadcasting of obscene material and bar the use of indecent material between 6 a.m. and 8 p.m.

Commercial TV, because it takes its over-the-air family-oriented function fairly seriously, is rarely castigated for obscenity or indecency. However, as noted earlier, in 1990 a Chicago station license was revoked because of adult movies. Also, in 1988, a Kansas City, Missouri, station was fined $2000 for airing *Private Lessons*, a film about an older woman trying to seduce a young boy.

Libel and Invasion of Privacy. Libel involves defaming someone's character in a way that affects that person's reputation or livelihood. Generally, the person who feels libeled takes the broadcaster to court. One of the idiosyncrasies of libel is that people who are considered public figures have to prove that the disparager engaged in actual malice. In other words, famous people have to prove that the journalist was out to get them and purposely broadcast false information. Honest mistakes don't count. But ordinary folks can win libel cases when something false and defamatory was reported about them, even if it was an honest mistake.

Invasion of privacy is related to libel in that it often leads to information that is damaging to a person. But the thrust here is on how the information is gathered. Invasion of privacy laws are the province of the individual states, and in most states invasion of privacy means that a person has a right to be left alone. Reporters who snoop around taking photos through bedroom windows are certainly setting themselves up for invasion of privacy charges in any state.

Libel and invasion of privacy are most likely to occur in the areas of news and documentary programming as well as with the more tabloid-oriented talk shows and with some of the reality-based shows. Sitcoms and totally fictional dramas have little to fear from these laws.

A person in charge of programming cannot check every fact or every method of gathering information that is supplied by a staff member. However, programming heads should be careful when hiring people who may be confronted with temptations to commit libel or invasion of privacy. These people should be honest, ethical, and knowledgeable about the laws.

Fairness. For many years stations, and hence networks, were required to make sure that all points of view were given on controversial issues. This requirement stemmed from the "fairness doctrine" that wasn't really a doctrine at all but rather a series of FCC statements, court hearings, and interpretations of the Communications Act and its amendments.

During the deregulatory 1980s, the doctrine underwent an interesting transformation that illustrates once again the interrelationship of the various parts of government when it comes to programming principles. In 1981, the FCC asked Congress to repeal the fairness doctrine, but Congress did nothing. In retrospect, this was a proper action. Congress probably didn't have the authority to repeal something that wasn't codified as a law.

In 1985 the FCC undertook a study of the fairness doctrine and concluded that it was not in the public interest. About the same time, a court decision said that the doctrine wasn't a law but rather was an FCC regulation, so the FCC had the authority to abolish this. This raised the awareness level of some of the senators who, confusing fairness with Section 315, were afraid their television advertising time would be impaired. Such was not the case because none of the documents related to the fairness doctrine dealt directly with the issue of politicians; they dealt only with controversial issues.

But the senators, some of whom honestly believed in the concepts espoused by the fairness doctrine and some of whom were still confused, passed a 1987 bill to codify the provisions of fairness. However, President Ronald Reagan vetoed this bill. Shortly thereafter the FCC stopped enforcing fairness, in effect abolishing the concept. And that's where it stands except for occasional senatorial remarks about drafting another bill.

Meanwhile, most station programmers still attempt to be fair in their treatment of controversial subjects, but the decisions on what constitutes fairness can be made in-house without fear of government interpretation.

Other Government Laws and Regulations. There are a myriad of other laws, regulations, and court decisions that affect the day-to-day operation of a programming department. Stations that editorialize (stations are not required to do this) must be sure they make a positive effort to find people with opposing viewpoints and give them a chance to reply.

Copyright laws and court rulings can affect the content and production of programs. The whole financial interest–domestic syndication issue discussed in previous chapters stems from an FCC ruling that is presently being challenged in the courts. Changes in this could drastically affect the amount of programming undertaken by the networks. Changes in primetime access (another FCC ruling) could affect networks, stations, and syndicators.

Keeping up on all of this is difficult. That's why there are attorneys who specialize in communications or entertainment law. Programmers should consult these lawyers whenever necessary—preferably before the fact so that damage can be prevented. Programmers who need lawyers to bail them out every few months will discover they need to find some other form of employment.

Commercial broadcasters are not alone in being affected by government and other influential groups. The spheres of influence also extend to cable, radio, and public broadcasting, as we shall see in the next chapter.

NOTES

1. Ken Auletta, *Three Blind Mice* (New York: Random House, 1991): 93.
2. "Littlefield Exhorts Ad Execs: Stand Up to Watchdog Groups," *Variety* (January 24, 1991): F-4.
3. Stanley R. Greenberg, "Missiles of October" script, ABC Television Network, January 1974.
4. "'Most Biased' TV Shows Listed by Right-wing Group," *The Hollywood Reporter* (June 7, 1991): 3.
5. "Viewpoint: Censorship Today," *Forum* (Spring/Summer 1991): 2.
6. George Comstock et al., *Television and Social Behavior: A Technical Report to the Surgeon General's Scientific Advisory Committee on Television and Social Behavior* (Washington, DC: Government Printing Office, 1972).
7. Max Gunther, "All That TV Violence: Why Do We Love/Hate It?" *TV Guide (November 6, 1976): 6–10.*
8. Public Law No. 416, June 19, 1934, 73rd Congress (Washington, DC: Government Printing Office, 1934): Section 307(a).
9. Ibid., Section 315.

10 Influences on Programming: Part II

CABLE TELEVISION

Cable TV programming is subject to much less outside influence than programming on commercial TV. One reason for this stems from cable's narrowcasting philosophy. Pressure groups that want specific types of programming have difficulty pushing their demands on a network with a very limited concept. ACT would be silly to go to CNN and demand children's programming. The CNN programmers could answer, "We don't do that." Nickelodeon and Disney are channels devoted to children's programs, and ACT has been very pleased with those two services.

In addition, cable came to fruition during the deregulatory era of the 1980s. During a time when the government was rescinding regulations regarding broadcasting, it was not about to create them for cable. In fact, the appearance of cable helped commercial broadcasters throw off some government shackles. Regulation of broadcasting was based largely on the "scarcity theory." Because the airwaves allowed for only four or five channels in most cities, very few people could own commercial stations. Therefore, the government felt obligated to protect this rare commodity and make sure it was operated in the public interest. But when cable came along, just about anyone could start a network, so scarcity was no longer a viable idea. The fractionalization of programming and audience, along with the philosophy of deregulation, led to a limiting of controls on cable programming and a lessening of controls on broadcasters.

Government

Cable TV, however, has not been totally free of government influence. In its early days (the late 1940s through the mid-1970s), the FCC was reluctant to regulate cable because the signals did not go through the public airwaves but rather through wires. The FCC was not sure whether or not it had jurisdiction over this wired television and, as a result, stayed at arm's length. However, it did make rulings that affected cable in areas where cable and broadcasting overlapped—or clashed. For the most part, these regulations were made in an attempt to protect broadcasting interests from competition that might be created by cable. This led to rules regarding distant signal importation, syndicated exclusivity, and must carry—all of which are still in the regulation hopper today.

Distant Signal Importation and Syndicated Exclusivity. The principles related to distant signal importation and syndicated exclusivity rules were major sore points between broadcasters and cablecasters early in television history. Many of the first cable TV systems were established in areas that had little or no TV. They imported signals from distant cities so that their subscribers would have something to watch. This distant signal importation didn't sit well with TV stations in the area—they didn't much care for the competition. To make matters worse, sometimes the same program would be playing on an imported station that was playing on a local station. WXXX, which was licensed to Ourtown, would find that the "I Love Lucy" rerun it was playing was also being shown on the imported station, WZZZ, from Theirtown. If WXXX lost half its audience to WZZZ, it could only charge half as much for its commercial time. In addition, WXXX had bought what it thought were exclusive rights to program the syndicated version of "I Love Lucy" in Ourtown. It wanted to maintain this syndicated exclusivity.

The stations went pounding on the FCC's door to prevent distant signal importation or at least to make the cable systems black out the duplicate programs so that stations could have syndicated exclusivity. In 1965 rulings were passed that favored broadcasters in that they limited the number of stations cable systems could import and also required the cable systems to black out duplicate programs. Broadcasters were happy; cablecasters weren't. When superstations were introduced, they were considered distant imports. Therefore, the rulings on distant signal importation and syndicated exclusivity applied to WTBS, WOR, WGN, and the other nationally distributed stations.

As the cable industry grew and gained clout, it appealed the rulings related to distant signal importation and syndicated exclusivity and sometimes won. Broadcasters would then initiate their own appeals. Throughout the years, these regulations have undergone numerous rulings, re-rulings, reinstatements, and revocations.

As of this writing a very complicated formula exists for distant signal importation that limits the number of stations a system can import based on its market size and the number of stations in the market. Syndicated exclusivity is being enforced, so if a superstation or other distant import is programming something within thirty days of when a local station plans to program it, the program on the superstation must be blacked out or replaced with something else. This is a logistical pain to programmers at cable systems, but the superstations have tried to help by assuming the burden of trying to make sure they don't show programs that might be aired on local stations.

Must Carry. In general, must-carry rules state that cable TV systems have to allocate channel space to the local broadcast stations. Must carry was part of the same 1965 regulations that governed distant signal importation and syndicated exclusivity, but no one was concerned about carrying local stations because cable didn't have much other programming to place on its systems.

But when all the cable networks arose in the early 1980s, many cable systems, especially those in large cities with many VHF and UHF stations, found themselves hard up for channel space to accommodate all the local channels and the cable networks. Why, they reasoned, should they have to carry three PBS stations that programmed virtually the same thing and four different religious stations that hardly anyone in their franchise area watched? Wouldn't they be of more service (and, by the way, earn more money) if they could use some of those channels to add pay services such as Cinemax or The Movie Channel?

So, egged on by Ted Turner, who wanted channel space for all his many services, the cable system operators besieged the FCC to convince the commissioners that must carry should be repealed. Of course, the broadcasters were fast on their heels to convince the FCC to the contrary. For the better part of a decade the FCC repealed must carry and then the courts upheld it—or vice versa. Sometimes must carry, distant signal importation, and syndicated exclusivity would all be on the FCC docket at the same time, and the FCC, cable systems, and stations would try to work up compromises that were not always logically connected—"I'll swap you must carry for three distant signal importations."

As of the present, stations and cable systems are busily negotiating compensation arrangements. See Chapter 3 for more detailed explanations.

The First Amendment. In addition to the above rulings that are specifically cable related, cable television has found itself influenced by other more general government policies. Certain individual channels are subject to more involvement than others. CNN, for example, has encountered many issues related to the First Amendment.

In 1990 it became embroiled in a controversy over prior censorship. CNN had obtained tapes of conversations between deposed Panamanian dictator Manuel Noriega and his lawyers. It began airing these tapes, but a Miami judge, fearing that the exposure would jeopardize the ability to conduct a fair trial related to Noriega's drug charges, ordered CNN to stop cablecasting the tapes and turn them over to him. He wanted to listen to the material and then decide whether or not the tapes could be played. CNN retorted that this was prior censorship, which, as stated earlier, is outlawed by the Communications Act. The network took the issue to the Supreme Court, but it sided with the Miami judge. The high court's decision was based on the fact that the judge was not imposing prior censorship because his ban was only temporary in order to decide whether a permanent ban was needed. CNN called this "a loss of a battle in a continuing war against censorship."[1] Eventually the network was allowed to play the tapes, but the fear this decision left in members of the press was that more judges would issue "temporary" restraints.

FIGURE 10.1 Of all the cable channels, CNN is most likely to have problems with journalistic pursuits versus the First Amendment. Photo courtesy of and © Cable News Network. All rights reserved.

Obscenity and Indecency. Obscenity and indecency have been issues for several cable channels. Playboy has been threatened with legal action related to these subjects. But because people must pay separately for the Playboy channel in order for it to enter their homes and because the programming has never been deemed obscene, the channel still exists. However, many systems do not carry Playboy due to the public outcry against it in their communities. The pay movie channels have also come under obscenity-indecency-related attacks because they do not edit movies with explicit sex scenes and/or graphic violence. But, as with Playboy, those who are offended are free to cancel their subscriptions.

Legal action has closed down one explicit sex-oriented channel. XXXtasy was a scrambled channel that could be obtained only by subscription. But in 1990, several enterprising (and lawbreaking) teenagers in Alabama broke its code and taped its programs and then sold the tapes to fellow students. An Alabama district attorney charged XXXtasy with obscenity, and a grand jury indicted the company. Rather than fight the case, XXXtasy closed up shop.

The basic channels can be faced with issues of obscenity, too. A&E, for example, has a policy of editing out frontal nudity from films that it cablecasts. However, Woody Allen has a clause in his contract that prevents altering his movies in any way. So when A&E is faced with a Woody Allen movie that has a slight bit of frontal nudity, it must decide whether to shelve the film or disregard its own policy.

Top Management

The main external influence on cable programming comes from within the cable structure—the multiple system owners (MSOs). Cable TV has a great deal of vertical integration wherein large companies have ownership roles in organizations that produce programs, networks that distribute programs to systems, and systems that exhibit networks to the consumer. For example Viacom, which is a program syndicator and producer, is the third largest cable system owner. In addition, it is sole owner of six cable networks and part owner of three more.

The result is that MSOs frequently have a great deal of say as to which cable networks their cable systems carry. They quite naturally want their cable systems to offer all those networks in which they have a financial interest. Doing so fills the MSOs, coffers. For this reason some cable networks seek financial ownership on the part of the MSOs; in that way they can be included in the "family" of channels chosen by the MSO. The downside, of course, is that "he who pays the piper calls the tune." These MSOs will want to have some say in programming philosophy.

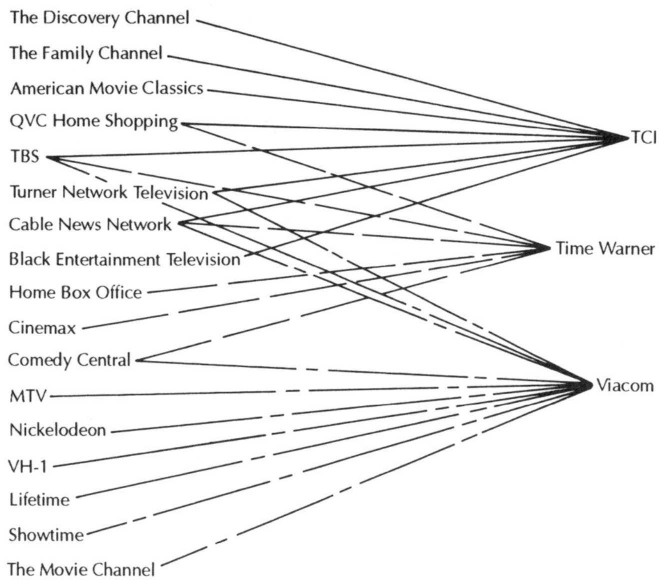

FIGURE 10.2
This chart shows three of the largest cable system owners (TCI, Time Warner, and Viacom) and the cable networks in which they hold a financial interest. Sources: "The Cable Network Programming Universe," *Broadcasting* (January 7, 1991): 94–95; and "National Cable Programming Services," *Variety* (November 15, 1991): C-10–C-12.

Some MSOs like to select virtually all the network services their systems should carry—even ones that they don't own. The degree of MSO control depends primarily on the philosophy of top management. TCI, for example, makes most of the channel carriage decisions for its systems. ATC (owned by Time Warner) gives much more local autonomy to its systems.

Arguments can be made for both approaches. Centralization produces economies of scale. If the MSO is undertaking all of the buying for all its systems, it can sometimes negotiate better deals from the networks than individual systems can. This centralized approach also cuts down on the need for local system programmers, a plus from the balance sheet point of view (but not from the point of view of those who want jobs in programming). On the other hand, such a policy is not likely to be responsive to local needs. Not all communities are demographically similar. Some skew older or more rural than others. Because so many cable channels narrowcast, the services desired by the citizens of one community might be quite different from those of another. It would be a disservice to the subscribers if a system loaded up on all the children's services available in a community consisting almost totally of adults.

COMMERCIAL RADIO

The influences on radio are more similar in form and intensity to commercial television than are those of cable TV. But they are not as intense as those exerted against TV, mainly because radio is more fragmented and less visible (no pun intended) than TV.

Stations, networks, and syndicators have somewhat of a symbiotic relationship, but they do not really influence each other a great deal. Stations do not live or die by their network or syndicated programming. Networks affiliate with so many different stations that the ties to any one station are not that strong. Likewise for syndicators.

When a radio station is selecting or changing a format, it is subject to a great deal of influence from owners, station representatives, the sales department, and others who have something to gain (or lose) from the station's overall prosperity.

But salespeople, accountants, group owners and the like rarely express opinions on individual aspects of programming. If a controversial disc jockey is making it hard for the salespeople to sell time, they might complain, but so much of radio involves selling spots that appear within various time segments that advertisers do not usually react to having their ads near certain musical selections—which, of course, come and go very quickly. No broadcast standards department breathes down the neck of radio programmers, but the program directors themselves are expected to police what is aired and keep it in line with community standards. Records are reviewed by critics, but radio programming itself is not subject to much critical review.

Although there are academic studies on the effects of radio programming, they are certainly not as profuse as those related to TV. In fact, the National Association of Broadcasters has a grant program designed to encourage people to undertake research studies involving radio; it has few takers.

Pressure groups have a valid outlet on radio. Many talk shows are there simply to be a forum for public opinion, so, unless pressure groups become bothersome by tying up the lines, they are welcomed.

Government

Radio has had a longer history than television with government and has lived through several ups and downs of regulatory philosophy, all of which have affected programming.

The Federal Communications Commission. During the early years of radio when the Federal Radio Commission (the forerunner of the Federal Communications Commission) oversaw radio, a number of licenses were revoked because of improper programming. One of these was for a station run by Dr. J. R. Brinkley, who prescribed medical treatments over the radio. Usually his listeners' health problems could be cured only by special prescriptions obtainable from druggists who belonged to a pharmaceutical association that Dr. Brinkley owned. One of his most creative cures involved goat gland treatments to improve male virility. The FRC took a jaundiced view of most of these medical treatments, but Dr. Brinkley was crafty, so it took some fancy legal maneuvers to actually remove him from the airwaves.

A few other early stations had licenses revoked for programming policies, mainly those of engaging in bitter attacks against individuals or groups—politicians, Jews, prostitutes, judges. But overall, the number of station revocations was small.

In 1946, the FCC issued an eighty-page document detailing what stations should do in order to have their licenses renewed. This included numerous details about airing local public affairs programs, keeping commercials limited, and maintaining a well-balanced programming schedule. Radio broadcasters immediately dubbed this document the "Blue Book," in part because of its blue cover but more because blue penciling is associated with censorship. The reaction to the Blue Book was so negative that the FCC never really implemented its provisions.

In the 1960s and 1970s radio stations, and their TV counterparts, underwent elaborate license renewal that involved a process called ascertainment. Station personnel had to interview community leaders about problems in the community and then propose program ideas to deal with these problems. They would make promises concerning airing these programs in documents submitted to the FCC for their license renewal. Then when their next renewal came up (at that time station licenses were up for renewal every three years), they had to prove that they had aired the programs on community problems that they had promised in their last renewal. This process was called "promise versus performance."

Pressure groups of local citizens were also very involved in license renewal during this period. They would demand that stations air certain types of programs (more children's programs, more Chicano programs, more programs about community organizations). If the stations did not respond to their liking, they would write nasty letters to be included in the stations' file at the FCC, submit formal petitions to deny the station its license renewal, or even ask to take over the station themselves.

In general, license renewal during this period was a very complicated, paper-intensive process. Stations usually sent large boxes of documents to Washington where they were supposedly read by FCC staffers. The license renewal process placed great burdens on the programming departments in terms of interviewing community leaders about problems, dealing with pressure groups, and preparing much of the mass of paper that was submitted.

And then came deregulation. Now radio station licenses are looked at only every seven years (TV is five years) rather than every three. Ascertainment is gone, as is promise versus performance. Radio stations don't even have to program news or public affairs if they don't want to. And license renewal can be obtained by filling out a postcard-sized form. The programming department is hardly involved.

Obscenity and Indecency. Obscenity and indecency charges arise most often in relationship to talk show topics and music lyrics. For awhile during the 1980s, a phenomenon known as "topless radio" spread around the country. People would call in and recount to a talk show host (and the listening audience) their explicit sexual experiences. After the

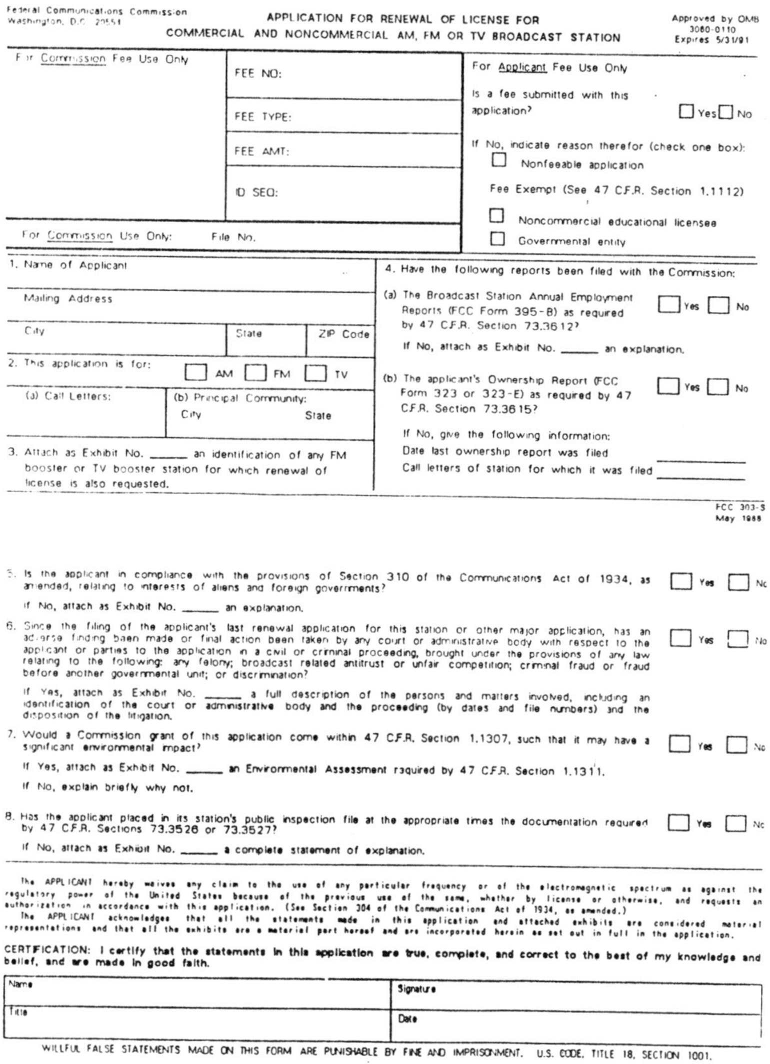

Federal Communications Commission
Washington, D.C. 20554

APPLICATION FOR RENEWAL OF LICENSE FOR
COMMERCIAL AND NONCOMMERCIAL AM, FM OR TV BROADCAST STATION

Approved by OMB
3060-0110
Expires 5/31/91

For Commission Fee Use Only

FEE NO:

FEE TYPE:

FEE AMT:

ID SEQ:

For Applicant Fee Use Only

Is a fee submitted with this application? ☐ Yes ☐ No

If No, indicate reason therefor (check one box):
☐ Nonfeeable application
Fee Exempt (See 47 C.F.R. Section 1.1112)
☐ Noncommercial educational licensee
☐ Governmental entity

For Commission Use Only: File No.

1. Name of Applicant

Mailing Address

City | State | ZIP Code

2. This application is for: ☐ AM ☐ FM ☐ TV

(a) Call Letters:

(b) Principal Community:
City State

3. Attach as Exhibit No. ______ an identification of any FM booster or TV booster station for which renewal of license is also requested.

4. Have the following reports been filed with the Commission:

(a) The Broadcast Station Annual Employment Reports (FCC Form 395-B) as required by 47 C.F.R. Section 73.3612? ☐ Yes ☐ No

If No, attach as Exhibit No. ______ an explanation.

(b) The applicant's Ownership Report (FCC Form 323 or 323-E) as required by 47 C.F.R. Section 73.3615? ☐ Yes ☐ No

If No, give the following information:
Date last ownership report was filed ______
Call letters of station for which it was filed ______

FCC 303-S
May 1988

5. Is the applicant in compliance with the provisions of Section 310 of the Communications Act of 1934, as amended, relating to interests of aliens and foreign governments? ☐ Yes ☐ No

If No, attach as Exhibit No. ______ an explanation.

6. Since the filing of the applicant's last renewal application for this station or other major application, has an adverse finding been made or final action been taken by any court or administrative body with respect to the applicant or parties to the application in a civil or criminal proceeding, brought under the provisions of any law relating to the following: any felony; broadcast related antitrust or unfair competition; criminal fraud or fraud before another governmental unit; or discrimination? ☐ Yes ☐ No

If Yes, attach as Exhibit No. ______ a full description of the persons and matters involved, including an identification of the court or administrative body and the proceeding (by dates and file numbers) and the disposition of the litigation.

7. Would a Commission grant of this application come within 47 C.F.R. Section 1.1307, such that it may have a significant environmental impact? ☐ Yes ☐ No

If Yes, attach as Exhibit No. ______ an Environmental Assessment required by 47 C.F.R. Section 1.1311.

If No, explain briefly why not.

8. Has the applicant placed in its station's public inspection file at the appropriate times the documentation required by 47 C.F.R. Sections 73.3526 or 73.3527? ☐ Yes ☐ No

If No, attach as Exhibit No. ______ a complete statement of explanation.

The APPLICANT hereby waives any claim to the use of any particular frequency or of the electromagnetic spectrum as against the regulatory power of the United States because of the previous use of the same, whether by license or otherwise, and requests an authorization in accordance with this application. (See Section 304 of the Communications Act of 1934, as amended.)

The APPLICANT acknowledges that all the statements made in this application and attached exhibits are considered material representations and that all the exhibits are a material part hereof and are incorporated herein as set out in full in the application.

CERTIFICATION: I certify that the statements in this application are true, complete, and correct to the best of my knowledge and belief, and are made in good faith.

Name	Signature
Title	Date

WILLFUL FALSE STATEMENTS MADE ON THIS FORM ARE PUNISHABLE BY FINE AND IMPRISONMENT. U.S. CODE, TITLE 18, SECTION 1001.

FIGURE 10.3
The front and back of the postcard-sized license renewal form required by the FCC.

FCC fined one station for this and the fine was upheld in the courts, "topless radio" stopped. Of course, people still call radio talk shows and talk about sexual experiences, but not as explicitly as they were encouraged to do for the brief era of "topless radio." Call screeners, station employees who talk to callers before they are allowed on the air, have been instructed to screen out anyone who might talk too explicitly.

Another indecency case is presently wending its way through the courts. Group station owner Infinity Broadcasting has been charged with broadcasting indecent language used by air personality Howard Stern. Stern, whose programs receive high ratings in markets where he is aired, dwells on sexual matters, including sexual intercourse, orgasm, masturbation, lesbianism, homosexuality, breasts, and nudity of male and female genitalia. The FCC fines levied against Infinity total $600,000. The company is appealing through the courts. Stay tuned.

Problems arise with music lyrics when they contain words used to discuss sexual or excretory acts. Once again this is a gray area in terms of what is appropriate from one time to another and from one community to another. During the 1980s, a group of wives of Congressmen made a strong case against sexually oriented lyrics, suggesting that records with such lyrics be so labeled and that such material be kept off radio. About all they succeeded in doing, as far as radio was concerned, was taming things down for awhile.

For the most part, station programmers make individual decisions on what their particular station should or should not air. For example, the rock group Guns N' Roses is known for its boundary-testing language. Some stations won't play the group at all; some make their decisions on a song-by-song basis; and some play virtually everything they record. As a program director for a station that plays most of the Guns N' Roses cuts explained, "It's the language of our audience. We have an aggressive, 18- to 24-year-old male audience and a lot of these are people like Axl Rose, angry young men. They use colorful words."[2]

Sometimes radio station program directors make decisions about whether or not to play a song based on how understandable the lyrics are. A rock station deejay noted about one such track: "Unless you read the lyric sheet or are an incredibly huge fan of the band, it's hard to discern what they're saying."[3] Then, too, if rock groups fear their music might not be played because of certain offensive words, they will change the words on a recording that is sent to radio stations but leave the original words on records and CDs sold in stores.

Other Government Laws and Regulations. Government regulations other than those related to license renewal and obscenity also affect radio programmers. In the 1950s Congress held hearings on payola (the accepting of gifts in exchange for playing a record on the air) and amended the Communications Act to prevent this practice. Payola still occurs on occasion, but now it is patently illegal. Plugola (promoting a certain restaurant, concert, music store, etc. in exchange for favors) is also a station no-no.

The U.S. Criminal Code says radio and TV stations cannot hold lotteries. A contest is deemed a lottery if people have to pay to enter, if chance is involved, and if a prize is given. Radio stations usually avoid lotteries by ensuring that people do not have to pay to enter their contests.

Of course, radio stations, like TV stations, are subject to equal time, libel, invasion of privacy, and other government laws and regulations. Issues related to these are more likely to come up for networks than stations because decisions are likely to involve news, and, except for all-news stations, radio stations do not concentrate on gathering news—they obtain it from networks.

PUBLIC BROADCASTING

The influences and interrelationships between public broadcasting stations and networks are quite strong. In both public TV and public radio, stations pay the networks for programs. They are the networks' primary customers and the frontline informants. If a significant number of NPR affiliates feel that the host on a music show talks too much, NPR programmers will tell the host to edit the material. When stations need time within programs to conduct their money-raising activities, the network makes sure the programs are organized in such a way that the stations can cut away as needed.

Sometimes the stations send different signals, however, making it difficult for the network to decide how to treat certain programming material. For example, in July of 1991, 174 of PBS's 284 stations refused to carry "Tongues Untied," an episode in the documentary series, "P.O.V." The show described a highly personal account about the search for black gay male identity. It contained graphic language, some nudity, and one scene showing two men kissing. Only three stations in the top twenty markets dropped the program, but twelve of the eighteen statewide systems that cover smaller markets turned it down. Marc Weiss, executive producer of "P.O.V.," explained the difference in attitude: "Statewide systems are supported by state legislative funds, which makes them more vulnerable to angry local citizens. If a station gets fifteen calls, that alarms them because they are so dependent on fund raising."[4]

Although the influences from within the public broadcasting family can be significant, those from without tend to be even stronger. Part of this is because public broadcasting is "public." It is funded, to a great degree, by the government through taxes, and that in itself makes it subject to influence from the government and pressure groups. Added to that is the fact that another significant portion of its income comes from public donations, leading the public to feel they have all the more right to speak their singular or collective minds. The other main source of income is corporate underwriters, and although they are not supposed to be involved with program content, such is not always the case. In fact, on occasion, they have caused more commotion than commercial broadcasting's advertisers.

Another reason public broadcasting is subject to influence and attack is that its programming tends to be more controversial than commercial broadcasting. Public broadcasting's mission is to deal with problems and artistic breakthrough; these are by nature controversial. Commercial TV's goal is to make money; controversy is an anathema. As NPR's Peter Pennycamp has put it, "Public radio and TV programs have edges; in commercial broadcasting, the edges have all been rounded off."

Pressure Groups

Both public radio and public TV tackle controversial subjects and frequently take stands on those issues. This engenders resentment on

FIGURE 10.4 "Spirits of the Present: The Legacy of Native America" had an advisory board from Native American communities that suggested, among other things, that the series offer an exploration of different topics as told from the viewpoint of contemporary Native American people. The series was a co-production of Radio Smithsonian and the Native American Public Broadcasting Consortium and was aired by American Public Radio. (Photo courtesy of American Public Radio; Susan Braine, photographer.)

the part of various pressure groups, but, for the most part, public broadcasters see this as a normal part of doing business. Public radio and TV are set up to be responsive to those who feel disenfranchised. PBS's head programmer meets regularly with minorities, not to discuss specific programs but to hear about their overall needs. Some of the many advisory boards involved in public broadcasting take into account viewpoints of various pressure groups.

After public broadcasters decide to air something, however, they usually hold firm regarding their decision. When NPR broadcasts material favoring gun control, it can count on hearing from the National Rifle Association. The complaints are vociferous at first but eventually blow over—until the next report favoring gun control.

If public broadcasters feel that members of the public have a point, they do alter their programming. APR received complaints from gays in San Francisco who objected to the way the Christian Science Monitor covered the subject of homosexuality. APR relayed these complaints to the producers of "Monitor," and the program underwent an attitudinal change in regard to the subject.

When PBS aired the docudrama "Death of a Princess" in 1980, the Saudi Arabian government acted like a pressure group, complaining bitterly. The program depicted the confined life of a Saudi Arabian princess who, along with her commoner lover, was executed for adultery. The Saudis complained that the program was an attack on the Islamic religion and demanded an apology. When none was forthcoming, the furor eventually died down.

Sometimes it is difficult to tell whether or not the pressure groups have been successful. For example, the National Innholders Association, a lobbying group of cattle raisers, loggers, and others who own private property within or near national parks and forests, took umbrage at a documentary entitled "The New Range Wars" about cattle grazing in public parks. The program was one in a series of environmental specials produced for PBS and Turner Network Television by the National Audubon Society. Almost 50 percent of its funding came from General Electric. The Innholders Association waged a nationwide campaign to boycott G.E. products because, as its director Charles Cushman claimed, "The company is funding propaganda pieces."[5] At the end of the season, G.E. announced it would no longer underwrite the series because of "recession-dictated cutbacks." Mr. Cushman believes his group's pressure played a major role in the decision.

Of course, there is always the fear that pressure groups affect programming indirectly. Although they may not dictate the particular decision of the moment, programmers may be reluctant to stick their necks out on certain issues in the future. No one keeps track of ideas that were never proposed because of fear of reaction from pressure groups.

Underwriters

Public broadcasting was designed to be commercial-free so that advertisers could not influence program content. But as government funding has decreased percentage-wise, both public radio and TV have had to rely more and more on corporate largess in the form of underwriting.

Public broadcasting's experience with underwriting is somewhat akin to commercial broadcasting's approach to advertisers. It tries whenever possible not to mix program decision-making and financial support. At one point Tom and Ray Magliozzi of NPR's "Car Talk" made some very critical remarks about Chevrolet, a major underwriter of another NPR program, "Performance Today." Chevrolet said it would pull out its money unless Tom and Ray apologized on the air. NPR programmers did not ask Tom and Ray to apologize; in fact, they did not even tell the two of the controversy. Perspiration poured for a few days, but eventually Chevrolet decided to continue supporting "Performance Today."

However, another NPR program, "Heat," had to be taken off the air because it could not find underwriters. "Heat" was a program with an unusual format that delved into the area of overlap between public affairs and culture. NPR programmers believed in the concept and poured $800,000 into the program, recouping only $400,000, most of it from subscribing stations. With a $400,000 deficit and no takers among the underwriters, the program was forced off the air.

Even government agencies that give grants to public broadcasting can try to influence program content. In the late 1980s PBS aired "The Africans," a miniseries about the influence of Western countries on the African continent. The series looked at the subject from the point of view of an African and was not at all flattering toward the United States and England. The National Endowment for the Humanities, which had contributed money for the production, labeled the series an "anti-Western diatribe" and demanded that its credit be removed from the tapes.

Even if underwriters don't alter the specific content of a program, they can influence what is or is not aired through the use of restrictive grants. In other words, the money is given for particular types of stories or programs. The Lilly Endowment provides grants for the broadcast of issues affecting children and youth; the United States–Japan Foundation contributes funds for coverage of Japan. Public broadcasters don't like these limitations. Although the grants do not influence the content of the shows, they do affect whether something is covered at all. Japan might receive more frequent treatment than Cambodia, not because its news stories are more important but because the money exists to cover Japan.

Government

Theoretically, the Corporation for Public Broadcasting was set up to insulate public broadcasting from the government. The money allocated by Congress is forwarded to CPB, which then distributes it to stations and networks. CPB's role is to make sure that the government does not pass along its politics as it passes along its money. But such has not always been the case.

The President. During the Richard Nixon years, the president showed even more rancor toward public broadcasting than toward commercial broadcasting. Public broadcasting, with regularly scheduled programs such as "Washington Week in Review," was fairly consistently against Nixon and his policies.

The administration took the position that public broadcasting should emphasize local not national issues; let the commercial networks, who were better equipped, cover national events. The administration also promulgated the addition of more cultural programming, which, of course, would cut down on the time available for the more damaging news, documentary, and public affairs programs. When public broadcasting did not acquiesce, Nixon tried to use financial power. When Congress passed public broadcasting's 1972–74 budget bill, Nixon vetoed it. This angered several people on the CPB board, who resigned. Nixon appointed replacements more in tune with his ideas, causing an enormous schism between CPB, which was now Nixon-oriented, and the rest of the public broadcasting family, which was largely anti-Nixon. The Watergate hearings and Nixon's downfall, as well as a reorganization of both the CPB and PBS boards, brought the level of the conflict down enough notches that peace could once again reign.

Obscenity and Indecency. Public radio has the dubious distinction of causing the most famous of the obscenity cases. It arose in 1973 when WBAI, a Pacifica station in New York, broadcast a monologue by comedian George Carlin in which he spoofed seven dirty words that could not be said on the public airwaves. In the course of the spoofing, he mentioned the terms 106 times in 12 minutes. The program was aired at 2:00 p.m., and a father driving in a car with his young son complained to the FCC. This led to a lengthy court case that ended in the Supreme Court. The court sided with the FCC but in a rather muddy manner. It did not specifically say whether the WBAI program was obscene or even indecent; it said the program should not have been aired in a time period when children were likely to be in the audience. This led to the FCC's establishment of 6 a.m. to 8 p.m. as hours when indecent material should not be aired.

PBS frequently wrestles with language and nudity. Vice president John Grant says, "Individual stations make decisions about language and nudity, but we try to act as a gatekeeper. We evaluate for the stations, but some stations may have standards that are more stringent than ours."

• • • • • • • • • • • • • • • •

Influences on all the forms of electronic media have their downsides for those who are charged with developing programming. Rules, regulations, and opinions often get in the way of what appears to be logical programming and scheduling. But there are many positive points to be made for the existence of these inside and outside forces. Programmers, with the power to influence the minds of millions of listeners and viewers, should not work unchecked. Power corrupts and absolute power corrupts absolutely. The necessity of thinking in terms of how others may react often leads to more responsible actions.

So far, we have talked about where program ideas come from, how they get developed and tested, what the essential ingredients are, and who, besides the program department, has an influence on content. However, all too frequently programmers who have been mindful of all the foregoing destroy their efforts by paying too little attention to where, when, and how the show is scheduled. We now move on to discussion of this vital process.

• • • • • • • • • • • • • • • •

THE ISSUE OF "STOP THE CHURCH"

One of the Public Broadcasting System's series in the early 1990s was "P.O.V." ("Point of View"). Each episode consisted of a short film on a matter of social or political consequence. Usually shot on a shoestring by nonprofessionals, the pictures were more passionate presentations of positions than balanced documentaries. They were designed to raise consciousness on subjects the producers believed were underexposed and/or misunderstood.

In July of 1991, one of the productions ran into a firestorm of criticism. The show, "Tongues Untied," dealt with black male homosexuality and included frank language and some nudity. Over a hundred of PBS stations refused to air the program and others ran it at off-hours to minimize its audience.

On August 12 PBS announced the cancellation of "P.O.V.'s" next scheduled episode, "Stop the Church." Produced by ACT UP, a gay activist organization, the film depicted the preparation for and execution of a 1989 demonstration at St. Patrick's Cathedral in New York City against the Catholic Church's stance on AIDS. The satirical song "The Vatican Rag" was played as background music to scenes of Catholic worship, and ACT UP members were shown disrupting a mass being said by Cardinal John O'Connor. During the services activists were heard yelling at O'Connor, "Stop killing us, stop killing us," presumably because of the Church's objection to the promotion and distribution of condoms.

David Davis, president of "P.O.V.," acknowledged the earlier disturbance over "Tongues Untied" and explained that he was pulling "Stop the Church" because he felt that "another controversy at this time would break [stations'] backs and undermine their confidence in 'P.O.V.' "[6] But PBS spokesperson Mary Jane McKinven denied the linkage with the earlier show. She stated that it "has a pervasive tone of ridicule which rather overwhelms its critique of church policy."[7] Her comment was supported by John Grant, PBS's vice president for scheduling: "Our decision has nothing to do with the attack on the church; . . . it has nothing to do with the fact that it was made by an AIDS activist group. It was the tone and the ridicule that we found inappropriate for broadcast."[8]

Although the Public Broadcasting System officially removed the program from its schedule, individual stations were still free to obtain a print and play it. Three chose to do so: KCET in Los Angeles, WGBH in Boston, and KQED in San Francisco.

KCET's decision to run the program caused a tumultuous brouhaha in Los Angeles. Cardinal Roger M. Mahony accused the station of surrendering to "blackmail" by gay activists and called on southern Californians to consider withholding contributions from KCET. According to Mahony, gays and AIDS activists had threatened to withdraw financial support and jam the switchboard during the station's August pledge drive to make it impossible for donors to register their contributions.

At his press conference on September 5, two days before the scheduled broadcast, the Cardinal stated, "This absence of responsible leadership at KCET leads me to believe that we should hold the station morally, and possibly legally, responsible for every future act of terrorism against churches, temples, and synagogues because KCET has told potential perpetrators of such hate crimes that not only is such activity acceptable, it is worthy of televised documentaries celebrating and glorifying it." He went on to say that "until the station builds bridges instead of driving wedges between various groups, I do not feel that Channel 28 is worthy of financial support from the people of southern California and their businesses."[9]

Despite this blistering attack, KCET held to its decision to air the show. At his own news conference held later on the same day, KCET President William Kobin said that he was "distressed by Mahony's actions. KCET believes strongly that its viewers deserve the same opportunity as Cardinal Mahony to view this film and make up their own minds regarding this controversy."[10]

The film appeared the following night followed by a discussion of such relevant issues as First Amendment rights, taste, judgment, censorship, and the role of television in the presentation of controversial subjects. Participating on the panel were representatives of gay groups and the Catholic Church, but Cardinal Mahony, who was invited to attend, chose not to appear.

The dispute represents a rich area of discussion for present and future broadcast decision-makers. The collision was inescapable. It stemmed from the diverse attitudes of the two parties on homosexuality. The Catholic Church holds that same-gender sexual activity is morally objectionable. In his press conference Mahony repeated the church's position that people with AIDS contract the disease through their choice of life-style and the only way to prevent the disease's spread is abstinence. The producer of "Stop the Church," Robert Hilferty, claimed that it was this very stance by Catholicism that prompted the demonstration at St. Patrick's and the making of the film. "My film is not a hate film. It was not intended to ridicule Catholicism or the Catholic Church," he stated. "It was a protest against the fundamental position which, in ACT UP's judgment, was resulting in the death of gays."

Consider now the unhappy choices facing the KCET programmers who had just gone through a bloodbath with "Tongues Untied" and who now faced another highly visible battle, again involving homosexuality. The station was committed to "P.O.V.," a series dedicated to the presentation of controversial subject matter. But if "Stop the Church" were to be broadcast, a donnybrook would surely ensue, possibly with severe financial repercussions. And just as certainly, if it wasn't aired, there would be similar aftereffects.

A surface solution would be to honor the judgment of the network officials and withhold the program. However, gay organizations are well aware that individual PBS stations operate with considerable autonomy and are quite capable of broadcasting shows that are disapproved, or discouraged, by the network. What to do? Either way the station would alienate a broad constituency and run the risk of revenue loss in a period of recession when contributions were hard to come by. KCET decided to schedule the program and attempted to moderate Catholic fury by scheduling an after-show panel session in which church advocates had an opportunity to rebut the charges. Might they have taken another course? If so, what?

The program did have some financial consequences. KCET estimated it lost $130,000, most of it due to one local businessman who pulled back a large pledge. But the station claimed this experience would not change its programming philosophy. Barbara Goen, KCET's vice president for public information, said, "Any attempt at curtailing our income source, or threatening it, is very serious. It probably has greater significance in hard times than in flush times. But does it mean we change the way we make program decisions? No, it doesn't."[11]

It is interesting to speculate on what advice a TV program consultant, had one been retained, might have offered to the Cardinal. "Stop the Church" registered a 4.5 rating and a 9.0 share. For the five preceding weeks the time period averaged a 1.7 rating and a 3.5 share; for the five weeks following the figures were 1.2 and 2.3. In short, the heated controversy over the show generated an audience approximately three times greater than normal. Perhaps the Cardinal felt that no matter what the audience results, he could hardly let the broadcast take place without protest. But on a purely statistical basis, an airing without any controversy would surely have resulted in a smaller audience, an outcome he certainly would have preferred.

NOTES

1. "Justices Refuse to Allow Airing of Noriega Tapes," *Los Angeles Times* (November 19, 1990): A-20.
2. "Radio 'X,'" *Los Angeles Times* (October 17, 1991): F-3.
3. Ibid.
4. "PBS Network Hit by Charges of Censorship," *Los Angeles Times* (August 14, 1991): F-1.
5. "Specials: An Endangered Species," *Los Angeles Times* (December 12, 1991): F-1.
6. "PBS Network Hit," F-1.
7. Ibid.
8. Ibid.
9. "KCET Unworthy of Public Support, Mahony Declares," *Los Angeles Times* (September 6, 1991): F-1.
10. Ibid.
11. "KCET Pays Price in Flap with Church," *Los Angeles Times (October 10, 1991): F-1.*

11 Scheduling Strategies: Part I

To construct a successful programming lineup, programmers must do more than just fill the time periods. Many TV shows or radio program segments that seemed to contain all the right ingredients for a long and profitable life have had short and painful histories for reasons quite apart from their inherent merit. Programs not only have to be developed, they have to be nurtured. Too many productions have simply been tossed on the air with no plan, no promotion, no lead-in, and, therefore, no hope.

Once programmers have produced a promising show, they must be equally adept at its placement and treatment on the schedule. The time period, the competition, the receptivity of the audience are all factors to be assessed before the show is committed to the schedule.

COMMERCIAL TELEVISION

Scheduling strategies apply primarily to networks and stations. Syndicators sell to stations and rarely, at least initially, are able to dictate time periods. After a syndicated show has been on the air for awhile and has developed a large following, the distributor may be strong enough to demand a specific position on the schedule. And, of course, certain shows are designed for a particular daypart and can't be scheduled effectively anywhere else. But, for the most part, the syndicators sell the shows, and the stations place them wherever they wish.

Ideally, programmers seek a large audience with the lead-off show and structure the programs that follow in such a way that the audience will flow continuously throughout the schedule. This isn't always possible. Sometimes a competitor's opening program will be an established blockbuster that makes it impossible for others to start off effectively. Other times the competitor's strength may be in the middle of the schedule. Then the only strategy is to ride out the "bad" period and attempt to rebuild when the powerblock is over and the audience released.

When putting together a schedule, programmers should consider the following principles.

Fitting the Show to the Available Audience

As we have shown earlier, most programs have a primary appeal to a particular audience. For example, action-adventure appeals principally to men, sitcoms to women, contemporary music to teens, serials to younger women, talk shows and game shows to older women. Occasionally, a particular show will attract an almost universal audience. For example, more than 90 percent of all the homes in the country watched at least one episode of "Roots" when it was initially aired in 1977. Often something of a current nature, such as the coverage of a war or disaster, will also have wide all-viewer appeal. But generally the primary appeal of a show is to a specific demographic group.

Therefore, a program must be placed in a time period in which its core viewers are available. It makes no sense to schedule a show with a predominant teen appeal on a Saturday night. That audience isn't home. They're at movie theaters, or basketball games, or anywhere other than in front of the set with Mom and Dad.

ABC relearned this lesson in the 1990–91 season. The network bucked the conventional wisdom and scheduled a nightlong lineup for the 18- to 34-year-olds: "The Young Riders," "Twin Peaks," and "China Beach" (replaced in mid-season by "Under Cover"). ABC hoped that if it built the franchise, the young viewers would come. They didn't. In the last week of their Saturday telecasts, "Twin Peaks" and "Under Cover" were tied for 85th place among 89 programs rated by the A. C. Nielsen Company. "We've all known that the true 'Twin Peaks' fan is a party animal and not home on Saturday night, and they [ABC] have sorely missed the opportunity to have a party on a weeknight," wryly observed the show's executive producer, David Lynch. The only comment from the network's president of entertainment, Robert Iger, was that "Saturday has not served 'Twin Peaks' as well as we had hoped."[1]

But late night Saturday? That's something different. Many young viewers have returned home by then and are eager for entertainment, as "Saturday Night Live" has impressively proved over three decades.

When buying syndicated shows, station programmers look for series that are appropriate to the time period: Oprah in the afternoon, Arsenio late at night, "Wheel" in primetime access. Whenever possible, a network affiliate will try to buy a syndicated show that coordinates with a network lead-in (assuming the lead-in is strong) in order to continue the audience flow. Many independents schedule off-network sitcoms in access, not only to counterprogram affiliates but also to take advantage of the large audience of teens and women during this time period. (The primetime access rule prohibits network affiliates in the top fifty markets from running off-network programs in the access period.)

Before locking in a program, programmers must study the time period's demographic history. If the target audience is underrepresented, a more favorable position should be sought. No matter how strong the show, if the key viewers are not available, the project will fail.

Dayparting

Closely related to the principle of fitting the show to the available audience is the concept of "dayparting." People's needs, activities, and moods change throughout the day, and dayparting takes this into account by changing what is presented and how it is presented throughout the various stages of the day.

In the morning, when people first wake up, they often want information to help them plan the day—weather, traffic reports, important news. Because most people must go to work, they do not have a great deal of time to spend with media. That is why many of the morning shows are divided into short segments. Someone can digest, along with the toast, an interview with Tom Selleck but not an hour-long gripping drama.

As the morning wears on, people who are able to spend time with television are able to do so in a more leisurely fashion. As a result, game shows, soap operas, and talk shows dominate daytime TV.

In the mid-afternoon students return home from school. Many television stations change programming fare entirely in order to appeal to young children.

As people begin arriving home from work, TV stations switch from programming oriented toward children and homemakers to news. This allows adults, both male and female, to catch up on the day's events.

The evening is the time for the most leisurely TV viewing of all—the time when all categories of viewers are relaxed and able to spend several hours with one or more programs. Comedy and drama are the primary fare during these hours.

From 11:30 p.m. until 12:30 or 1:00 a.m., the talk/variety form presented in segments has been most popular for two basic reasons: 1) people who are preparing for sleep generally do not want their minds to be challenged or overstimulated (doing so may cause them to be restless for hours), and 2) viewers turn off the lights at varying times; segmented shows permit them to switch off the set after an interview has been completed. In this way a programmer is able to attract a large number of viewers for at least part of a show. If a two-hour movie were scheduled, the viewer would be tempted to say, "I can't watch all of it; therefore, I won't watch any of it."

Stations that stay on through the early morning hours as a service to insomniacs or night workers who don't want to retire as soon as they come home usually fill the time with inexpensive movies. This is more for reasons of economy than daypart strategy. Of late, networks have been supplying news to their affiliates during the wee hours, and a small but devoted audience of news junkies have tuned in.

On the weekend, programming changes dramatically to accommodate the life-style of the audience. The daytime lineups are studded with sporting events to reach the male viewers who are available in abundance. Saturday morning is often devoted to children's programming to accommodate early-rising children and their late-rising parents. Saturday evening schedules are canted toward the older audience in recognition of the absence of teens and young adults. (We have already seen in the case of "Twin Peaks" what happens when this social fact is ignored.) Sunday morning on the networks is wall-to-wall news and public affairs programming. This is partly to take advantage of the availability of adults who have the time to digest the news in deeper, more thoughtful quantities, but it is also a way to pay off public affairs obligations in a commercially insensitive time period. Whenever their community-consciousness is called into question, networks and stations can point with pride to these estimable service shows without having to sacrifice large profits in the process.

Launching the Show: The First Strategy

For a program to be a success, it must first be sampled. If viewers are unaware of a show, the early ratings will be weak and the series will, in all likelihood, languish quickly. In a soft economic environment, buyers are disinclined to stay with programs that do not instantly indicate promise. Programmers have devised two strategies that can greatly hasten the sampling process: 1) introduce the show in a quiet time, and 2) place the premiere in a time period where a large audience is virtually guaranteed.

Introducing the Show in a Quiet Time. For all practical purposes, programmers are stuck with a pattern of originals from fall to mid-spring and reruns thereafter. So how can they achieve their goal of launching shows away from the clamor of competing promotions? The networks, notably Fox, are experimenting with some new approaches.

Since Fox is the newest network and, therefore, the one least rigidified in tradition, it is the one most inclined to innovation. In the summer of 1991, Peter Chernin, president of Fox Entertainment, declared that his company had fallen into the same "self-destructive" practice of launching all new shows in the fall. He noted that the previous season the four networks had "introduced 34 new series as if it was a massacre. Seventy-five percent failed. No other business debuts its product like that. No longer at Fox. We will roll out new series in every month of the year."[2]

His new plan went into effect in July of 1991 with the premiere of "Beverly Hills 90210." "In the first week," Chernin claimed, "it showed an 84 percent improvement over its '90–'91 average."[3] By mid-fall it was second in its time period, topped only by the formidable veteran "Cheers." The economic impact of the Fox strategy is still unclear. The summer premiere enabled the show to get off to an excellent start and build momentum before the promotional cannonading began in September, but the strategy forced them to play repeats in some of the highest-rated months in the year. The rating and financial effects will be studied carefully by all the networks.

Robert Iger, ABC Entertainment president, dismissed the Fox plan to premiere shows on a fifty-two-week basis as "nothing new." He cited CBS's debut of "Northern Exposure" the previous summer and ABC's late winter and spring launchings of such shows as "Wonder Years," "Twin Peaks," and "China Beach." Iger favored a more traditional strategy in which the bulk of the programs would be rolled out over a three-week period, with six series to be held out for introduction later in the season.

Placing the Premiere in a Hit Time Period. A favorite launching strategy is to schedule a newcomer after a monster event, the Super Bowl being the ideal choice. Hardly a year goes by that the network carrying the game doesn't introduce a new show behind it in the hope that the audience of around 40 million will like what it sees and an instant hit will be born. Sometimes it works, witness "Wonder Years," and sometimes it doesn't, witness "Tenspeed and Brown Shoe." Although the results vary, the strategy is fundamentally sound. There is no better way to get a program sampled than to place it immediately following the event that is the highest-rated program virtually every year.

A variation of this strategy is to slot a newcomer into a time period that follows a powerhouse series such as "Roseanne" or "Cheers." Although it is not the period the show will occupy in series form, it is a way to gain immediate exposure. This greater familiarity will help the show become established when it moves to its permanent position.

In 1990 NBC attempted another version of this device. Dubbed the "double pump," the technique called for a new program to be exhibited twice in one week in strong time periods. The following week the series assumed its official slot. The plan created massive consumer confusion and was acknowledged by network executives to have been a failure.

Alan Wurtzel, ABC's research and marketing chief, claims that his network's research over the years 1988 to 1990 revealed that 50 percent of all households that will ever watch a program will get hooked by one of the first two episodes. Obviously, the sooner they sample these shows the better. This is why broadcasters strive so tirelessly to develop new techniques to bring viewers into the tent for the first show.

Launching Syndicated Series: Some Special Problems. The traditional fall kick-off has been particularly difficult for syndicated shows. The promotional fusillade for network series is so thunderous that it's virtually impossible for syndicated programs to create any viewer awareness. Many of them remain unknown until well into the autumn. With overnight ratings available in all major and most midsize markets, a syndicated show's failure to grow can be spotted immediately by station executives. Some of the more hair-triggered among them will pull the program before it has had a reasonable opportunity to develop even a modest sampling.

This disturbing condition has led a number of distributors to begin experimenting with non-fall premieres. "We no longer look at September as the only time to introduce a new show,"[4] said Shelly Schwab, president of MCA Universal TV, which successfully launched the weekly series "Harry and the Hendersons" and "Roggin's Heroes" in January. Viacom's Michael Gerber echoed those sentiments and announced the debuts of three new shows in January. Dick Kurlander, an analyst at Petry Television, a station representative firm, believes that the ratio of fall launches to nontraditional debuts will go from the current 95-5 to 80-20 by the mid-1990s.

There are two major factors militating against any sort of wholesale departure from fall premieres for syndicated shows. Summer HUT (households using television) levels are considerably lower than the fall–spring period. Even though a new show may achieve a higher share of audience in July than September, the pie is so much smaller that the total audience may still fall short of a conventional launch.

Secondly, January premieres run into station clearance difficulties. Station managers who commit to programs in the fall are frequently locked into pay-or-play contracts for at least twenty-six weeks. For purely economic reasons, they may wish to stay with the failing show until the full commitment is satisfied, particularly during a recessionary period. Programs launched in January may have to struggle for many months with subpar lineups until additional time periods break open.

To repeat, the first objective of a launch is to get the program sampled. Any technique of timing that enables the program to be introduced in a period of relative uncompetitiveness is desirable. However, this advantage must be weighed against the less attractive factors of starting in a period of low set usage, being forced to schedule reruns in high HUT months when opponents will be running original episodes, and offering a show for sale at a time when many buyers can't accommodate it.

Tentpoling

We have just seen that one of the best ways to gain sampling for a show is to introduce it behind a big hit and then move it to its regular time period. Even better is to place it behind a hit and leave it there. As Isaac Newton noted a few centuries ago, a body at rest tends to remain at rest. Viewers who know that the program they are watching will be followed by one with the same appeals will be inclined to stay tuned. This inertia works to the benefit of the new show and enables it to obtain the desired sampling. The same principle can also be applied to programs that are struggling elsewhere on the schedule. By being moved next to hits, they can often be revived and enjoy healthy, lengthy runs.

"The technique of placing new shows behind hit shows is probably as old as broadcasting itself," says Tony Barr, a longtime program executive with both ABC and CBS. "In January 1976 'Happy Days' was a half-hour smash at 8:00 p.m. on Tuesdays. Paramount spun off 'Laverne and Shirley,' a sitcom of exactly the same type of humor and broad-based appeal, and ABC scheduled it right behind 'Happy Days' at 8:30 p.m. In effect, 'Happy Days' was converted from a half-hour show into an hour show. There was virtually no drop-off in audience and one hit became two hits."

Sometimes a powerhouse show is strong enough to hold up programs both before and after it. If a new show is scheduled right before a very popular program, people tuning in early in anticipation of the hit will sample the end of the new entry. Hopefully, they will be intrigued and tune in the next week for the entire show. Placing an established show before a hit is not as strong a programming strategy as placing it after, but the method can work.

This concept of scheduling weak or new programs around a strong show is referred to as "tentpoling"—the pole in the middle holds up the two weaker shows, especially the one that follows.

The Fox network made excellent use of tentpoling when it expanded the number of evenings it broadcast. It had developed a strong Sunday lineup that included "The Simpsons," "America's Most Wanted," and "Married . . . With Children." When it expanded to more nights of programming in 1990, it used these popular programs as anchors for the other nights—"The Simpsons" on Thursday, "America's Most Wanted" on Friday, and "Married . . . With Children" on Sunday. "We got the tentpoles and they are performing admirably," said Fox scheduling executive Sandy Grushow.[5]

Of course, tentpoling was far more effective in the precable era when program choices were fewer and the remote control unit had not yet been invented. The long walk from the couch to the set was more exercise than most viewers cared for, and given a reasonable excuse to stay put, would. Today they can whiz through thirty channels before the next show's opening credits have rolled and the pass-along strategy is less assured.

Nevertheless, it continues its usefulness. "A Different World," a comedy spin-off of the "The Cosby Show," was introduced in September 1987 in the time period immediately following the parent program, which at that time was the number one show on television. The derivative production was an instant success and remained in the top ten for four years. Its compatibility with the lead-in, coupled with the enormous popularity of Cosby, resulted in an overpowering hour at the start of every Thursday evening.

Station managers use this same strategy with blockbuster syndication series. When "Jeopardy" was launched in its third incarnation in 1984, it was placed behind the triumphant "Wheel of Fortune" in many markets. It quickly became an enormous success and helped form an invulnerable hour of dominance.

Hammocking

Probably the surest way to generate an audience for a new program is to slot it between two established shows, the so-called "hammock" principle. With power in front and power behind, the new program benefits from both the pass-along viewing from the preceding show and the anticipated entertainment from the following program. The audience tendency towards inertia is almost irresistible.

The case of "A Different World" was cited earlier. Not only did this series follow the megahit "The Cosby Show," it also preceded "Cheers," a multiyear blockbuster. In its privileged hammock, it had every opportunity to succeed, and it took full advantage of its prized position.

Not every attempt at hammocking has produced a new success. In the early 1970s, CBS presented an evening lineup on Saturday night that was probably the strongest schedule in the history of the medium. The night led off at 8:00 p.m. with "All in the Family," a series that consistently reached 50+ share levels. At 9:00 p.m. was the much-admired Mary Tyler Moore program whose share scores ranged in the mid-40s. Bob Newhart at 9:30 p.m. and Carol Burnett from 10:00 to 11:00 closed out the night with high-30 to low-40 shares. It was an awesome array, but it had one weak slot, 8:30–9:00 p.m. The network tried at least one new show every year for three years and never did solve the problem. All of the hammocked shows scored in the 30+ share range, but considering the potential of the time period, the performance was rightly considered unsatisfactory.

A story (probably apocryphal) is told about ABC president Leonard Goldenson who, in the early days of the network, observed that his competitors were separating two hits and sliding a new show in between. Impressed by this strategy, he supposedly asked his programming chief, "Why don't we do that?" "Because, Leonard," came the response, "we don't have two hits."

Hammocking in the local program schedules of affiliates is more difficult than at independents. The spaces between the network-supplied blocks are either too brief to permit the technique, or the time periods are committed to particular programs such as local news or a popular talk show host. Independents that have to fill the entire broadcast day can and do use the hammock principle. It is frequently seen in cartoon blocks, in which a newcomer is sandwiched between two established series.

FIGURE 11.1 The three-network primetime schedule for Tuesday nights in 1990–91 (Fox did not provide programming). Note how the networks counterprogram each other.

Primetime Programming for Tuesday Nights, 1990-1991

Time	ABC	CBS	NBC
8:00	Full House	Rescue 911	I'll Fly Away
8:30	Home Improvement	" "	" "
9:00	Roseanne	Movie	Heat of the Night
9:30	Coach	"	" "
10:00	Homefront	"	Law and Order
10:30	" "	"	" "

Counterprogramming

No program, no matter how popular, can satisfy all viewers. The neglected or dissatisfied audience becomes a good target for competing shows. The tactic of filling a time period with a program whose appeal is dissimilar to an opponent's is called "counterprogramming."

The 1991 three-network primetime lineup on Tuesday night provides a good example of this device. From 8:00 to 9:00 p.m. ABC scheduled two situation comedies, the second of which, "Home Improvement," was a promising newcomer hammocked comfortably between the well-established "Full House" and the number one rated series, "Roseanne." CBS countered with "Rescue 911," a fast-paced "actuality" adventure, and NBC introduced "I'll Fly Away," a soft, family-values, dramatic series. In the 9:00 to 10:00 hour ABC continued its comedy skein with the aforementioned "Roseanne" and "Coach," a show that had blossomed into a hit in its enviable time period. Once again, the two competitors had to seek an alternative audience. CBS elected to present feature films and NBC offered "In the Heat of the Night," a law enforcement adventure series. To close out the night ABC decided to try a new family drama, "Homefront," in the belief that the lineup's tremendous momentum would pass along a large audience to the newcomer and give it a good chance to succeed. CBS, of course, played the second half of its movie, and NBC followed "In the Heat of the Night" with another crime-and-punishment series, "Law and Order." A nightlong effort such as this on the part of each broadcaster to carve away an audience on which the others are not concentrating is classic counterprogramming strategy. However, this technique may not always be the best way to thwart a competitor.

In those cases where an opponent has tried to stake a claim on a particular audience but has done so with a relatively weak entry, the best strategy may be to go after that audience with a stronger version of the same genre. If the move is successful, the challenger will capture the viewers and seriously erode the opponent's audience base, not only for that time period, but possibly the ensuing ones as well.

The question arises: What do you do against a monster hit? Network executives have wrestled with this problem from "I Love Lucy" in the 1950s through the "The Cosby Show" in the 1980s. The temptation is to throw up a test pattern, abandon the time period, and hope no one will notice. A pleasant fantasy, but unrealistic. At times programmers have decided to cut their losses and fill the period with the most inexpensive programming available. A case can be made for this fiscal prudence, but it is basically unacceptable.

For one thing, stations have to sell commercials around the period, and the spots will be worthless if the program is dismal. Also, such a lack of effort says all the wrong things to the company's many constituents (advertisers, investors, etc.) Most executives feel it is best to try something experimental. There isn't much to lose, and, who knows, one might just get lucky. Even if it doesn't work, critics and commentators will applaud the innovative effort and this will help ease the pain of a stricken time period.

Sylvester L. "Pat" Weaver, president of NBC-TV in the early 1950s, recalls: "Whenever the other guys came up with a big winner, I knew it was time to call in my zanies, the guys with the off-the-wall ideas. I knew no conventional show could make it, but maybe something really unusual might click."

ABC scored strongly with this strategy in 1990. CBS's "Murder, She Wrote," preceded by the powerhouse "60 Minutes," had been delivering hammer blows to the opposition since 1984. ABC decided to take a flyer on a program that featured home videotapes shot by ordinary citizens. The novel concept caught on and "America's Funniest Home Videos" became a top ten series virtually overnight.

There are times when it is excellent strategy to take on a blockbuster hit. The timing must be exquisite, but if it's right, the colossus can be toppled. Two conditions are required: the incumbent hit must be in a waning state, and the challenger must be a promising show in the early stages of its ascendancy. These are two critical and delicate judgments for the programmer, and if either is wrong the venture will fail.

At the beginning of the 1990–91 season, executives at Fox believed that "The Cosby Show," in its seventh season, had softened to the point where it could be taken. They also felt that "The Simpsons" in just one year had become such a national phenomenon that it could successfully challenge the champion. It did. The two ran neck and neck in the ratings for the entire year. At the end of the season, "The Cosby Show" retired.

Counterprogramming is a way of life for independent stations. They must always examine the schedules of their network-affiliated opponents and go after the audience that is left unattended. The primetime and late-night strategies of the four Los Angeles independent TV stations in 1992 are instructive. At 10:00 p.m., when the networks were presenting ambitious entertainment programs, the four indies were all playing local news hours. At 11:00 p.m. when the affiliates swung into their local news, the indies offered "Cheers" on channel 5, the "Addams Family" on 9, "Studs" on 11, and "Arsenio Hall" on 13. One might note that Arsenio's 11:00 p.m. time period also gave him a half hour headstart over the similarly formatted "Tonight" show (see "Bridging," below). This kind of artful dodging is the route to survival and profit for the indies.

Bridging

The less opportunity the audience has to sample your adversary's program, the better chance you have for success. If you have something compelling unfolding at the time the competition is about to start, you distract the viewer from the temptation to stray.

The most effective use of this device is to have a program under way well in advance of the competitor's start time (as we have just seen in the case of Arsenio Hall). Networks and stations have oftentimes scheduled movies to begin one hour before the start of a powerful opposing show. Viewers who have invested a full hour in a film are unlikely to break away for something else.

A variation on the strategy is to edit a long-form program so that the competitor's start time is spanned by entertainment. Take, for example, a 9:00–10:00 p.m. program that is opposed by two half-hours. If the producer of the hour-long program schedules a two-minute commercial from 9:25 to 9:27, a strong piece of action can be unfolding at 9:30 p.m. when the second of the two half-hours is about to begin.

As with many other programming strategies, bridging has been affected by the invention of the remote tuner. During commercial breaks, hyperactive TV viewers may very well zip to five or six different channels to find other fare or just satisfy their curiosity as to what else is playing. If the program they have been watching has been intriguing enough, they will return. But if something else catches their fancy, even though it may be almost over, they may become lost viewers.

Blunting

A cousin to bridging is "blunting." The goal is the same: minimize the competitor's opportunity to be sampled. Except under the most dire emergencies, the premieres of all programs are announced many weeks in advance. The need for notification to advertisers and affiliated stations, plus the time to rev up a promotion campaign makes last-minute introductions impractical. This long lead period enables rivals to fashion competitive strategies.

Strong episodes with marquee guest stars can be scheduled against the newcomer. So can a special program that combines hot celebrities with other high-appeal elements. A major motion picture featuring a recently announced Academy Award nominee can receive a "world TV premiere" on the competitor's premiere night. Of course, either the special or movie can (should) bridge the newcomer's start time. If there is sufficient lead time and the program inventory permits, a series of extraordinary events and programming can be placed in the time period for weeks leading up to the competition's premiere. The idea is to build a viewing habit so binding that a viewer won't even be tempted to seek programming elsewhere.

Blunting requires vigilance and imagination, but the rewards can be exceptional. No programmer should supinely accept the premiere of an opponent's show, especially one that occupies a key position in the schedule. Obviously, no company has the resources to attempt the blunting of every competitive move. But if there is a lot riding on the outcome of a new show's launch, every effort should be made to capture the viewer with even greater attractions.

The putative founder of the blunting technique in network television is Fred Silverman. When he became program chief of CBS in 1970, he instituted a policy of yearlong vigilance. Large sheets of paper were designed in which space was provided for the programs of the three networks, half hour by half hour for all seven nights of the week. A sheet was maintained for every week in the year. Each week key CBS network strategists met to exchange information and update the charts. As soon as one of the other networks made a program announcement, the plotting began. Rarely was there an ABC or NBC program launched that wasn't harassed by CBS diversions and enticements.

The system contributed significantly to CBS's number-one status for all five years of Silverman's program leadership. He then moved to ABC-TV, where he used the technique to score more winning years. However, by the time he arrived at NBC, both his competitors were schooled in the process and the originator's advantage had vanished.

Stacking

Programmers attempt to develop an audience flow by assembling programs of similar appeal in order to sweep the viewer along from one time period to the next. A classic example of this strategy was CBS's 1991 Monday night schedule of four sitcoms capped by a humorous, easy-to-take hour. The lineup dominated through the year and on a number of occasions was victorious in every half hour.

Before embarking on such a strategy, a programmer should carefully evaluate two considerations: 1) is there a powerful show to begin the schedule, and 2) is there a weak link in the chain? Without a strong leadoff program, the lineup will be unable to develop the momentum necessary to start the audience flow. If a competitor takes a commanding lead at the beginning of the race, it will be very difficult for the "stacked" lineup to gain pass-along benefits. Similarly, if there is an especially weak program in the string, viewers will drift to other offerings and the flow will be interrupted. If there is some doubt about the appeal of one of the shows, programmers would be well advised to place it at the end of the lineup.

Stacking strategy calls for the most popular programs to be placed at the start of the schedule (for reasons just explained) and at the hour positions, when all stations, except those running movies or special events, switch to new programs. A "magnet" show at that time can not only hold on to the station's audience but can also attract the "released" viewers from the competition.

CBS Monday Night Schedule

8:00–8:30 — Evening Shade
8:30–9:00 — Major Dad
9:00–9:30 — Murphy Brown
9:30–10:00 — Designing Women
10:00–11:00 — Northern Exposure

FIGURE 11.2
The CBS 1991 Monday night schedule shows how programs can be stacked.

It is interesting to note that the other two networks counterprogrammed CBS's comedy night by scheduling movies[6] from 9:00–11:00 p.m. However, CBS placed its strongest sitcom, "Murphy Brown," at 9:00 p.m., which enabled the network to maintain the audience flow for the remainder of the evening.

Stacking to maintain audience flow is a standard procedure in virtually every daypart. In weekly daytime programming the traditional strategy is to schedule games or other nonserial forms in the morning and serials in the afternoon. For decades the networks have presented an uninterrupted string of animated shows on Saturday morning to reach the 2- to 12-year-old audience. Research indicates that the youngest children watch the earliest shows, and, as the morning moves along, the audience tends to get older. In theory programmers select the order of the cartoons to reflect this development, although how they determine an advancing intellectual content in these programs is a mystery.

Stunting

Sometimes programs must take extraordinary measures to maintain their audience levels. The competition might be picking up momentum; a series of preemptions might have caused a loss of viewing habit; a big sweeps week's performance might be required to obtain a contract renewal. Whatever the reason, there are occasions when a show needs a major injection of audience appeal. As a short-term solution, programmers frequently resort to "stunting," the insertion of entertainment elements not normally associated with the series, to obtain a ratings spike.

One of the most popular ploys is to construct an episode around a movie personality, or an athlete, or a celebrity whose recent activities commanded national attention. The appearance of these "names" provides powerful promotional opportunities and allows the program to exploit the public's curiosity about prominent figures.

Another traditional device is the development of a multiparter. Episodes that are usually complete in one showing are extended to two or more presentations because "they're just too big, too important to be told in their usual length." Presumably viewers will be so gripped by the first segment, they wouldn't dare miss the remaining episodes. Needless to say, the first chapter better be very strong or there could be a rating disaster for the conclusion.

In the spring of 1992, the stunt of choice was weddings. No fewer than twelve network primetime series produced one or more episodes around a marriage ceremony. The ABC research department estimated that a wedding episode can score three points higher than a series average rating, a very substantial boost for a show in a sweeps week.

The mother of all stunts was built into CBS's primetime soap opera, "Dallas." In the closing minutes of the last original episode of the 1979–80 season, J. R. Ewing, the power-mad, unscrupulous oil magnate, was shot by an unknown assailant and rushed to the hospital. Throughout the summer and into the first two months of the following season, viewers around the world (the series was an international phenomenon by then) speculated about "Who shot J. R.?" On November 21, 1980, the identity of the mysterious attacker was revealed to an estimated worldwide audience of over 300 million. In the United States more people viewed "Dallas" that evening than voted in the presidential election a few weeks earlier. It is still TV's second highest-rated episode, topped only by the 1983 finale of "M*A*S*H."

Stunting is a short-term device that is used for an immediate and specific purpose. No audience can be hyped indefinitely. Soon the novelty diminishes and interest declines. Also, stunts are frequently over-budget efforts and are financially unfeasible for any extended period. However, used judiciously and sparingly, they can be extremely effective.

One particular downside of stunting should be cited. In order to accommodate the stunt, producers frequently have to modify the basic idea of the show. A storyline that plays up the special appeals of a celebrity guest must necessarily shift the focus away from the core ensemble. It can be a chancy undertaking. If viewers enjoy the guest's performance but don't like, or understand, the relationships of the regulars, the goal of the sampling has been lost. On all stunting extremes, caution is advised.

Crossprogramming

"Crossprogramming" is the interconnection of two shows for mutual benefit. In its most sophisticated form, a story is started on one program and is completed on another. It can be highly productive, but because it requires an unusual combination of circumstances, it is rarely employed.

One example occurred in the early 1970s. "Marcus Welby, M.D.," a weekly drama about a Los Angeles general practitioner, was one of the highest-rated shows in network television. It appeared on ABC on Tuesday at 10:00 p.m. On the following night in the same time period, ABC presented "Owen Marshall, Counsellor at Law," a series dealing with the practice of a Santa Barbara attorney. The production company for both programs was MCA Universal. A story was developed in which Welby handled a serious medical matter with great skill and saved the patient. However, as a result of his efforts, he became embroiled in a law suit, which threatened his medical career. The legal case was played out the following night on "Owen Marshall." Marshall made an appearance at the end of the Welby hour to set the scene for the second half of the story, and Welby appeared at the trial to complete the connection between the shows. It worked superbly, and the ratings of both programs enjoyed significant increases.

One of NBC's more ambitious crossprogramming efforts occurred on Saturday, November 9, 1991. The night opened with a special one-hour version of "The Golden Girls" in which a hurricane is about to hit Miami, the location of the series. In the next half hour (9:00–9:30 p.m.) Golden Girl Sophia Petrillo (Estelle Getty) dropped in on the cast of "Empty Nest" to advise them of the storm. Finally, Rose Nyland (Betty White) volunteered her "help" at the local hospital on "Nurses" (9:30–10:00 p.m.) All three shows were produced by Witt/Thomas/Harris Productions, which made possible the interconnection of the stories.

Tie-ins of this magnitude require substantial advance planning, compatible formats, the good will of all, and a common production company. Such a combination is rare, which accounts for the infrequent use of elaborate crossprogramming. But when it can be used, it is a powerful programming tool.

Less ambitious versions, which produce more modest results, can also be developed. Appearances by stars on each other's programs and references on one program to events on another are two of the more familiar usages. This harkens back to the early days of radio, when Gracie Allen, the comedienne of "Burns and Allen," visited a large number of other radio shows looking for her brother.

Crossprogramming works because the avid viewers of the first show can be added to the core audience of the second, thereby producing incremental ratings, primarily for the latter. Obviously, the opening episode should be scheduled on the more popular of the two shows. It is easy to see how this maneuvering can cause resentment among the staff and cast of the higher-rated program. "The production company is just using us to beef up the numbers of the other program," is an opinion a programmer should be braced to hear. It calls for some nimble diplomacy, but then again, nothing in broadcasting is easy.

Theming

A much-used and very effective device among local stations is the combining of movies into theme weeks. If a station has a large inventory of films, the possibilities are endless: "Elvis Week," "Monster Week," "Mystery Week," etc., etc. The promotion becomes centralized around the theme, and constant repetition of the common element makes the whole seem greater than the sum of the parts. Many ordinary pictures have turned out to be solid ratings successes because they were part of a unifying concept.

The device can also be used for off-network series that contain a greater-than-average number of episodes. With a little imagination and some steadfast digging, one can always develop a week of "Fonzie's Greatest Romances," or "Matlock's Toughest Cases."

Stripping

Another strategy used by local stations involves "stripping"—placing the same show in the same time period every weekday, and

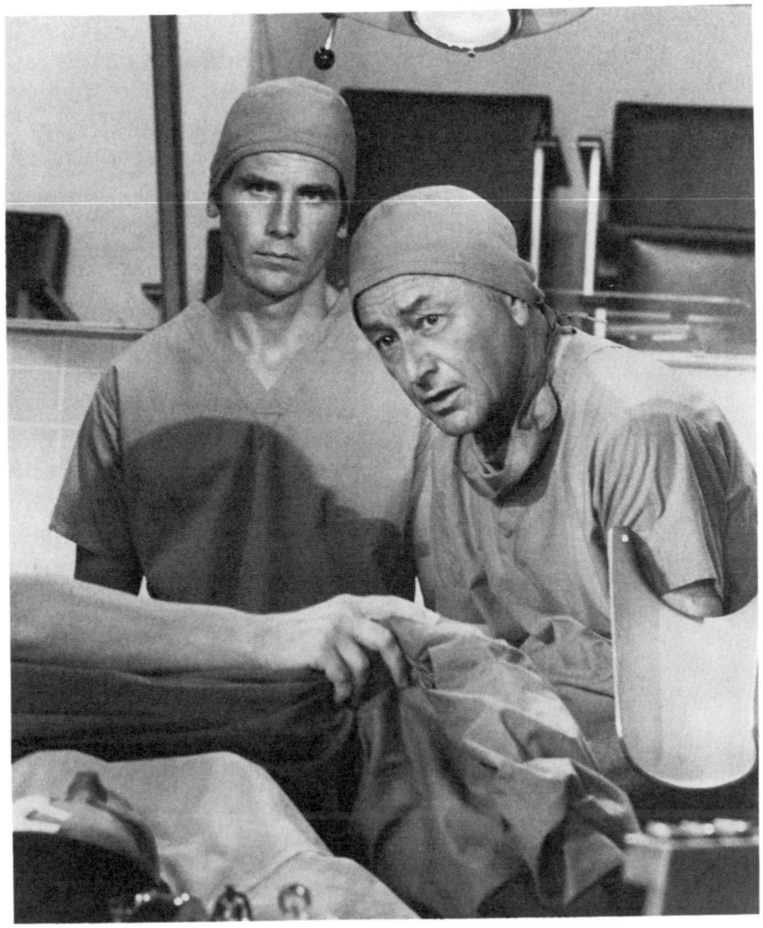

FIGURE 11.3 "Marcus Welby, M.D.," starring Robert Young (right) and James Brolin, was a top five series in the early 1970s. When a two-part story was begun on this show, the overwhelming percentage of viewers tuned in the next night to see the plot completed on "Owen Marshall: Counselor at Law." (Photo courtesy of MCA Television Group.)

sometimes even on Saturday and Sunday. It is done partly for economic prudence and partly because it is a sound audience strategy. As we have seen earlier, stations that buy a syndicated package of off-network shows have a fixed period within which the agreed-upon number of plays must be run off. Usually, the only way to meet this requirement is to schedule the series every day, Monday through Friday.

However, the placement of a show in a predictable time and place is also a solid strategy. Viewers who are fans of the program like to know exactly where and when they can find it. Nothing irritates and frustrates a viewer more than a lengthy search through a program guidebook to locate the air time of a favorite series. Even if it were possible for a station to meet its pay-or-play obligations on several series by alternating them night after night, it would not be sensible scheduling. Fans of one show might not like the other, and they could quickly get out of the habit of watching the time period at all.

Rerunning

In commercial TV, rerunning is more a financial necessity than a programming strategy. However, there are right times and wrong times to schedule reruns, and therein lies some strategy. But first, some background.

FIGURE 11.4
"I Love Lucy," starring Lucille Ball, pioneered the use of film for weekly TV series. The technique enabled the episodes to be rerun with high quality. The programs are still syndicated today, some forty years after they were first produced. (Photo courtesy of the Academy of Television Arts and Sciences.)

In the early days of television, very few programs were rerun—for technical reasons. Videotape had not been invented and film was considered much too expensive for this fledgling medium. Most programs were aired live—mistakes and all. A process called kinescope recording was developed so that some programs could be saved or run on the West Coast at the same "clock hours" as their East Coast origination. But these "kines," which were films made off a TV monitor, were of very poor quality.

From their beginnings in the early 1950s, Lucille Ball and Desi Arnaz filmed "I Love Lucy," and everyone thought they were a little crazy—until the foreign market opened up. Then Lucy and Desi began raking in the money from selling their shows to other countries, and a few fellow producers decided to follow suit. It wasn't long before American broadcasting realized it could play these films, too. And thus was born the rerun.

A basic annual pattern of program scheduling set in. Programs were introduced in the fall, ended their season in midspring, went into reruns for the summer, and emerged with new product the following fall. This came about through a combination of reality, surmise, social habits, and economics.

The key reality was the drop-off in viewing over the summer months. From June through August, daylight remains well into the evening. People stay outside longer, and that's why (according to conventional wisdom) the TV set stands dark and unattended. Also, during the summer a substantial percentage of viewers are on vacation and are unavailable for, or uninterested in, watching TV. Down through the decades a vociferous executive minority has claimed that this assessment is largely an incorrect surmise, grown into a mythic delusion. Give the people "Roots" or the answer to "Who shot J. R.?" in the summer and, they say, the theory would be demolished. As we will indicate shortly, the myth-debunkers are gradually testing their thesis. Defenders of the traditional cycle are quick to point out two other important reasons for the summer hiatus—

cast and staff demand a rest, and networks need the revenue from reruns.

It is certainly true that a weekly series is physically exhausting for cast and staff. The hours are long, the pressure relentless, and the opportunity to unwind almost nonexistent. A break is necessary so that they don't kill themselves, or each other. Scoffers downplay the fatigue factor and point to the round-the-calendar schedules of daytime serials that have been produced without loss of life or audience over four decades. Even NBC's erstwhile programming guru, Brandon Tartikoff, has stated: "If you can make 265 episodes each year of "Days of Our Lives," I've got to believe you can make 40–45 episodes of "Knots Landing."[7]

Viewers would undoubtedly welcome such increased production, but it is not likely to happen. It is far easier to camouflage the absence of a lead player in a soap opera, where the cast is huge and the subplots numerous, than it is to hide the fact that Blake Carrington is nowhere to be found for two or three weeks on "Dynasty." Tartikoff counters with: "Start some of your newer storylines and newer characters when you hit the month of June and build them up."[8] Although that approach may keep the show on the air, it seems predictable that without the stars, the ratings would drop dramatically. In addition to requiring a rest from the daily grind, series leads also want some free time to capitalize on their popularity by making movies and high-paying public appearances.

Probably the most compelling argument against doing "40–45 episodes a season" is the economics of a series. Most network deals for primetime programs call for two plays of each episode plus an option for a third. For a typical half-hour sitcom, the network will pay a license fee of $550,000 to the production company. If the network sells each of the six 30-second spots in the show for $75,000, the gross from the first play will total $450,000. However, from that amount the network must deduct a 15 percent commission to the advertising agencies that placed the commercials, as well as compensation to the stations carrying the show and other network expenses. The loss to the network on the initial run of each episode is substantial. A season of nothing but original episodes for all primetime shows would break a network within a year.

But when a repeat of the episode is scheduled, the cost of payments to performers and other guild members ($50,000–$75,000) is the only production expense. Although the value of the commercials is somewhat reduced because of the audience fall-off in the summer and the diminished attraction of seeing an episode twice in the same season, the revenue on a reasonably popular program is more than enough to offset costs and send the program into profit. Despite complaints of viewers and the frustration of some program and production executives, an annual cycle that incorporates a rerun of virtually every original episode is a strong possibility for the discernible future.

Given the fact that reruns are a necessity, programmers must decide when is the best time to schedule them; or, to be more accurate, when is the least harmful time to run them off. There are a few general dos and don'ts:

1. Don't schedule a rerun for the first ten or twelve weeks of a new series unless forced to by production realities. New series take time to become established and a repeat episode early in a show's run may drive away viewers who have not yet become devoted fans.
2. Don't waste a new episode against a blockbuster special. When the opposition schedules a surefire smash against you, burn off a repeat—nobody's going to be watching anyhow. You can save the new show for a more advantageous time.
3. Do hold in reserve fresh episodes for sweeps weeks, including the May sweep. There is a temptation to run off all new episodes during the traditional season (September to April) and forget that financially the May sweep is just as important as the others.
4. Don't waste an original on a night when viewing is traditionally low (for example, Christmas Eve and Christmas Day).
5. Don't run a repeat as a lead-in to the premiere of a new series. Plan ahead to be sure the new show will have the best possible chance to achieve sampling.

6. Do give production commitments to producers sufficiently in advance for them to deliver episodes without rushing. Late pick-ups result not only in inferior episodes because of haste-makes-waste production, but also in the too-early placement of repeats.

Boosting the Audience in Sweep Periods

Although every day in television is important, some are more important than others. Four times a year for thirty-day periods (February, May, July, and November), Nielsen and Arbitron conduct special "sweeps" of audience viewing habits in every market in the country. These monitorings are vital to the economic health of each station. The statistical data form the basis of the advertising rates until the results of the next sweep are available. Programmers are under great pressure to build the highest possible ratings for each time period.

In bygone times, stations and networks would go to extraordinary lengths to spike their numbers. Contests, giveaways, and assorted other viewing inducements were commonplace. Eventually the ratings services notified the stations that such "briberies" would disqualify the show from evaluation.

Many programs—local, network, and syndicated—then shifted to sensational material. For thirty days waves of crime, debauchery, and prostitution were observed throughout the land. A return to virtue was usually detected the day after the sweep ended. The scheduling of lurid and degenerate topics during sweep periods has been decried by journalists and civic leaders, but the practice continues with no discernible reduction.

Programmers frequently reach new heights of inventiveness during sweep periods. One station featured a week-long series about Nielsen families in its local news show. The goal was to attract the Nielsen homes in the area to watch a show about themselves and, of course, swell the ratings for the program. In another instance, the Carsey-Werner Distribution Company placed an ad in a New York newspaper for its badly underperforming "You Bet Your Life" that read "Nielsen Families . . . Meter Check Tonight!!! Watch Bill Cosby . . . tonight on 'You Bet Your Life' 7:00 p.m., Channel 2." Nielsen took a dim view of both these schemes. In the first instance it eliminated the news program from its sweeps report, and in the second it eliminated the entire time period.

So desperate do some stations become, they resort to out-and-out fraudulence. One station blatantly offered to pay diary-keepers to list its call letters in their reports. Another asked members of the sample to send in their diaries to the station in exchange for a prize. Stunts such as these are unconscionable and rarely go undetected (rival stations are quick to report such shenanigans to the ratings services). When the dirty tricks are verified, the services invalidate the figures and the appropriate time periods are stricken from the station's report.

However, it is entirely proper during sweep periods for stations and networks to employ the many audience-building techniques that have been described in this chapter. The battle for viewers (and therefore dollars) is crucial, and spiking devices are a legitimate part of the competitive system. Thus, if you notice an unusual number of specials, or a flurry of big-name appearances, or a string of blockbuster movies all occurring over a brief span, chances are one of the sweeps is in full swing.

Patience

One of the most important strategies in programming is one that is too rarely used: patience. Most shows don't instantly burst forth and blossom. They take some time to find themselves and their audience. What looked so good in the pilot or the story projections just doesn't come alive when the show first gets under way. Changes are necessary; people have to be replaced; emphases adjusted. Slowly the elements gel and a potential winner begins to take shape, if it's given the chance.

On the other hand, in broadcasting, the stakes are high. Advertisers are guaranteed a certain number of viewers. If these viewers are not delivered, the distributor must make good by giving the advertisers free commercials in other programs, a costly procedure. Furthermore, producers whose shows follow a stumbling entry become upset. They seldom suffer in silence. Top management wants to know what the programmer is going to do about it: "We can't just sit there, you know."

The overpowering temptation is to quiet all the clamor by announcing a change. It shows an awareness of the problem and a willingness to act. Nobody can fault aggressiveness, can they? The trouble is, it may be the totally wrong thing to do.

Some programs look good, feel good, and play well; yet they don't immediately draw an audience. These are the ones a programmer should stay with for awhile.

The history of television is laden with stories of slow starters that were inches away from extinction and eventually went on to become huge successes. The original Dick Van Dyke show, "Cheers," "Barney Miller," and "Hill Street Blues" are just a few of the scores of hits that owe their long runs to the patience of programmers who had faith.

In addition to professional judgment and "feel," there are other signs of incipient success that decision-makers should look for:

1. Exceptional mail response. Some broadcasters deride the importance of mail because the correspondence represents such a small portion of the viewing universe. But seasoned programmers recognize that an unusually high mail count could indicate something "is going on out there." It suggests the show should stay on until there is a fair opportunity to see if cards and letters will correlate to viewers.

2. Strong demographics. The demographics of new programs should be monitored very closely. Young viewers are much more volatile than the mature audience. If a new program presents especially young demographics, it might be a signal that older viewers will eventually be attracted. It's worth the wait.

3. Word of mouth. It's hard to quantify, but once in a while there is a buzz about a program that hints at the possibility of future success. A savvy programmer, one who knows how to filter out the flattering words of his or her mother or accountant, is able to detect a special quality in what he or she reads and hears. It should encourage a go-slow policy.

Most of all, programmers should trust their instincts. If the program is soundly conceived and well executed, pressures for an early dismissal should be resisted. There's a good chance the judgment will be vindicated. But even if it's not, the programmer will gain praise for steadfastness, and more than likely will obtain a first look at a grateful production company's next program concept.

There may be times when a show deserves to be canceled but simply can't be dropped because there are no replacements on the shelf. Employment-conscious programmers who have improvidently left themselves in this condition frequently use the time-honored strategy of stepping forth boldly to declare their faith in the struggling show. Better to be criticized for an excess of patience than a shortage of preparation.

These strategies of audience enhancement are most visible in the highly competitive arena of commercial television. But they are also very evident in the other forms of broadcasting, as we will see in the next chapter.

TWO FREDS GO HEAD TO HEAD

One of the most fascinating competitions in the history of network television occurred during the 1979–80 season. It contained the stuff of novels: high stakes, breathtaking gambles, and personal animosities. For students of broadcast strategies, it is a veritable treasure-house.

The story actually began in the early spring of 1975. Fred Pierce, recently appointed president of ABC Television, was searching for a president of ABC Entertainment, the chief programming position at the network. The company, although showing some substantial rating strength in its primetime lineup, was still in third place, and Pierce's mission was to improve its standing.

He pursued, and eventually signed, Fred Silverman, who had led CBS to the number one position in primetime ratings for five consecutive seasons. Silverman had been feeling undervalued at CBS and saw the ABC offer as an opportunity to be both successful and appreciated. He arrived in June 1975, and within nine months ABC vaulted to the top, a position it maintained through the next two seasons under his guidance.

Although the achievement was the product of many skilled executives and suppliers, the two principal architects were undoubtedly the two Freds. However, because of his more assertive personality and his dramatic transition from another network, Silverman received greater public applause, a condition his boss tolerated in the interests of corporate harmony. But Silverman grew restless and, as his three-year contract drew to a close, he sought new challenges. His request to head up the ABC News Department as well as Entertainment was denied.

In January 1978 he announced that he would become the president of the National Broadcasting Company upon the termination of his ABC contract in June. Pierce was enraged. He claimed that Silverman had assured him that, if he were to leave ABC, he would not go to another network. Pierce so informed the ABC board of directors, and the betrayal left him embarrassed and vengeful. Thus the scene was set for the remarkable events of the next twenty months.

Because he was still technically under contract to ABC, Silverman was enjoined from participating in any broadcast activities until June, an injunction that ABC carefully monitored. Since all three networks had to set their fall program schedules by May (to allow sufficient time for the production and sales of the shows), Silverman was effectively eliminated from all influence on NBC's initial 1978–79 lineup. Although he juggled a few shows in the fall and prematurely rushed some others onto the schedule in January, he was unable to derail the ABC programming juggernaut, which he had largely designed. ABC completed the season with a 2.2 rating point lead over CBS and a 3.8 victory over NBC.

The 1979–80 season was to be an entirely different matter. From the beginning Silverman would be involved in all phases of the competition: the development of shows, the setting of the schedule, the night-by-night, spot-by-spot promotion. His history of success, number one every full year he had been program chief, was a daunting record for his opponents to ponder.

But a still-seething Fred Pierce was not about to be intimidated. He vowed that ABC would be as aggressive in first place as it had been in third. If his bold moves led

FIGURE 11.5 This shows the network primetime schedules for the 1979–1980 season. Source: Alex McNeil, *Total Television* (New York: Penguin Books, 1984): 820.

1979–1980

Day	7	7:30	8	8:30	9	9:30	10	10:30	11 PM
SUN	**OUT OF THE BLUE**	**A NEW KIND OF FAMILY**	MORK & MINDY 2	**THE ASSOCIATES**	MOVIE				ABC
SUN	60 MINUTES 12		ARCHIE BUNKER'S PLACE[1] 10	ONE DAY AT A TIME 5	ALICE 4	THE JEFFERSONS 6	**TRAPPER JOHN, M.D.**		CBS
SUN	DISNEY'S WONDERFUL WORLD 26		THE BIG EVENT 3				**PRIME TIME SUNDAY**		NBC
MON			**240-ROBERT**		NFL MONDAY NIGHT FOOTBALL 10				ABC
MON			THE WHITE SHADOW 2		M*A*S*H 8	WKRP IN CINCINNATI 2	LOU GRANT 3		CBS
MON			LITTLE HOUSE ON THE PRAIRIE 6		MOVIE				NBC
TUE			HAPPY DAYS 7	**ANGIE** 2	THREE'S COMPANY 4	TAXI 2	**THE LAZARUS SYNDROME**		ABC
TUE			**CALIFORNIA FEVER**		MOVIE				CBS
TUE			**THE MISADVENTURES OF SHERIFF LOBO**		MOVIE				NBC
WED			EIGHT IS ENOUGH 4		CHARLIE'S ANGELS 4		VEGA$ 2		ABC
WED			**THE LAST RESORT**	**STRUCK BY LIGHTNING**	MOVIE				CBS
WED			REAL PEOPLE 2		DIFF'RENT STROKES 2	**HELLO, LARRY** 2	**THE BEST OF SATURDAY NIGHT LIVE[2]**		NBC
THU			LAVERNE AND SHIRLEY 5	**BENSON**	BARNEY MILLER 6	SOAP 3	20/20 2		ABC
THU			THE WALTONS 8		**HAWAII FIVE-O** 12		**BARNABY JONES** 8		CBS
THU			**BUCK ROGERS IN THE 25TH CENTURY**		QUINCY 4		**KATE LOVES A MYSTERY** 2		NBC
FRI			FANTASY ISLAND 3		MOVIE				ABC
FRI			THE INCREDIBLE HULK 3		THE DUKES OF HAZZARD 2		DALLAS 3		CBS
FRI			**SHIRLEY**		**THE ROCKFORD FILES** 6		**EISCHIED**		NBC
SAT			**THE ROPERS** 2	**DETECTIVE SCHOOL**	THE LOVE BOAT 3		**HART TO HART**		ABC
SAT			**WORKING STIFFS**	**THE BAD NEWS BEARS** 2	**BIG SHAMUS, LITTLE SHAMUS**		**PARIS**		CBS
SAT			CHiPS 3		BJ AND THE BEAR 2		**A MAN CALLED SLOANE**		NBC

NOTES: [1]Formerly *All in the Family* [2]Reruns of *NBC's Saturday Night Live* to victory and the industry concluded he was the key Fred all along, well, that would be a happy by-product.

On April 23, 1979, ABC revealed its daring, aggressive schedule for the 1979–80 season. To the astonishment of the industry, the lineup called for three hit shows to leave their traditional periods and move to new positions. "Laverne and Shirley," for three years a smash success in tandem with "Happy Days" (Tuesday, 8:00 to 9:00 p.m.), was relocated to Thursday at 8:00 p.m. This was truly a surprise to industry analysts who were aware that the program's qualitative research (the measurement of a show's inherent strength rather than its rating) was, at best, average.

"Mork and Mindy," starring Robin Williams, was moved out of Thursday, 8:00 p.m., where it had burst into top ten status the previous season, and was relocated at Sunday, 8:00 p.m. ABC's goal was to wrest the audience from CBS at the completion of the high-rated "60 Minutes." Replacing Mork on Friday was "Fantasy Island," heretofore a runaway winner on Saturday at 10:00 p.m.

A week later NBC announced its schedule. The industry (and presumably Pierce) was dumbfounded. The feared Silverman had introduced just six new shows. Almost 75 percent of a third-place schedule was being returned. The rumors were that this surprisingly low-key schedule was a failure of program development rather than a rush of conservatism in the normally hard-charging Silverman.

CBS was similarly unadventuresome. Seven new shows were added to the lineup and only one returning program, "The Jeffersons," had been moved to another time period.

In the first three weeks of the season, NBC stunted on virtually every night and stepped into the lead with a 19.7 rating vs. 18.8 for ABC and 17.0 for CBS. A concerned Bob Daly, president of CBS Entertainment, stoically announced that "it is premature to move before we get more ratings." By the end of October, NBC, now out of special events and other stunts, had slipped into third and were no longer a factor in the competetive race.

ABC's coverage of the baseball playoffs and World Series, always high-rated entries, boosted the network into first place. Relying heavily on its basic schedule, CBS eased into second. Although there was the usual flurry of Christmas specials, the three networks generally settled into their standard program lineups for the remainder of 1979.

Two weeks into 1980, CBS caught ABC, causing Daly to comment, "ABC made some mistakes; that's quite obvious. They left an opening for us when they moved 'Laverne and Shirley.' The new show, 'Hart in San Francisco,' left ABC vulnerable to counterprogramming. They moved three shows, and it hasn't helped them; we moved 'The Jeffersons' and it helped our night." As a result of a number-one finish for the week of January 21, CBS took over first place in season-to-date ratings for the first time since April 19, 1976.

But ABC was far from finished. For two full weeks in the latter part of February, the network's primetime schedule was devoted to Winter Olympics coverage. The games were a huge attraction and averaged an overall rating of 24.5 with a 38 share. It shot them back to first, a position they held by a 0.1 rating point going into the last week of the season.

Now the heavy thinkers got serious. Pierce and ABC had two aces to play, "The Academy Awards," always one of the top-rated programs of the year, and a repeat of *The Sting*, starring Paul Newman and Robert Redford, which had scored a 31.9 rating in its first TV appearance. Daly and CBS countered with a two-part movie-of-the-week entitled *Guyana Tragedy*, a dramatic recreation of the events surrounding the mass suicide of 909 followers of American cult leader the Reverend James Jones, and a special two-hour episode of "Dukes of Hazzard," a broad, good-ole-boy, hour-long comedy series that had premiered a year earlier and developed a large and devoted audience.

The Oscars, as expected, performed superbly with a 33.7 rating. But the impact was virtually neutralized on the next two evenings when *Guyana Tragedy* averaged a 30.3 rating. And thus it came down to the last day of the season. ABC needed a prodigious rating from *The Sting* to pull out the victory. The "Dukes" special was formidable, but would it be strong enough? It was. "Dukes" won all four of their head-to-head half hours, and CBS took the week by 1.3

rating points. That was sufficient to overcome the deficit and give them the season's championship. The near-legendary Silverman, whose dazzling track record had triggered ABC's ill-advised moves, limped home a distant third.

It is significant to note that before the season had ended, all three relocated ABC shows ("Laverne and Shirley," "Fantasy Island," and "Mork and Mindy") had been returned to their original time periods. "The Jeffersons" stayed put for the entire season, and two more after.

The hero of the drama, Bob Daly, stated: "What we learned this season is not to change the schedule so fast, not to panic when our new fall entries fail to meet expectations, to be prepared with backup series and not to rely on theatrical movies alone. Balance is essential."

Students of programming strategy will also have observed:

• Although programming is an intensely competitive business, it is best to minimize personal conflicts; emotionalism can cloud judgment.

• Stunting is a short-term solution; it is not a substitute for a solid schedule.

• Tinkering with a successful lineup can be perilous; well enough should be left alone. Relocated programs do not always reclaim their full audiences when restored to their original positions.

• An incumbent has an advantage over a transplanted show; it is the difference between an audience that's already there versus an audience that *might come* there.

• Shows with the same demographic profile are not necessarily interchangeable. The mood, style, and pace of one program may be far more suited to a particular time period than another show with the same apparent appeal.

• And then, of course, just in case the point has been missed, there's patience. Frequently the best thing to do is to do nothing; just give your original judgment a chance to prove itself.

NOTES

1. "ABC Pulls 'Peaks', Skeds Saturday Pics," *Hollywood Reporter* (February 19, 1991): 1.
2. "Fox: No More Fall Launch Fray," *Broadcasting* (July 22, 1991): 15.
3. Ibid.
4. "Syndicators Falling Away from Traditional Debut Skeds," *The Hollywood Reporter* (April 29, 1991): 1.
5. "Sandy Grushow: Charting Fox's Future," *Broadcasting* (April 1, 1991): 41.
6. ABC presented NFL football throughout the fall and switched to movies for the rest of the year.
7. "The Sorry State of Network TV," *Los Angeles Times* (May 6, 1991): F-1.
8. Ibid.

12 Scheduling Strategies: Part II

CABLE TELEVISION

Because there are so many different types of cable TV services, most of which engage in narrowcasting, it is difficult to discuss generalized programming strategies. What works for The Disney Channel does not work for CNN, or MTV, or even HBO. But overall, some of the same strategies used in commercial television can be found in the cable world. The main element that differs is the emphasis and, to some extent, the execution. In addition, cable has some unique characteristics that affect scheduling.

Conventional Scheduling Techniques

A few of broadcasting's conventional techniques are essentially useless in cable. Audience shares for cable networks are so tiny that tentpoles and hammocks do little good. Two programs with regular shares of 3.2 might be able to bring up a program in between them from a 3.0 to a 3.1, but that really is of little significance. Likewise, the small share size does not allow cable networks to be very effective with blunting, stunting, or crossprogramming. Counterprogramming is fairly useless in cable TV because no network can counter all the other cable networks, let alone the broadcast networks. With so many services available, two or more services will almost always be showing similar material. For example, two intellectually oriented channels, The Discovery Channel and The Learning Channel, differentiate themselves in that the former deals with the experiential, physical world while the latter deals with the world of ideas. But that's splitting hairs. USA network is the only cable service that considers itself in direct competition with the broadcast networks and, as a result, does try to counterprogram them.

Fitting the Show to the Available Audience. Cable networks try to program for the available audience, but, given the narrowcasting and fragmentation of cable, sometimes there aren't many viewers available. Peter Hansen, vice president of programming for A&E, admits, "Realistically our upperscale target audience just isn't sitting around watching TV during the daytime."

Dayparting. Some cable networks daypart and others do not. The Disney Channel is very specific about its dayparting—the early day is for young children; the late afternoon is for older children; the evening is for the whole family; and the period after 9:00 p.m. is for adults. The Learning Channel has three distinct dayparts and even has specific names for them—"Classroom America" takes up the morning with programming that can be used by teachers in the classroom; "Pursuits," in the afternoon, is dedicated to programming about personal growth, such as career advice, exercise, and cooking; and "Renaissance Spirit," during primetime, consists of dramatic productions and documentaries on humanistic topics. But to other channels, such as The Weather Channel and CNN, dayparting isn't a valid concept. The weather changes from morning to evening, but the method of talking about it doesn't.

Quite a few of the services make changes in their weekend schedule to take into account the life-style changes that take place then. Bravo, for example, programs three hours of children's cultural fare on Saturday and Sunday mornings, but the rest of the schedule is not specifically dayparted. Even CNN has lighter, more featurelike news on the weekends than during the week.

Bridging. The cable TV movie channels have used bridging very effectively, not always by design but often because the movies that they show uncut and without commercials last a little over an hour and a half. This can bridge across three or four sitcoms or two dramas that are showing on network TV. Audience members never have a commercial break during which to sample other programs.

Stacking. Cable networks do try to program for audience flow. They, like the commercial networks, want to hold an audience once they have captured it. Sometimes they do this by stacking programs of a similar nature one after the other. However, because most cable networks narrowcast, stacking is really not a necessary concept. All of the programs presented on the network are intended to attract the same well-defined audience, so, in essence, the whole network is stacked.

Theming. Theme building (Tuesday Night of Terror, a week of Elizabeth Taylor films) is very common in cable, especially on the movie services. It makes for excellent promotional opportunities that can distinguish one cable system from another.

Stripping. Stripping occurs on many of the networks. Ones that play syndicated programming, such as USA and The Family Channel, strip just like the independent stations. The superstations also strip—they are independent stations. Even HBO strips several children's programs early in the morning.

Patience. Cable networks have a great deal more patience than commercial networks—in part because they have such a need for material. The problem is more in acquiring than removing. Also, the industry has two sources of revenue—advertisers and subscribers. The advertising rates are nowhere near as high for cable networks as for commercial networks and the ratings for individual networks are low. So cable operators do not need to be as switch-happy as commercial broadcasters. They can let something grow and develop because the stakes on any particular program are not that high.

Rerunning. The rerunning of programs on cable needs to be given special attention because, within the cable business, rerunning has been developed to an art (or a science, depending on your point of view). It has not been thought of as an economic cop-out, but rather as a way of life. Instead of emphasizing that the show should fit the available audience, cable networks assume there is an available audience, however small, at various times. The philosophy is that by showing a program at six distinctly different times, everyone who wants to will be able to catch it. Rerunning, which has a negative connotation because of its commercial television roots, has been euphemized by cable TV; networks refer to a "rotating schedule" and reruns themselves are called "encores" to distinguish them from first showings, which are called "premieres."

The pay cable services, in particular, are very scientific about how they rerun movies and programs made specially for them. They try to make sure each program or movie is run in as many different time periods as possible. Most of the twenty to fifty movies and programs scheduled each month are shown four to six different times, and many are carried over from one month to another.

Pay-per-view services rerun even more often. Sometimes they run the same movie several times a day or even over and over again for an entire day. The goal is to make each movie accessible to everyone who wants to order it.

Sometimes outside factors affect the rotating (rerun) schedule of a cable network. For example, the contract they sign to obtain a particular movie may contain a clause limiting the number of showings. The window process determines when the movies can premiere and how long they can run. As mentioned earlier (see Chapter 4), after movies are shown in theaters, they are released first to pay-per-view, then to videocassettes, then to pay cable, then to network TV, and then to independent stations and basic cable. If the movie studio has decided to withhold a movie for a lengthy period before releasing it to any TV forms, the flick may lose some of its appeal and the cable systems may decide not to rerun it as often as they might otherwise. Sometimes studios do this with a movie so that they can give it a second theatrical run before turning it over to video.

The basic cable services also rerun and rerun and rerun. Because their rating numbers are small, they want to build up a large cumulative audience so that they can encourage advertisers to buy within the different showings of a program. But they, like TV stations, are also dealing with hard economics. Contracts with syndicators or production companies usually stipulate that they can run the program several times for a set price. For the most part, the cablers try to get their full money's worth.

Rerunning of programs is also very common on the local origination and public access channels operated by cable systems. This is due primarily to lack of material to fill twenty-four hours and lack of funds to purchase outside programming. Many local systems program written announcements that run throughout over and over again on the screen, usually accompanied by music or a local radio station.

At some point in the near future, rerunning may be taken to its logical extreme—the same program over and over all day. This scheduling philosophy is tried on occasion by the pay-per-view services. They program the same movies throughout the day so subscribers can tune in whenever it is most convenient. This is similar to the "scheduling" of a movie theater. If technology allows the number of cable channels to grow into the hundreds, this type of repeat-repeat-repeat scheduling may become quite common. A number of companies are exploring this option.

Unique Scheduling Techniques

In addition to using some of the scheduling techniques employed by commercial television (albeit in a somewhat altered form), cable television has some of its own unique scheduling strategies.

July 1992

Mon.	Tues.	Wed.	Thurs.	Fri.	Sat.	Sun.
		1	2 3:00 P.M.: Big	3	4 7:30 P.M.: Naked Gun 2 1/2	5
6	7 11:00 A.M.: Naked Gun 2 1/2	8	9	10	11	12 4:30 P.M.: Naked Gun 2 1/2 6:00 P.M: Big
13	14	15	16	17 9:00 A.M.: Big	18	19
20	21	22 8:00 P.M.: Big 11:30 P.M.: Naked Gun 2 1/2	23 9:00 A.M.: Naked Gun 2 1/2	24	25 6:00 A.M.: Big	26
27	28 12:15 P.M.: Naked Gun 2 1/2	29				

FIGURE 12.1
This shows the multiple cablecastings of two movies, *Big* (1988) and *Naked Gun 2 1/2* (1991), on HBO during July 1992. Source: *Cableview* (July 1992): passim.

Monthly Balance. One of the scheduling concerns facing cable but not broadcasting is a need for monthly balance. Subscribers lay out their money at the beginning of each month. If they do not feel they received their money's worth last month, they may not subscribe for next month. Therefore, cablers feel the need to please as many different people as possible during the month to maximize re-ups. A movie service will make sure that during the month it has movies that appeal to men as well as women, rural audiences as well as urban, young people as well as old. The Movie Channel, for example, would not feature war movies for an entire month because these would appeal primarily to men and might alienate women. The cable networks also try to have two or three blockbuster premieres per month—either big hits from the movie studios or something they have funded and produced themselves that can be easily promoted.

First Come, First Served. First come, first served is a programming strategy that occurs on local cable access channels, and nowhere else. Usually when members of the community produce programs, they can choose the time they want the programs to air, provided no one else from the community has previously requested that time. If disputes arise about schedule time, someone from the cable system needs to arbitrate.

Channel Assignments. The concept of

scheduling at a local cable system is very different from the concept at any network or station. Aside from local origination and access channels, cable system scheduling consists of deciding what program services to pass through to subscribers and what channel number to assign to each of these services. Often the sales force has more input into this decision than the people involved with programming because what is really important to the cable system is that it has a lineup of channels that will make people want to subscribe.

The following is the cable channel lineup for Palmer Cablevision in Naples, Florida.[1] Note the grouping of the various channels.

Basic Service
- 2—WBBH NBC
- 3—WSFP PBS/Fort Myers
- 4—FTX Fox
- 5—WINK CBS
- 6—WNPL Independent/Naples
- 7—WEVU ABC
- 8—C-Span I House of Representatives
- 9—WGN Independent/Chicago Superstation
- 10—Local Origination/ClassiVision
- 11—City and County Government/Sunshine Network
- 12—TBS Independent/Atlanta Superstation
- 13—WPBT PBS/Miami

Pay Channels
- 14—Cinemax
- 15—Showtime
- 16—HBO
- 17—The Disney Channel
- 18—The Movie Channel

Preferred Basic
- 19—Home Shopping Network
- 20—The Family Channel
- 21—Cable News Network
- 22—USA
- 23—Headline News Network
- 24—Lifetime
- 25—Univision Spanish
- 26—ESPN
- 27—The Nashville Network
- 28—Turner Network Television
- 29—American Movie Classics
- 30—Arts and Entertainment Network
- 31—The Discovery Channel
- 32—The Weather Channel
- 33—Nickelodeon
- 34—MTV
- 35—Consumer News and Business Channel
- 36—VH-1 and Comedy Central
- 37—Prevue Guide Electronic Programming Information
- 38—QVC Home Shopping
- 39—Visitors Television
- 40—Comedy Central
- 41—C-SPAN II Senate
- 42—Mind Extension University
- 43—The Learning Channel
- 44—Bravo
- 45—SportsChannel Florida
- 46—Weather Radar
- 47—The Monitor Channel
- 48—E! TV
- 49—Black Entertainment Television
- 50—The Travel Channel
- 51—Court TV
- 52—Palmer Pay-per-view Information
- 53—Request TV
- 54—Collier County School Board
- 55—Eternal Word Television Network

However, some interesting pressures have come to bear in the area of channel assignment. VHF stations want the same channel number on a cable system as they have for their over-the-air signal. Channel 5 wants to be channel 5 on every cable system in town in order to promote its programs easily. Technically, it is very easy for a cable system to transfer channel 5's over-the-air signal so that it shows up on channel 6 (or channel 106) in the subscriber's home. But usually the cable systems are happy to accommodate the VHF stations because they are popular with viewers.

The problems come with the UHF stations that usually don't want to be on the same channel as their over-the-air number. They want a lower number—down there where people are always flipping through channels to get from one VHF station to another. The cable systems are not willing to accommodate the UHFs. They want those lower numbers for cable network services—the ones that bring them revenue. This squabble is constant between cable systems and UHF stations.

a

FIGURE 12.2 Nickelodeon, which programs for youth with series such as "Roundhouse" (a), sometimes shares a cable channel with Arts and Entertainment, which programs cultural material such as "Afternoon Z Afternoon" (b), one of the dramas seen on "American Playwrights Theatre." (Photos courtesy of MTV Networks (a); Arts and Entertainment/Linda Alaniz (b).)

Both have complained about each other to whatever bodies within the government will listen, but nothing concrete has been decided.

The cable networks have tried to apply pressure on the cable systems for same-channel carriage. Again this is for national promotional purposes. CNN would like to be able to advertise, "Watch our news on channel 10." Of course, it can't because on many cable systems, channel 10 is a local station. The cable networks' campaign for consistency in channel numbers has been only moderately successful. In some individual markets, where all the local VHFs have the same channel number, the various cable systems have cooperated to give most of the cable networks the same channel.

But, of course, not all systems have the same number of channels, so this introduces a fly into the ointment. And not all systems carry the same services. While Uptown Cable may carry The Family Channel and not American Movie Classics, Downtown Cable may do just the opposite.

Some cable systems, especially those with limited channel capacity, schedule two cable networks on one channel, e.g., Financial News Network during the day and Nostalgia

b

Television at night, or Nickelodeon during the day and Arts and Entertainment at night. This is a bit confusing to subscribers, but it gives them more variety of choice than they would otherwise have.

Launching

Cable's philosophy on launching is not consistent. As you have no doubt learned to expect by now, it varies from network to network. Some believe in the traditional fall launch to accommodate media advertising. "If TV magazines have a 'fall issue,' why not be in it?" says The Discovery Channel's Fred Moyer.[2] Others want to avoid the fall for fear they won't be heard in all the din. Lifetime's Pat Fili notes, "We premiered our original primetime series for Tuesday nights in July to get a jump and break through the clutter . . . personally I don't want to premiere any primetime series in mid-September to early October."[3]

Some cable networks, following Fox's lead (or leading Fox, depending on whose side of the story you want to believe), launch shows on a year-round basis. Others prefer January when the largest number of people are indoors, available to watch TV.

Pay-per-view has a unique situation in that it needs constant launching. Each time it cablecasts a new movie or boxing bout or other sports spectacular, it must find a way to let the potential audience know it's there. The PPV networks have promotional staffs that keep busy on a constant basis trying to make the public in general, and PPV subscribers in particular, aware of the new programming.

Overall, cable scheduling is a complex process, largely because of the variety of disparate services available. It promises to become even more complex as channel capacity and fragmentation increase.

COMMERCIAL RADIO

Radio, because it deals primarily with formats rather than individual programs, has an entirely different basis for scheduling. Terms such as tentpoling, hammocking, bridging, blunting, stacking, and stripping (all of which refer to the manipulation of individual programs) do not apply for most of radio programming. Counterprogramming is inherent within the format itself; a country music radio station is always counterprogramming an album-oriented rock station.

All-talk formats make use of some of the television-oriented scheduling strategies because they tend to be comprised of individual programs hosted by specific personalities. But even talk shows tend to be long (usually about three hours), so there is little need for considering the interrelationships among them.

The Clock

The main scheduling instrument of most commercial radio stations and some of the networks is what is referred to as a "clock" or "wheel."

This is a circle that represents an hour broken down into segments. Stations generally program each hour approximately the same to enable audience members to know what they will hear at a certain time. For example,

FIGURE 12.3
A typical clock for an all-news station.

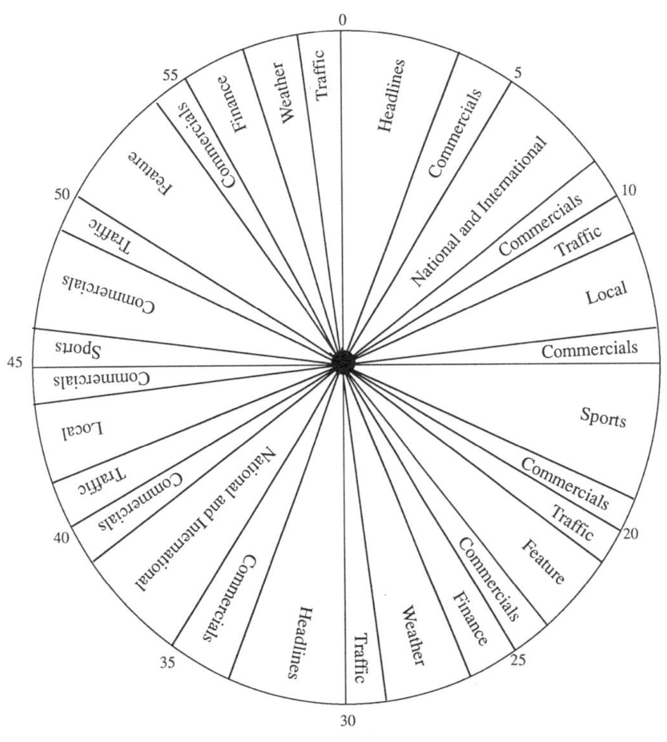

with the all-news station clock shown in Figure 12.3, listeners could depend on hearing traffic every ten minutes and sports at quarter after and quarter to the hour.

Music station clocks indicate the placement of news, commercials, and other verbal elements, but they also usually indicate the types of music that will be played at various times. A great deal of thought and planning goes into the actual scheduling of music both in terms of what songs to play and what order to play them.

Scheduling Music. A station that concentrates on current hits must constantly be on the lookout for music that has the potential to become popular. Program directors (PDs) and disc jockeys receive copies of new releases from record companies, listen to them, and select for airing the ones they think their audience members will enjoy. Each format and, to some extent, each station, has its own sound; as a result, PDs do not all choose the same selections.

To help confirm their decisions, program directors often consult the charts in trade magazines such as *Billboard* and *Radio and Records*. These list the top-selling songs in a variety of categories—adult contemporary, country, rock, soul, jazz. However, popularity of a record is often a result of its airplay. Airplay leads to sales and listings in the trades, and these, in turn, lead to more airplay.

Not all songs played on the radio are current hits. Many are songs from the past that can be roughly characterized as follows:

- Oldies—music from thirty or forty years ago. If the song was at one time in the top ten and sold about a million copies, it is called a golden oldie; if it was number one and sold several million copies, it is supergold.
- Classics—music from ten or twenty years ago. These, too, can be golden or supergold.
- Recurrents—music that is no longer current but is not yet considered old.

A note of caution. These terms are not totally standardized within the industry. Some programmers consider anything over three

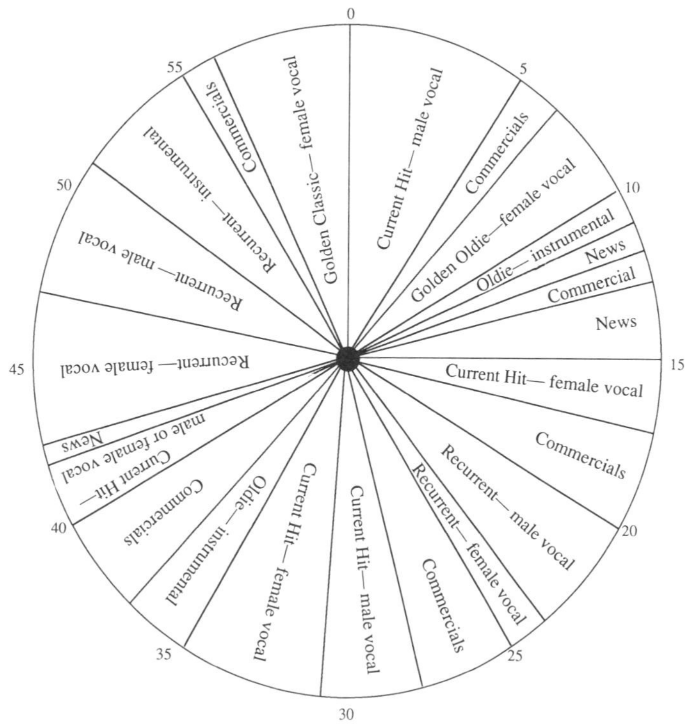

FIGURE 12.4
A typical clock for a music station.

years old to be an oldie. In some stations a recurrent is a fading hit while at others it is music that is at least two years old. Also the terms differ. For example, supergold may be referred to as power gold. Mike O'Reilly, program director at WPDC in Elizabethtown, Pennsylvania, divides his music into five groupings—new, familiar, recent, gold, and oldies. Other PDs have their own designations.

A radio station programmer must decide what proportion of current hits, recurrents, classics, and oldies to play. Sometimes the format decides this. An oldies format, for example, will consist only of music that is old enough to qualify. But many formats have a variety of music from different periods.

Some stations use the charts, not to determine when they should be playing something but rather when they should not be playing it. Stations that are trying to break new hits want to air songs before they hit the charts. Rachael Donahue, who is a programmer for such a station in Santa Monica, California, says, "We stop playing something when it makes the charts and everyone else starts playing it." Other stations want to start playing music as it is falling off the charts. Phil Wilson of KGGO, a contemporary format station in Des Moines, says: "Familiarity is very important to my audience. I add music when it is on the way down."

There is no magic formula for deciding how many songs of different ages to play. The decision is an intuitive one made by the programmer.

Other factors besides the age of a song go into the makeup of a wheel. Many stations like to vary vocal and instrumental, or alternate male and female vocalists. Others take into account the overall volume level of the music, perhaps making sure that each quarter hour contains at least one soft song. Others group songs by the nature of their lyrics.

Thought must also be given to the transition from one piece of music to another. If a fast song immediately follows a slow song, the result can be jarring. Usually a station will program a jingle or some commercials between the two. The same is true when going from soft music to loud music or sometimes from a minor key to a major key.

All of these strategies are considered by stations programming music and by networks or syndication companies that supply music services. Knowledge of these elements can make the music programming more successful, but, as with television, the best strategies usually arise from a competent programmer's instincts.

Scheduling Other Program Elements. Aspects of programming other than music must also be considered in the makeup of a clock. The placement of news, weather, traffic, sports, contests, and even commercials can be quite important. People tend to be more aware of the time placements of these items than they are of the subtleties of music positioning. They are much more likely to know that they can tune in five minutes after the hour to hear the latest sports scores than they are to realize that a station never plays more than two female vocal selections in a row.

All-news stations adhere quite strictly to their clock because listeners tend to tune in and tune out. If people caught in traffic know that a particular station broadcasts traffic reports at :15, :30, and :45 after the hour, they are likely to switch to the station at one of those times for the update. Music-oriented stations often have news breaks on the hour or half hour, and people become accustomed to tuning in at that time if they want to catch up on the day's events. A station that programs to people whose overriding interest is music can arrange its breaks so that they do not come at the hour or half hour. In that way, it can attract listeners who push their radio buttons as soon as the news comes on another station. Other stations vary the clock over the course of the day. They program news in the morning when it really seems to matter and then don't program news at all for the rest of the day.

The number of commercials, their frequency, and their position in the clock are all very important. Stations that program fewer overall commercials than the competition are likely to attract listeners. However, if they cannot sell their fewer commercials at a higher price, they may not be able to balance the budget. For this reason, the number of commercials often becomes more of a sales and management decision than a programming one. Frequent commercial breaks seem less jarring on stations with talk formats than those with music formats. Music stations often try to hold their audience members through a commercial break with a "tease"—an announcement of a great new hit coming right up.

Radio stations often tout commercial-free periods when they play straight music for twenty or thirty minutes, or even an hour. Obviously, in these cases the normal clock changes. Again, such a practice might be good for the audience but not the bottom line. A programmer who proposes this idea should have in mind a method for recovering the lost revenue.

The placement of commercials within the clock is also important. If your station has a format similar to another station in town, it is wise to schedule your commercial breaks at a time different from the competing station. In that way you catch the button pushers who are searching for something other than commercials. Of course, this means the competing station will be programming music over your commercials and may grab your listeners. Such is the gamesmanship of radio scheduling strategies.

The disc jockey is another element that figures into the clock. If the disc jockey is popular and is the main element attracting listeners to the station, then more time should be allowed within the hour for comments from the deejay. Often this chatter involves contests that are likely to intrigue a substantial portion of the listening audience and that need time for their instruction and execution. Many early morning shows have very little music and concentrate on the antics of the host. If the deejay is just a facilitator of transitions from one element of programming to another, however, the clock should be more tightly packed.

Variations in Clocks. Nothing is sacred about a clock. It can be fine tuned as much as necessary to develop the best results. However, refining it too much and too often shows a lack of patience and has the disadvantages already mentioned in regard to patience and television programming. Fiddling with the wheel without giving it a chance to find its audience is not advisable.

Some stations place almost fanatical requirements on adhering to the clock—usually to keep otherwise freewheeling disc jockeys in line. Some PDs list the specific songs that should be played throughout the entire day, giving the deejay absolutely no flexibility. Most, however, use the clock as a guideline that can be ignored for numerous reasons—there are not enough popular instrumentals available at the moment to rotate them as often as indicated; several current hits are longer than the average musical selections and will absorb more space on the clock; some major news event has occurred and needs to be updated frequently; the winner of a contest is an intriguing personality who should be interviewed at some length.

And then there are stations that don't use a clock. KSRF in Santa Monica, California, for example, works from a music list. The PD makes a list of songs and the deejays are free to play anything from that list whenever they want. Usually there are one or two songs at the top of the list that need to be played more often than others, and those may be more specifically slotted.

Networks with music formats have clocks, just as stations do. But many networks and almost all syndicators simply supply material for stations to use when they wish. In these cases, the concept of clock does not apply.

Dayparting

Radio scheduling strategies rely quite heavily on dayparting. People's activities, moods, rhythms, and uses of radio change greatly over the course of a day. Radio must take people's biological clocks into account to enable programming to progress in a consistent, successful manner.

Early in the morning, people want to know what has been happening (or not happening) overnight. Many listeners wake up to clock radios, so stations that feature primarily music are quite likely to have more news and information segments in the morning than at other times of day. Early morning disc jockeys are more "talky" than their afternoon counterparts. All-news stations have more format competition in the morning than during other periods.

As the day wears on, people engage in other activities (caring for children, driving the car to a sales meeting), and radio becomes a background to these activities. Music becomes more predominant and talk less predominant on many radio stations. Those stations with news and/or talk formats slide into feature-type material or call-in programs.

In the mid-afternoon, when students are out of school, teen-oriented radio stations pick up the pace and tempo of the music, often featuring their most popular disc jockeys in the late afternoon.

As people leave work and begin the commute home, radio once again gives them information, but usually in a gentler way than in the morning. People are heading home to relax and do not need the up-to-date facts that they might need before the work day. In fact, some stations will play more soothing music at rush hour than at other times of day in order to help people relax as they head home.

In the evening, radio is used primarily as

background to accompany reading, studying, or other activities. The use of radio is greater by young people than by older audience members, who are more likely to be tuned to television. In fact, some of radio's most exciting, and outlandish, moments occur late at night when the audience is primarily young adults receptive to innovation. Many recording groups make several versions of their song, some longer than others and some containing more borderline language than others. Late at night is when the longer, more racy cuts of music are likely to be played. As Rachael Donahue of KSRF says: "We'll put on the three-minute cut of a song in the morning. But at night, when we mostly just play music, we'll use the eight-minute cut."

Networks are conscious of dayparting, too, and try to program material to fit the station's dayparting philosophies. In fact, Westwood One requires that their material run in specific time blocks that they refer to as their "air window." Of course, networks have total daypart control of call-in request shows. These programs, such as Unistar's "Solid Gold Saturday Night," must be aired live to permit the audience to participate.

A radio programmer must take these daypart characteristics into account and merge them, along with format and wheel considerations, into the best possible scheduling strategies.

Launching

The time when launching is most important to a radio station is when it changes format. The new format must be launched in such a way that a new potential audience is made aware of the station. This usually involves using media other than radio to get across the message. Often a station that is changing format will change call letters. It's better to sound like the new guy in town than to have the image of an easy listening format lingering over a contemporary hit format. Besides, the old call letters may not be appropriate. WROC would be a negative for a country station. A bus billboard or TV commercial that announces a new station with a new format is often quite effective.

All-talk stations launch individual hosts or programs and use some of the more traditional TV-type launching techniques. Radio does not deal with seasons, however. If launches are necessary, they are made when the occasion arises and not necessarily in the fall or any other particular time of year.

PUBLIC BROADCASTING

Programming strategies aren't as important within public broadcasting as they are in the commercial world. As NPR's Bill Buzenberg puts it: "Public broadcasting needs to think first about content and less about format and strategies. That's one big way that we differ from commercial broadcasting." But public broadcasting does find some of the typical programming strategies useful from time to time.

Public Television

As with cable, public television's small numbers (usually a rating of 2 or 3) render strategies such as bridging, blunting, and crossprogramming virtually useless. PBS is more likely to program defensively than offensively—it will try to avoid being bridged as opposed to trying to set up a bridge against the networks. "If we know the networks are stunting something big, we won't put on a program for which we want a large viewership," says John Grant, PBS's scheduling chief. Occasionally public television attempts its own stunts, particularly when stations are trying to raise money from viewers during pledge weeks. Stripping occurs mainly with the children's programs that almost always appear at the same time every day. Stacking is also occasionally used—a set of three contiguous cooking shows on Saturday afternoons. But none of these strategies is undertaken with the vigor that they are by the commercial networks.

Counterprogramming. One of the reasons public television does not need to stress strategies is that it consists of only one network. By nature, PBS counterprograms other networks. Even if it is programming a drama at the same time that CBS is programming a drama, the pace and content of the shows will be different enough that the two will not directly challenge each other.

Pillars. PBS schedules around its long-lasting, popular programs, which it refers to as "pillars." In a way, these pillars (e.g., "Masterpiece Theatre," "Nova," "Great Performances") act as tentpoles for other programming, but that is not really PBS's intent. Rather these are old reliables that are rarely moved from their traditional program slots.

In recent years, PBS has begun scheduling limited series such as "The Civil War," "Land of the Eagles," and "The Race to Save the Planet." These are run over a period of several consecutive nights and preempt regular programs—even the pillars.

Rerunning. Many public television programs are rerun, especially if they have obtained popularity. "Sesame Street" is rerun ad infinitum because of audience demand. On many public TV stations, much of the weekend schedule is devoted to reruns of programming shown during the week. John Grant explains: "In its early days public TV reran many programs because of lack of product. Then when we got more product, most of the rerunning stopped. Now we rerun not because of lack of product but because we want to give everyone a chance to see the program. We realize that since the advent of cable TV, the viewer has more choices and may not want to watch our program when it is first aired. So we rerun."

Dayparting. Public television stations determine their own dayparting to a much greater degree than the commercial stations affiliated with NBC, CBS, ABC, and Fox. The contract signed between commercial networks and affiliates penalizes a station if it does not run a program when the network wants it to run. Such is not the case with PBS, although in recent years informal pressure has been placed on the major PBS affiliates to run most of the programs at set times. The reason behind this change is that PBS wants to develop national promotional campaigns that tell all viewers exactly when they can expect to see a particular program on their local PBS station. But stations often have minds of their own regarding what time of day certain programs should air in their communities.

FIGURE 12.5
"Sesame Street" is both stripped and rerun (and rerun and rerun) on most public television stations. (Photo courtesy of Children's Television Workshop.)

Although procedures differ from station to station, most stations do daypart. A common pattern among the major stations is:

- Early morning—PBS programs for young children, such as "Sesame Street" and "Mister Rogers' Neighborhood."
- Late morning—Programs designed for teachers to use in their classes as part of their instruction. Some of these may be produced by the station but most will be PBS children's programs or programs purchased from organizations that specialize in this type of material.
- Noon—A college credit course programmed at this time so that people can watch at work during their lunch hour.
- Early afternoon—Programming to be used in school, reruns of PBS programs from the night before, and/or programming from one of the regional networks.
- Late afternoon—PBS programs for older children, such as "Square One TV" and "Reading Rainbow."
- Early evening—News from a regional network and PBS's "The MacNeil/Lehrer Newshour."
- Evening—PBS's lineup of primetime programs, mostly drama, documentaries, and cultural programming.

- Late night—Programming selected by the local program director: locally produced documentaries, foreign programs, repeats requested by viewers, etc. Some stations go off the air after PBS's evening shows.

The following was the schedule for public TV station WQED in Pittsburgh on Friday, August 14, 1992.[4] A very definite dayparting philosophy can be seen. Mornings are for children; an early afternoon block (when children are likely to be napping) is directed toward women; late afternoon is for children; and the evening is geared for the general adult audience.

7:00 a.m.—Sesame Street
8:00 a.m.—Mister Rogers' Neighborhood
8:30 a.m.—Sesame Street
9:30 a.m.—Lamb Chop's Play Along
10:00 a.m.—Barney and Friends
10:30 a.m.—Reading Rainbow
11:00 a.m.—Where in the World Is Carmen Sandiego?
11:30 a.m.—Mister Rogers' Neighborhood
12:00 noon—Sesame Street
1:00 p.m.—Prime Time in the Daytime
2:00 p.m.—Gourmet Cooking
2:30 p.m.—13 Painting
3:00 p.m.—Sesame Street
4:00 p.m.—Shining Time Station
4:30 p.m.—Lamb Chop's Play Along
5:00 p.m.—Reading Rainbow
5:30 p.m.—Square One TV
6:00 p.m.—Where in the World Is Carmen Sandiego?
6:30 p.m.—Nightly Business Report
7:00 p.m.—The MacNeil-Lehrer Newshour
8:00 p.m.—Washington Week in Review
8:30 p.m.—Wall $treet Week
9:00 p.m.—Talking with David Frost
10:00 p.m.—The Magic of Bing Crosby
11:00 p.m.—Viewer's Choice *
*viewers have voted as to which program should be rerun

The weekends are usually filled with children's shows, self-help programs, movies, local public affairs programs, and PBS reruns. PBS does not distribute shows on Saturday night, but it has its regular primetime lineup on Sunday evenings.

Many variations of program scheduling exist among public TV stations, however. For example, WGBH in Boston devotes its entire daytime schedule to children and bills its evening PBS programming as "Excellence in TV." Its sister station WGBX (a subsidiary station with a limited use discount) schedules blocks of programs—how-tos, college credit courses, public affairs, the Massachusetts state legislature. "We don't expect anyone to watch for a long time," says Dan Everett, director of broadcasting, "but the programs are there for those who want them."

Launching. Public television is in there launching with everyone else—in the early fall. Sometimes it uses a high-profile series such as "The Civil War" to launch its season. Although PBS does not try to counterprogram the intellectually oriented cable services such as A&E and The Discovery Channel, it has increased its promotional efforts to try to counter the erosion that those services might have on its audience.

Public Radio

Public radio has even more local autonomy than public television. APR and NPR affiliates are free to select the network programs they want and intermix them however they wish with locally produced material.

Dayparting. However, both NPR and APR program with dayparting in mind and make suggestions to stations as to when certain programs would play best. NPR's "Morning Edition," for example, would sound ridiculous if it aired at 3:00 p.m. Likewise, most stations that program NPR's evening newscast "All Things Considered" do so in the early evening hours when people are likely to be in their cars commuting home.

NPR presents news differently in the morning than in the evening to take into account the fact that people are in a hurry in the morning. "Morning Edition" is more modular than "All Things Considered." In the morning no one story lasts longer that eight minutes while in the evening a single story can occupy as much as twenty minutes—both numbers unheard of by commercial standards.

NPR produces "Performance Today" with the idea that it will follow "Morning Edition." It does this by making the program more foreground (i.e., more talk-oriented) than other music programs. Although in NPR's mind, "Car Talk" is made to be a Saturday morning show, some stations place it elsewhere. APR suggests that Garrison Keillor's "The American Radio Company" be carried Saturday night and repeated on Sunday afternoon. Some stations do, some don't. "Whad'ya Know?" is a live call-in show and so must be aired when APR sends it over the satellite. Sometimes APR will suggest that one of their programs follow an NPR program. Says Dan Jensen of APR, "We've had a few shows that we thought would fit very nicely after 'All Things Considered.' "

FIGURE 12.6 APR suggests that Garrison Keillor's homespun music and philosophy program, "The American Radio Company" be aired Saturday night and repeated Sunday afternoon, but not all stations program it that way. (Photo courtesy of American Public Radio/ Fred Petters.)

The Clock. Clocks are very important for network public radio programs that may be aired live because stations must know the exact times that they can break in for local news or announcements. Provision must also be made so that stations that do not wish to program local items need not do so. NPR's "All Things Considered" handles this problem by building natural breaks into some stories. For example, it may program a five-minute story that could conclude at the end of three minutes. Those stations wishing to make local announcements can cut off the story after three minutes and return to "All Things Considered" two minutes later. Other stations let the story run for five minutes. NPR also uses short musical interludes called "zippers" that can be played in their entirety by some stations and cut away from by others.

Stations are very eclectic in their programming philosophies. Some, particularly those that are public affairs–oriented, schedule similar to television in that they air individual programs and pay a fair amount of attention to audience flow from one program to another. Others, mainly the ones that emphasize music, program more like commercial radio in that they use clocks. However, many of public radio's musical selections are so long that the clock has very few entries on it.

The following was the schedule for KCRW in Santa Monica on Monday, November 4,1991.[5] Note the variety of sources of the material aired.

- 6:00 a.m.—NPR's Morning Edition (NPR news)
- 9:00 a.m.—Morning Becomes Eclectic (local music show)
- 12:00 noon—Newstand (NPR, BBC, and local news)
- 12:30 p.m.—International Affairs (local)
- 1:00 p.m.—Arts/LA (local)
- 1:30 p.m.—Crossroads (NPR)
- 2:00 p.m.—The Count of Monte Cristo (BBC play)
- 3:00 p.m.—BBC Newshour (APR)
- 3:30 p.m.—Marketplace (APR)
- 4:00 p.m.—All Things Considered (NPR news)
- 6:55 p.m.—Art Talk (local)
- 7:00 p.m.—Evening Becomes Eclectic (local music show)
- 10:00 p.m.—Short Stories from Modern Mexico (local)
- 11:00 p.m.—Androcles and the Lion (BBC play)
- 12:00 midnight—BBC World Service News
- 12:05 a.m.—Late Night Notes (local alternative music show)

FIGURE 12.7 "Whad'ya Know?" hosted by Michael Feldman, is a comedy/game show that features banter with the audience, interviews with non-celebrity guests, and phone conversations with people from the "town of the week." Because of the phone calls, the show must be aired live. (Photo courtesy of American Public Radio.)

Launching. Public radio has been criticized for its lack of launching and, in fact, its general lack of promotion. Some consider it the country's best-kept secret. Although it is making increasing effort to make itself known, it has a long way to go to make a dent in the din of commercial promotion.

Overall, the development of public broadcasting scheduling strategies is more laissez-faire than in the commercial (or even most of the cable) world. People truly interested in the type of programming public broadcasting has to offer will seek it out and watch or listen to it with a fairly high degree of attention. It is not likely to be the programming people choose when they plunk down in the easy chair and start scanning with the remote control for something that is relaxing and non-taxing.

• • • • • • • • • • • • • • • •

After all the development, testing, and strategizing is completed, there remains the question—did it work? This brings us to the scorekeepers: Who are they? How do they get their numbers? And how accurate are they?

• • • • • • • • • • • • • • • •

NOTES

1. "ONTV," *Naples Daily News* (September 27, 1992): 5.
2. "Cable Networks Try Out Fall Premiere Strategy," *Electronic Media* (September 9, 1991): 1.
3. Ibid.
4. *Pittsburgh* (August 1992): 97 and 101.
5. *Splice, KCRW's Monthly Newsletter* (November 1991): 2–3.

13 Program Evaluation

Theoretically, electronic media programming could be evaluated any number of ways—the number of fan letters received, the artistic merit of the programs as rated by TV critics, the social good accomplished by airing the material, etc. Sometimes one or more of these methods is used, depending on the goal of the program (see Chapters 7 and 8). But the most prevalent form of evaluation is body count: how many and who. The procedure involved in counting the number of people or households watching or listening to particular programming is called "audience measurement." The common term used to refer to the numbers obtained through audience measurement is "ratings."

Based on these audience measurement or rating statistics, broadcasters and cablecasters set their commercial fees, from which they derive their revenue. Over the years buyers and sellers have agreed upon certain scorekeepers whom they believe are able to estimate audience sizes and demographics with reasonable accuracy. The most prominent of these systems are discussed in this chapter.

Since a programmer's job is to attract audiences that will produce a profit for the firm, he or she should know who is doing the counting and how the numbers are calculated. Sometimes programming executives feel uncomfortable with the mathematical orientation of research. But if they don't understand the underpinnings of ratings, they may find that people in the research department will be making decisions that should be the domain of programming.

The purpose of this chapter is to give programmers a basic overview of audience measurement in order to understand how it can (and should) be used. The chapter avoids any complex mathematics involved with the statistical computations. It offers guidelines that allow programmers to understand the basic strengths and weaknesses of the ratings process and enable them to communicate effectively with those who do the actual research.

COMMERCIAL TELEVISION

Commercial television uses two primary scorekeepers—Nielsen and Arbitron. Nielsen reports on networks and stations while Arbitron's primary concern is stations. In addition to these companies, there are many other specialized audience research firms that conduct in-depth research or use methodology different from that of Nielsen or Arbitron.

For example, a company called Strategy Research Corporation conducts face-to-face interviews in order to provide data concerning Hispanic viewing in families where only Spanish is spoken. These specialized firms are often hired by individual electronic media firms to fulfill particular needs. (A Spanish language station might hire Strategy Research Corporation.) But the two companies that operate on an extensive national basis are Nielsen and Arbitron.

Nielsen

The Nielsen company dates back to 1923, obviously long before the days of television audience research. A. C. Nielsen, Sr.'s main customer during the early days of his company was drug stores. He asked a number of these stores to save their invoices; then his company analyzed the invoices and sold the information to drug manufacturers so that they could predict their future sales.

Meters. In 1936, Nielsen bought a crude machine called an "audimeter" from two MIT professors. This device consisted of a moving roll of punched paper tape that could be attached to a radio in such a way that it would record when the radio was turned on and what station it was tuned to. Nielsen perfected this device and, in 1942, started using it to measure radio program audiences. The equipment was so complicated that specially trained technicians had to visit at least once a month each of the 1000 homes where Nielsen had placed the audimeter, to take off the old tape and put on a new one. But Nielsen was able to analyze this information and sell it to stations, advertisers, and others interested in knowing how many people were listening to particular radio programs.

Both Nielsen and the audimeter changed greatly over the years. The company started measuring TV in 1950, and then in 1964 dropped radio because the audimeter wasn't set up to handle the programming changes and audience fragmentation that radio was undergoing. Nielsen now handles only television.

The audimeter went through many stages, making it more user-friendly and allowing ordinary mortals to change the tape. Eventually it was connected directly through phone lines to the computers in Nielsen's Florida office. It recorded information as to when the TV set was turned on or off, and it registered what channels the TV was tuned to or switched to while it was on. People with audimeters also kept diaries that indicated which members of the family were watching which programs. These diaries were used to determine demographic information, such as age and sex, for the various programs.

FIGURE 13.1
This 1949 audimeter is one of the first that could be operated without a technician's assistance. People in the survey could remove a tape and mail it to Nielsen. (Photo courtesy of Nielsen Media Research.)

Then in the late 1980s Nielsen began using a new device called a "peoplemeter." In 1987 it replaced all its audimeters that were measuring national primetime programming with peoplemeters, and it has gradually introduced peoplemeters for calculations in large cities. But because of the expense involved in changing meters, many local markets are still measured by audimeters and diaries.

For a peoplemeter, each person in the family has a special button on a remote control device that he or she must push periodically while watching TV. (Buttons are also available for visitors.) When the set is turned on, all the people who are watching are to push their remote buttons. Anyone new entering the room to watch should push the appropriate button. When a person stops watching, he or she is to turn off the button. In this way, information is gathered not only about when the TV is on and what channel it is tuned to but also about who is watching. In other words, the peoplemeter gathers the information previously collected by both the audimeter and the diary. In the middle of the night, the data that have been recorded by the peoplemeter are automatically sent over phone lines to Nielsen's main computer.

New, more sophisticated measurements are under development. One of these uses an infrared device that senses heat. It would "learn" each family member's body heat and then register that person as viewing. People would not need to push buttons. The main bug in this system at present is that it often mistakes large dogs for people.

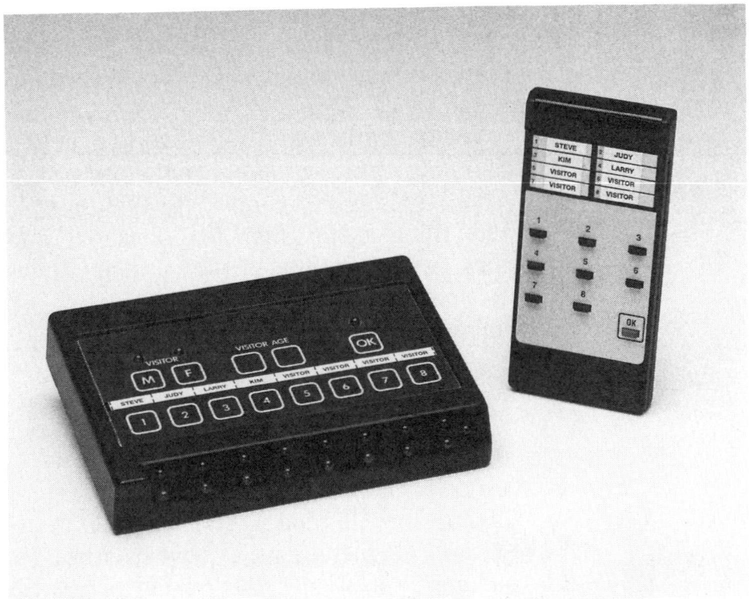

FIGURE 13.2
A Nielsen peoplemeter. (Photo courtesy of Nielsen Media Research.)

Sampling. So, how is a peoplemeter home selected? Certainly not by the residents going to the local TV station and volunteering. Nielsen chooses the households through random sampling. Theoretically almost anyone who can receive a particular TV program or service has a chance to be selected for a peoplemeter. (People who work in the TV industry, such as station and network programmers, are not allowed to participate, for obvious reasons.)

For national commercial TV network programs, the sample might include almost any place with a TV set. For many years only residential homes were metered, but a hue and cry arose in the late 1980s concerning all the viewers who didn't have a chance to be in the sample because they were in college dorms, bars, hotels, hospitals, and the like. As a result, Nielsen has begun to extend its sample base beyond homes.

For a local TV station program, the possible universe for the sample would include all places with TV sets in the particular market of the TV station. Nielsen divides the country into about 200 designated market areas (DMAs) in order to find samples for particular geographic areas. These DMAs are nonoverlapping areas where people generally view the same stations. Generally, they are cities—Boston, Cincinnati, Salt Lake City—but sometimes they are several cities that are close together—Albany-Schenectady-Troy, New York.

The sample universe for a syndicated show is all the places with TV sets that are in markets where the syndicated program is airing.

For its national sample that determines the ratings for the commercial TV network programs, Nielsen selects 4000 out of a total population of 93.1 million dwellings. This sample is about 0.004 percent of the nation's households, which makes someone's chances of becoming part of a Nielsen household extremely remote.

But Nielsen tries to make these 4000 dwellings as representative of the total United States public as it can. The goal is to have the sample mirror the interests and demographic characteristics of the total population. In order to do this, Nielsen attempts to locate every housing unit in the country. For the most part, it uses readily available government census data for finding households, but surveyors employed by Nielsen have found living quarters in railway cars, beer trucks, trees, and caves—all equipped with TV sets.

Once all the nation's housing units are uncovered, they are symbolically placed in a large hat from which 4000 are chosen. Actually a computer program randomly selects the sample group in such a way that each location has an equal chance of being selected. The computer program does make sure that no more than one housing unit per block is chosen. To do so would introduce bias since television usage in two homes in one block would be similar because of factors such as TV reception and the presence of cable TV.

Nielsen representatives then try to solicit the cooperation of the households that have been selected, by writing and phoning them. Generally about 55 percent of the people selected agree to have a peoplemeter installed in their home. This means Nielsen has to "dip into the hat" to find more volunteers in order to reach the number needed.

Once a sample member has agreed to participate, Nielsen sends a field representative to the home (or apartment, or dorm room, or vacation home). This person attaches the peoplemeter, taking into account the entire video configuration of the home—multiple TV sets, VCRs, satellite dishes, cable TV, and anything else that might be attached to the TV. The field representative also trains the family members in the use of the peoplemeter.

People participating as part of the Nielsen sample are paid—usually a grand total of fifty cents. Nielsen has found that paying people something makes them feel obligated, and they will therefore perform their duties (pushing their buttons) more faithfully. But it doesn't matter how much they are paid. People paid fifty cents perform just as well as those paid fifty dollars.

No one stays in the Nielsen sample for more than two years. Nielsen is trying to turn over the peoplemeter sample faster than two years because people tend to suffer from button-pushing fatigue. After they have had the peoplemeter for awhile, the novelty wears off and they no longer take the logging in and out seriously. But soliciting and changing sample members is expensive, so Nielsen must keep an eye to its budget in determining how to update its sample.

The procedures for obtaining sample members for local TV station surveys are similar to those for the national sample except that only the local area is used as the universe. The number of people in the sample depends on the size of the market. Large markets have 300 to 500 sample homes and smaller markets have less. Overall, Nielsen uses about 100,000 sample homes for all its local ratings.[1]

FIGURE 13.3
A theoretical ratings distribution.

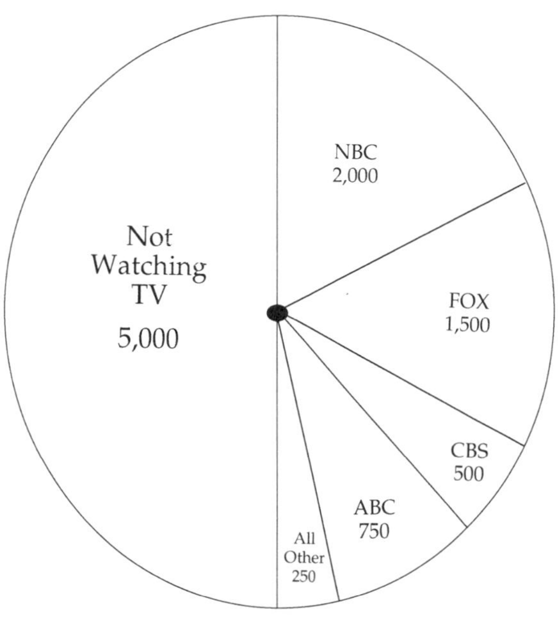

Calculations. Once the Nielsen computer has gathered all of the data from all of the peoplemeters, audimeters, and diaries, it undertakes the calculations that result in audience measurement data. A human being does look over the computer's shoulder as the data come in to check for obvious accuracy problems. If, for example, a peoplemeter reported that someone watched TV for 26 hours a day, something would be awry. Such data are unusable, and a field representative would be dispatched to the home to check on the health of the peoplemeter. A wide variety of statistics can be generated from the collected data, but the ones that are most important to commercial TV programming executives are the rating and the share.

A rating is simply the percentage of households tuned to a particular program out of all the TV households in that universe. In order to be counted, a household must view at least six minutes of a half-hour show. The universe for all network shows is all households in the United States. For purposes of mathematical simplicity, let's assume that there are a total of 10,000 households in the U.S. If 2000 of them are watching a show on NBC for at least six minutes, the rating for that show is 20 (2000/10,000 = 2/10 = 20 percent). If 1500 are watching a show on Fox, that rating is 15.

As we know, there are not 10,000 households in the United States; there are 93,100,000, and almost 100 percent of them have TVs. If the NBC show has a rating of 20, that means 18,620,000 households are watching. One rating point is worth 931,000 homes (1 percent of 93,100,000 homes), and 20 x 931,000 is 18,620,000.

But again, not all homes are actually measured; only 4000 are for the network ratings. A rating of 20 actually means that 800 of the 4000 Nielsen homes were watching the NBC show. However, because the Nielsen sample was randomly selected throughout the nation, it is assumed that 20 percent of all U.S. homes were watching NBC.

The ratings universe for a local station is the total number of homes in the designated market area. Thus a rating for a local show is the percentage of homes in the DMA that are tuned to that particular program for at least six minutes.

Historically, ratings have been tied to households, and when ratings are reported, they still usually refer to the number of households with a TV set tuned to the program. But TV viewing is becoming more and more of an individual activity as people within the same household watch different programs on different TVs and people watch more and more TV outside of their homes. Now that Nielsen has the peoplemeter, it can easily break down the data for individuals rather than households, and that is the trend for the future.

A share is the percentage of households viewing a particular show measured against all the TV households that have their sets on at that time. The rating is based on all households owning TVs, but the share is based just on sets that are being watched. So if, out of 10,000 homes owning TVs, only 5000 of them have sets turned on, then 2000 of them watching a particular program would give that program a share of 40.

If out of the 4000 Nielsen sample homes, only 3000 of them are watching TV, then 3000 is used as the base for determining the share. In order for a program to have a share of 33, 1000 of the homes would have to be watching that program (1000/3000 = 33 percent). That same program would have a rating of 25 (1000/4000 = 25). Shares will always be higher than ratings unless 100 percent of homes have a TV set turned on—a very improbable situation.

Shares, like ratings, are starting to be reported in terms of people rather than households. When shares are reported for households, they sometimes add up to more than 100 percent because of homes that have more than one TV set.

Another statistic that programmers sometimes deal with—or become nervous from—is the HUT (households using television) or the PUT (persons using television). This is simply the percentage of homes that turn the TV set on to anything or the percentage of people who watch TV. When HUTs and PUTS go down, everyone in the electronic media business experiences the jitters because most shares and ratings (and hence dollars of income) go down, too.

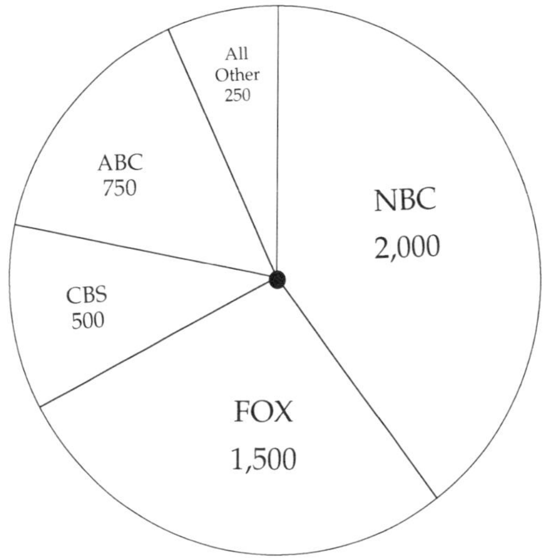

FIGURE 13.4
A theoretical distribution of shares.

One other of the many statistics used in audience measurement is the "cume." This term is short for "cumulative" and refers to the total audience impressions. For example, some syndicated shows run once during the week and once on the weekend. The production company combines the numbers from the two showings and sells a cume audience to the advertisers.[2]

Reports. From the data collected and calculated by the computer, Nielsen publishes rating reports. The reports on each night's primetime national network programming are usually ready about twelve hours after the data are collected. In other words, if the data are collected from the peoplemeters at 3:00 a.m., the finished reports on that day's network programming will be ready about 3:00 p.m. the next day. These reports are usually referred to as the overnights.

All of the reports dealing with national TV are referred to as the Nielsen Television Index (NTI). This NTI includes not only the overnights but a summary of households using television, reports that deal with daytime network programs, cumulative audiences over a period of time, and annual reports that show three-year trends.

The reports dealing with stations are not published as frequently as those for prime-

FIGURE 13.5 A sample page from the NTI that shows primetime. Note the demographic breakdowns. (Courtesy of Nielsen Media Research.)

74

PROGRAM AUDIENCE ESTIMATES (By Time Periods) SEP. 18–24, 1989

AVERAGE MINUTE AUDIENCE %

DAY / TIME / NETWORK PROGRAM NAME		HOUSEHOLDS	TOTAL PERS (2+)	WORKING WOMEN 18+	WORKING WOMEN 18-49	LOH 18-49 W/CH <3	WOMEN TOTAL	WOMEN 18-34	WOMEN 18-49	WOMEN 25-54	WOMEN 35-64	WOMEN 55+	MEN TOTAL	MEN 18-34	MEN 18-49	MEN 25-54	MEN 35-64	MEN 55+	TEENS TOT. 12-17	TEENS FEM. 12-17	CHILDREN TOT. 2-11	CHILDREN TOT. 6-11
THURSDAY EVENING																						
6.00- 6.30PM	TVU	47.8	27.5	23.5	21.5	28.6	33.4	24.1	24.8	27.6	31.8	49.3	25.2	19.5	18.8	20.0	23.5	42.3	19.1	21.3	22.2	21.9
6.30- 7.00PM	TVU	50.3	29.6	26.3	23.9	28.1	35.0	24.8	26.1	29.0	34.1	51.6	28.7	22.3	21.5	23.4	27.1	47.1	20.0	22.8	22.4	21.7
7.00- 7.30PM	TVU	51.4	30.8	28.4	25.6	31.1	36.5	26.3	27.2	30.1	35.6	53.7	29.8	23.3	23.0	25.1	28.7	47.1	20.6	23.8	23.2	23.1
7.30- 8.00PM	TVU	53.4	32.7	31.6	28.8	34.6	38.3	28.6	29.5	32.6	37.3	54.7	30.9	24.8	24.9	27.1	29.8	46.3	24.5	27.0	26.3	26.4
8.00- 8.30PM	TVU	59.4	38.6	37.0	34.5	44.3	43.8	35.3	36.1	39.4	42.3	58.4	36.6	31.7	31.4	33.7	34.5	50.2	31.9	37.5	33.6	34.7
A MISSION: IMPOSSIBLE		8.0	4.9	3.8	3.4	3.5^	5.3	2.5	3.6	4.8	5.8	8.0	5.5	3.6	4.1	4.8	5.4	9.2	3.0	2.6^	3.5	3.3
C 48 HOURS		12.6	7.9	7.7	6.3	6.7	10.4	5.1	6.3	7.5	10.3	18.4	7.8	5.0	5.6	6.6	8.1	13.7	2.8	2.4^	4.2	4.2
N BILL COSBY SHOW		26.1	16.7	18.4	18.4	24.3	19.7	20.8	19.3	19.7	18.0	20.6	12.3	14.1	12.5	12.0	10.1	12.8	18.5	23.8	18.4	20.2
8.30- 9.00PM	TVU	61.2	40.8	39.7	37.5	46.6	46.1	37.9	39.2	42.5	45.4	59.3	39.0	33.7	34.6	37.5	37.6	51.0	32.2	34.7	35.5	38.8
A MISSION: IMPOSSIBLE		9.3	6.0	4.6	4.3	4.6	5.9	3.8	4.8	5.8	6.5	7.8	6.9	5.7	6.0	6.5	6.6	9.3	4.5	3.3^	4.9	4.7
C 48 HOURS		15.0	9.4	9.5	8.0	10.5	12.1	7.3	8.3	9.7	12.2	19.3	9.4	6.3	7.5	8.5	9.9	14.4	3.9	4.2	5.3	5.5
N SISTER KATE 2ND PREVIEW(S)		23.0	14.7	16.7	16.6	20.2	17.9	19.0	17.3	17.3	16.2	19.6	10.2	11.7	10.1	10.0	8.3	11.5	16.0	19.8	16.6	19.0
9.00- 9.30PM	TVU	62.0	41.1	41.7	40.1	48.9	47.2	41.1	41.5	44.6	46.3	57.5	41.5	36.0	37.8	40.4	41.4	51.9	30.2	31.4	30.0	32.4
A YOUNG RIDERS		10.9	7.3	5.6	5.8	9.0	7.1	6.2	6.7	7.4	7.1	7.8	8.2	7.5	8.0	8.4	8.3	9.2	6.7	6.4	5.6	5.3
C TOP OF THE HILL		11.1	6.3	7.1	5.7	5.5	9.3	4.3	5.6	6.9	9.6	16.5	5.2	3.0	3.3	4.1	5.4	10.0	2.4	3.6^	3.2	3.6
N CHEERS		24.1	15.5	18.8	19.3	22.1	18.8	21.9	19.7	19.3	17.0	17.4	13.8	15.1	14.2	14.1	12.8	13.2	13.2	14.5	12.0	13.0
9.30-10.00PM	TVU	60.9	39.2	40.0	38.0	47.7	45.6	37.9	39.9	43.6	46.5	56.0	41.3	34.3	37.0	40.2	43.1	52.4	28.7	28.6	23.1	24.9
A ABC NEWSBRIEF-THU(SUS)>																						
A YOUNG RIDERS		12.8	8.3	6.3	6.2	12.8	8.5	7.5	7.8	8.8	8.6	9.2	9.3	8.5	8.8	9.6	9.2	11.4	7.6	7.4	5.7	5.6
C TOP OF THE HILL		11.1	6.2	7.3	5.8	5.3	9.4	4.3	5.9	7.1	10.1	16.6	5.0	2.6	3.0	3.8	5.5	10.2	2.3	3.5^	2.7	3.0
N HARDBALL PREVIEW(S)		18.8	11.5	14.6	15.3	17.7	13.8	15.6	15.2	14.9	13.6	11.4	11.5	12.7	12.1	12.0	11.2	9.7	9.8	9.5	6.2	7.1
10.00-10.30PM	TVU	59.3	37.5	40.0	37.2	46.6	44.9	36.2	39.1	43.2	47.1	55.2	40.5	32.9	35.6	39.2	42.8	52.4	22.7	21.6	18.6	19.6
A PRIMETIME LIVE		11.8	6.9	6.8	5.9	9.9	8.8	6.7	7.0	8.4	8.8	11.7	7.4	6.7	6.3	6.9	7.2	10.3	2.9	3.5^	3.2	2.8
C TOP OF THE HILL		11.1	6.3	7.4	6.4	6.9	9.1	4.5	6.2	7.2	9.9	15.3	5.6	2.8	3.6	4.5	6.6	10.9	2.1	3.1^	2.8	3.0
N HARDBALL PREVIEW(S)		17.7	10.8	13.3	13.4	15.8	12.6	13.9	13.9	14.1	13.4	10.0	11.5	12.7	11.8	12.0	10.9	10.2	8.6	8.4	5.3	5.7
10.30-11.00PM	TVU	56.2	35.2	37.9	34.8	43.6	42.5	34.9	37.4	41.3	44.7	51.3	39.0	32.4	34.8	38.1	41.4	49.3	20.7	19.4	14.5	15.4
A PRIMETIME LIVE		12.4	7.2	7.8	6.5	9.3	9.8	7.0	7.8	9.5	10.2	13.0	7.5	6.6	6.2	6.9	7.5	10.7	2.5	3.2^	2.2	2.2
C TOP OF THE HILL		10.7	6.2	7.0	6.1	6.7	8.6	3.9	5.7	6.9	9.6	14.3	5.9	2.8	3.8	4.8	7.1	11.4	2.4	3.1^	2.4	2.5
N HARDBALL PREVIEW(S)		17.0	10.3	12.6	12.7	13.7	12.1	13.1	13.4	13.5	12.9	9.8	11.5	12.4	11.8	12.2	11.2	10.2	7.8	6.9	4.2	4.3
11.00-11.30PM	TVU	47.7	28.5	31.0	28.0	33.0	35.5	27.8	30.6	34.7	37.6	43.7	32.4	26.7	28.8	31.4	33.8	41.3	13.4	11.7	8.9	8.9
11.30-12.00MD	TVU	35.3	20.3	23.7	21.7	25.1	25.7	21.8	23.0	25.6	26.5	29.5	22.9	21.7	21.8	22.6	22.4	25.7	9.4	7.7	5.5	5.2
N NBC NEWS SPEC. REPORT(SUS)>																						
12.00-12.30AM	TVU	27.3	14.9	17.2	15.3	19.5	18.7	16.3	16.7	18.5	19.0	21.2	17.6	17.8	17.3	16.8	16.4	19.1	6.2	5.6	3.2	2.8
12.30- 1.00AM	TVU	21.7	11.4	13.2	11.7	14.7	14.1	12.8	12.8	14.5	14.1	15.0	13.7	15.2	14.1	13.2	12.3	12.7	4.4	4.2	2.5	2.0
A ABC SPECIAL REPORT-12:58A(SUS)>																						
1.00- 1.30AM	TVU	17.3	8.8	10.6	9.9	11.0	11.0	10.8	10.5	12.1	11.2	9.9	10.6	12.7	11.1	10.1	9.1	8.7	3.2	3.2^	2.1	1.8
1.30- 2.00AM	TVU	13.4	6.8	8.5	8.0	8.5	8.4	8.8	8.5	9.4	8.6	7.1	8.0	9.7	8.3	7.3	6.7	6.9	2.5	2.7^	2.0	1.7
2.00- 2.30AM	TVU	10.4	5.2	6.4	6.2	6.3	6.3	6.6	6.5	7.2	6.7	5.0	6.0	6.9	6.2	5.5	5.3	5.7	2.2	2.7^	1.7	1.4^
2.30- 3.00AM	TVU	8.6	4.2	5.5	5.5	5.5	5.0	5.7	5.5	5.8	5.1	3.7	5.0	5.7	5.3	4.8	4.4	4.6	1.8	2.5^	1.4	.9^
3.00- 3.30AM	TVU	7.4	3.6	4.7	4.4	5.1	4.1	5.0	4.5	4.8	4.1	3.2	4.4	5.3	4.8	4.4	3.9	3.7	1.6^	2.1^	1.1	.7^
3.30- 4.00AM	TVU	6.6	3.1	4.1	3.8	4.0	3.6	4.2	4.0	4.3	3.6	2.8	3.9	4.5	4.3	3.9	3.7	3.0	1.5^	1.9^	.9^	.6^
4.00- 4.30AM	TVU	6.0	2.8	3.7	3.5	3.5^	3.2	3.5	3.4	3.8	3.4	2.6	3.5	3.8	3.7	3.6	3.5	3.1	1.3^	1.5^	.9^	.6^

75

FOR EXPLANATION OF SYMBOLS, SEE PAGE B.

time network TV. Generally station ratings are undertaken four times a year—November, February, May, and July—periods that are often referred to as sweeps. The weeks of the sweeps are very important to local stations because the numbers produced then determine advertising rates for the following three months. The report containing station data is called the Nielsen Station Index, and it, like the NTI, breaks the information down into various forms, including demographics.

Nielsen also publishes a special report for syndicated programming from data it obtains from its peoplemeters, audimeters, and diaries. Because of the increasing importance of the syndication market, Nielsen is working to publish reports about syndicated shows more quickly and more frequently than in the past. Nielsen Executive Vice President Andrew Faller says, "We want to make ratings of syndicated shows as timely as network ratings, because we want to make sure that there is no handicap for syndicated shows."[3]

The various Nielsen reports are purchased by networks, stations, advertisers, and any other companies with a need to know audience statistics. They are available in hard copy, on diskette, or on-line through a company's computer. The cost is usually \$4 million to \$5 million a year for networks and somewhat less to advertisers, stations, and others.

Arbitron

The other major audience measurement company, Arbitron, was formed in 1949. It analyzes only stations, although its data for affiliated stations can certainly be used to glean national ratings information.

In many ways its methodology is similar to that of Nielsen. It uses diaries and a meter that is similar to the audimeter. It has also experimented with a meter called Scan America. With this, people not only register their TV viewing, but they also run their groceries over the meter so that it can scan the bar code on the groceries and in that way correlate data concerning TV programs watched with that of purchases made.

The market areas that Arbitron surveys are similar to those of Nielsen, but they are called areas of dominant influence (ADIs) rather than DMAs.

Arbitron chooses its sample homes by first selecting 10,000 households in each market that are representative of the community in terms of such demographics as size of family, income, education, race, age of head of household, and access to cable TV. From this 10,000, Arbitron then selects the number of households needed for the survey, generally 300 to 500. The homes in the sample are changed gradually so that no one remains in the sample for longer than five years.

Arbitron relies more heavily on diaries than Nielsen because its meters do not gather the demographic information that peoplemeters collect. In addition, Arbitron uses diaries only (no meters) in some of the smaller markets. Approximately 40 percent of diaries are actually returned to the company. Arbitron attempts to make the fullest possible use of the information it obtains. In some cases, doing this involves filling in information that has not been supplied by the diary keeper. (Arbitron refers to this process as Monte Carlo Audience Ascription.) For example, if a diary indicates that a set was tuned to "Monday Night Football" but it does not note which members of the family watched it, someone at Arbitron may assume that it was viewed by the 35-year-old male household member and enter that data into the computer.

Another technique Arbitron uses is ethnic weighing. Some demographic groups return diaries at a higher rate than others. Arbitron often makes up for this by giving greater weight to some diaries than others. For example, if only 20 percent of the Hispanic sample returned diaries as compared to 40 percent of the Caucasian sample, the diaries from Hispanics are counted twice.

Both Monte Carlo Audience Ascription and

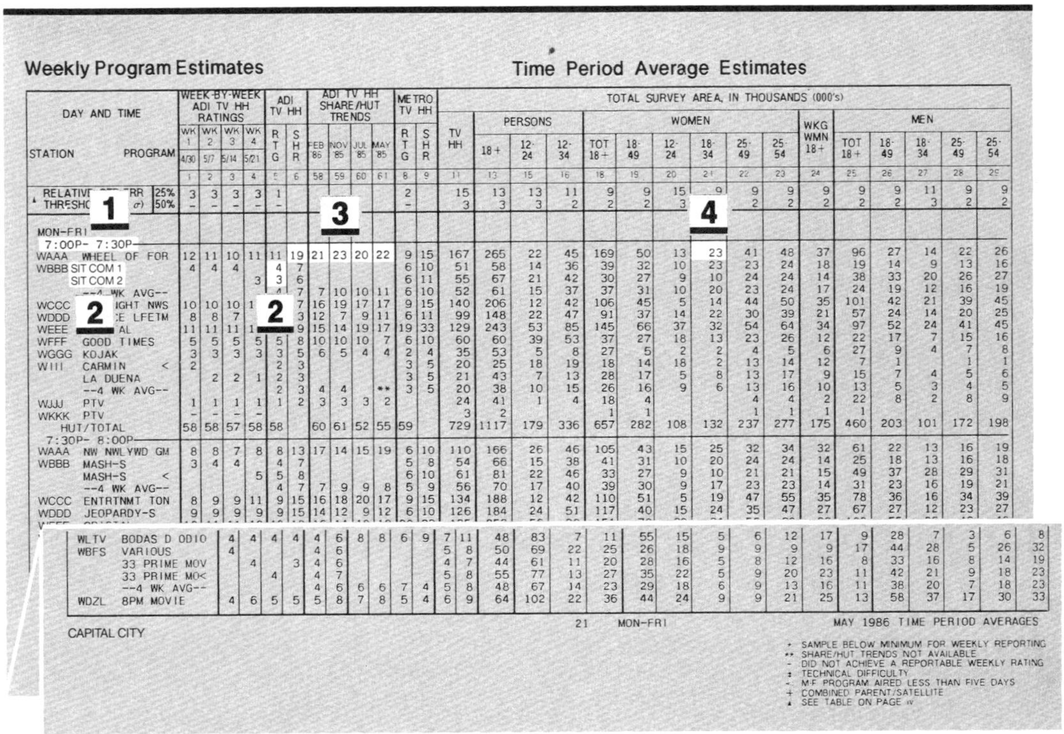

Weekly Program Estimates — Time Period Average Estimates

STATION	PROGRAM	WK 1 4/30	WK 2 5/7	WK 3 5/14	WK 4 5/21	ADI TV HH RTG	ADI TV HH SHR	FEB '86	NOV '85	JUL '85	MAY '85	METRO TV HH RTG	METRO TV HH SHR	TV HH	PERSONS 18+	PERSONS 12-24	PERSONS 12-34	WOMEN TOT 18+	WOMEN 18-49	WOMEN 12-24	WOMEN 18-34	WOMEN 25-49	WOMEN 25-54	WKG WMN 18+	MEN TOT 18+	MEN 18-49	MEN 18-34	MEN 25-49	MEN 25-54
		1	2	3	4	5	6	58	59	60	61	8	9	11	13	15	16	18	19	20	21	22	23	24	25	26	27	28	29
▲ RELATIVE STD ERR THRESHOLDS (1σ)	25%	3	3	3	3	1						2		15	13	13	11	9	9	15	9	9	9	9	9	9	11	9	9
	50%	–	–	–	–	–						–		3	3	3	2	2	2	3	[illegible]	2	2	2	2	2	3	2	2
MON-FRI 7:00P- 7:30P																													
WAAA	WHEEL OF FOR	12	11	10	11	11	19	21	23	20	22	9	15	167	265	22	45	169	50	13	23	41	48	37	96	27	14	22	26
WBBB	SIT COM 1	4	4	4		4	7					6	10	51	58	14	36	39	32	10	23	23	24	18	19	14	9	13	16
	SIT COM 2				3	3	6					6	11	55	67	21	42	30	27	9	10	24	24	14	38	33	20	26	27
	--4 WK AVG--					4	7	7	10	10	11	6	10	52	61	15	37	37	31	10	20	23	24	17	24	19	12	16	19
WCCC	…IGHT NWS	10	10	10	1	[illegible]	7	16	19	17	17	9	15	140	206	12	42	106	45	5	14	44	50	35	101	42	21	39	45
WDDD	…E LFETM	8	8	7		[illegible]	3	12	7	9	11	6	11	99	148	22	47	91	37	14	22	30	39	21	57	24	14	20	25
WEEE	…AL	11	11	11	1	[illegible]	9	15	14	19	17	19	33	129	243	53	85	145	66	37	32	54	64	34	97	52	24	41	45
WFFF	GOOD TIMES	5	5	5	5	5	8	10	10	10	7	6	10	60	60	39	53	37	27	18	13	23	26	12	22	17	7	15	16
WGGG	KOJAK	3	3	3	3	3	5	6	5	4	4	2	4	35	53	5	8	27	5	2	2	4	5	6	27	9	4	7	8
WIII	CARMIN <	2				2	3					3	5	20	25	18	19	18	14	18	2	13	14	12	7	1		1	1
	LA DUENA		2	2	1	2	3					3	5	21	43	7	13	28	17	5	8	13	17	9	15	7	4	5	6
	--4 WK AVG--					2	3	4	4		**	3	5	20	38	10	15	26	16	9	6	13	16	10	13	5	3	4	5
WJJJ	PTV	1	1	1	1	1	2	3	3	3	2			24	41	1	4	18	4			4	4	2	22	8	2	8	9
WKKK	PTV	–	–	–	–									3	2			1	1			1	1	1	1				
HUT/TOTAL		58	58	57	58	58		60	61	52	55	59		729	1117	179	336	657	282	108	132	237	277	175	460	203	101	172	198
7:30P- 8:00P																													
WAAA	NW NWLYWD GM	8	8	7	8	8	13	17	14	15	19	6	10	110	166	26	46	105	43	15	25	32	34	32	61	22	13	16	19
WBBB	MASH-S	3	4	4		4	7					5	8	54	66	15	38	41	31	10	20	24	24	14	25	18	13	16	18
	MASH-S <				5	5	8					6	10	61	81	22	46	33	27	9	10	21	21	15	49	37	28	29	31
	--4 WK AVG--					4	7	7	9	9	8	5	9	56	70	17	40	39	30	9	17	23	23	14	31	23	16	19	21
WCCC	ENTRTNMT TON	8	9	9	11	9	15	16	18	20	17	9	15	134	188	12	42	110	51	5	19	47	55	35	78	36	16	34	39
WDDD	JEOPARDY-S	9	9	9	9	9	15	14	12	9	12	6	10	126	184	24	51	117	40	15	24	35	47	27	67	27	12	23	27
WLTV	BODAS D ODIO	4	4	4	4	4	6	8	8	6	9	7	11	48	83	7	11	55	15	5	6	12	17	9	28	7	3	6	8
WBFS	VARIOUS	4				4	6					5	8	50	69	22	25	26	18	9	9	9	9	17	44	28	5	26	32
	33 PRIME MOV		4		3	4	6					4	7	44	61	11	20	28	16	5	8	12	16	8	33	16	8	14	19
	33 PRIME MO<			4		4	7					5	8	55	77	13	27	35	22	5	9	20	23	11	42	21	9	18	23
	--4 WK AVG--					4	6	6	6	7	4	5	8	48	67	14	23	29	18	6	9	13	16	11	38	20	7	18	23
WDZL	8PM MOVIE	4	6	5	5	5	8	7	8	5	4	6	9	64	102	22	36	44	24	9	9	21	25	13	58	37	17	30	33

CAPITAL CITY — 21 — MON-FRI — MAY 1986 TIME PERIOD AVERAGES

* SAMPLE BELOW MINIMUM FOR WEEKLY REPORTING
** SHARE/HUT TRENDS NOT AVAILABLE
\- DID NOT ACHIEVE A REPORTABLE WEEKLY RATING
‡ TECHNICAL DIFFICULTY
< M-F PROGRAM AIRED LESS THAN FIVE DAYS
\+ COMBINED PARENT/SATELLITE
▲ SEE TABLE ON PAGE iv

FIGURE 13.6 A sample page showing the format of an Arbitron television ratings report. Number 1 indicates that reports are given by the half hour. Number 2 shows how Arbitron indicates that a station ran different programs different weeks. Number 3 shows share trends over a year's time. Number 4 is a sample of demographics showing the number of women who viewed WAAA during the 7:00–7:30 time period. (Copyright 1992, The Arbitron Company. All rights reserved.)

ethnic weighing have been criticized for their possible inaccuracies. However, if only "perfect" diaries in a "perfect" proportion were used, the number of households involved would need to be much larger. This would, of course, raise the cost of audience measurement— something the clients do not seem willing to bear.

Arbitron, like Nielsen, has sweeps periods in November, February, May, and July. The reports that it publishes are very similar to those of the Nielsen Station Index in that they include various demographics and are published daily, weekly, monthly, quarterly, or yearly, depending on the type of information contained.[4]

CABLE TELEVISION

Cable television services are measured by Nielsen and Arbitron as they are measuring commercial TV networks and stations. In fact, cable TV viewing was measured by Nielsen as far back as 1972.

Problems Associated with Cable and Audience Measurement

The advent and growth of cable has caused numerous problems for the rating companies. For one thing, it has fractionalized the viewing audiences, making it harder to analyze everything. During the 1950s, 1960s, and 1970s, the three networks dominated and most cities had only four or five stations. Ratings for each network were generally in the 20s and shares were in the 30s. Now, not only have network ratings decreased into the low teens, but ratings such as 1 or 2 are common for TV stations and cable services. The probability of error on numbers this small is quite high.

Another problem with cable TV ratings revolves around ascertaining the program lineup on every channel on each cable system. When only networks and stations existed, Nielsen and Arbitron could keep track of channel numbers in each market and know, through published schedules, what was playing on each station. In addition, if there were doubts, company personnel could call local stations to verify programs and time periods. But when cable came along, people in one part of town might have WXXX on channel 6 while those down the street had it on channel 12, and those across town watched it on channel 37.

The cumbersomeness of dealing with these inconsistencies led Nielsen to develop Automated Measurement of Lineups (AMOL), which it introduced in 1982. Each network and syndicated program is coded invisibly in a manner similar to grocery store bar codes. Then as the program plays, Nielsen equipment tracks this code and in that way can determine exactly what program is shown on which channel at what time. This has been particularly helpful for calculating ratings for individual, syndicated programs that may be playing on innumerable stations and cable systems throughout the country.

VCRs, even more than cable, have presented difficulties for the rating companies. Generally, programs are counted as being watched if they are recorded. But many programs that are recorded are never watched, and some are watched several times. In addition, people playing back a tape often zip through commercials, destroying the whole purpose of ratings.

The rating companies' handling of cable TV and other newer media is somewhat in a state of flux. As the electronic media scene evolves, Nielsen and Arbitron try to devise solutions to the problems created in the wake, but for technical and financial reasons they cannot be totally up to date.

Reports and Statistics

Nielsen and Arbitron supply data on cable TV in some of the same reports that include network and station information. Figures for the major cable services, such as CNN and ESPN, are reported daily while those of the more minor services, such as FNN and The Weather Channel, are reported quarterly.

Nielsen publishes a special report called Nielsen Home Video that covers not only cable TV but also VCRs and home satellite dishes. Special statistics are often reported, such as the percentage of home-dish or VCR penetration in each market, the types of pro-

MON-FRI 8:00PM-11:00PM

NIELSEN CABLE ACTIVITY REPORT
SECTION 1: DUAL FEED

FIRST QUARTER 1992
DECEMBER 30, 1991 - MARCH 22, 1992

	CABLE NET COVERAGE AREA			TOTAL PAY CABLE			TOTAL CABLE			TOTAL SAMPLE			
	AA %	SHR %	AVG. WK CUME %	AA %	SHR %	AVG. WK CUME %	AA %	SHR %	AVG. WK CUME %	AA %	AA (000)	AVG. WK CUME %	AVG. WK CUME (000)
NETWORKS													
ABC				13.1	19	75.1	12.3	19	72.1	13.2	12138	71.4	65788
CBS				12.7	19	71.4	13.0	20	70.6	13.6	12493	69.6	64075
NBC				12.7	18	74.7	12.1	19	72.0	12.8	11791	70.8	65185
FOX				5.9	9	53.2	5.2	8	47.8	5.6	5125	46.3	42610
MAFF				<<		<<	<<		<<	<<		<<	
TOTAL NETWORKS				44.4		94.3	42.6		92.8	45.1	41548	91.7	84494
PAY CABLE													
CMX	5.4	8	46.4	1.7	2	14.2	0.7	1	6.3	0.5	450	4.1	3765
HBO	6.4	9	54.4	4.7	7	39.7	2.1	3	17.4	1.4	1256	11.3	10439
SHOW	4.7	7	45.7	1.6	2	15.6	0.7	1	6.9	0.5	436	4.5	4136
ALL OTHER PAY				1.2	2	12.5	0.5	1	5.5	0.3	319	3.7	3366
TOTAL PAY				9.1		60.4	4.1		26.5	2.7	2461	17.4	16008
SUPERSTATIONS													
TBS	2.2	3	24.6	2.2	3	24.6	2.1	3	23.4	1.4	1296	15.5	14255
WGN	1.5	2	19.3	0.8	1	11.2	0.7	1	10.0	0.6	534	7.3	6736
ALL OTHER SUPERS				1.3	2	13.1	1.2	2	12.2	1.3	1195	11.3	10400
TOTAL SUPER STNS				4.3		40.6	4.0		38.0	3.3	3025	29.1	26783
AD-SUPPORTED													
AEN	0.8	1	13.1	0.7	1	11.7	0.7	1	11.6	0.5	416	7.5	6922
BET	0.4	1	6.8	0.4	1	5.6	0.2		3.7	0.2	143	2.4	2202
CMDY	0.4	1	8.2	0.2		3.5	0.2		3.1	0.1	99	2.0	1861
CMT	0.3	1	5.3	0.1		1.4	0.1		1.3	0.1	52	0.9	827
CNBC	0.1		2.5	0.1		2.2	0.1		1.9	<<		1.2	1144
CNN	1.0	2	13.5	0.8	1	12.1	1.0	1	13.2	0.6	582	8.6	7946
DISC	1.1	2	16.1	1.0	1	15.9	1.0	2	15.2	0.7	613	9.9	9117
ENT	0.1		2.1	<<		0.7	<<		0.6	<<		0.4	387
ESPN	1.6	2	17.3	1.8	3	19.0	1.5	2	17.0	1.0	933	11.1	10226
FAM	1.0	2	11.6	0.9	1	10.5	1.0	1	10.6	0.6	569	6.9	6330
HLN	0.3	1	7.0	0.3		5.6	0.3		5.6	0.2	164	3.7	3397
LIF	1.2	2	14.6	1.1	2	13.6	1.1	2	13.3	0.7	629	8.6	7935
MTV	0.6	1	12.7	0.7	1	14.1	0.6	1	11.8	0.4	337	7.6	7000
NICK	1.0	2	13.2	1.0	1	13.6	1.0	2	12.5	0.6	580	8.1	7438
NTN	0.1		2.3	<<		0.4	<<		0.3	<<		0.2	211
PREV	0.3		5.3	0.2		3.0	0.1		2.0	0.1	70	1.3	1215
TLC	0.1		2.6	<<		0.7	<<		0.7	<<		0.5	416
TNN	1.4	2	10.0	0.7	1	7.7	1.2	2	9.1	0.8	739	5.9	5461
TNT	1.4	2	18.8	1.4	2	19.4	1.3	2	17.8	0.9	796	11.5	10592
TWC	0.2		5.4	0.2		4.9	0.2		4.6	0.1	99	3.0	2762
USA	2.6	4	26.8	2.7	4	27.6	2.5	4	25.8	1.7	1531	16.9	15593
VH1	0.3		7.1	0.3		6.2	0.2		5.3	0.1	129	3.4	3143
TOTAL				14.6		79.6	14.3		76.3	9.3	8571	49.8	45886
ALL OTHER CABLE ORIGINATION				3.4	5	34.9	3.2	5	32.1	2.1	1919	20.9	19287
TOTAL CABLE ORIGINATION				17.3	25	83.5	16.9	26	80.1	11.0	10164	52.3	48172
INDEPENDENTS				3.0	4	27.2	3.0	5	26.7	3.9	3555	27.8	25633
PBS				1.5	2	21.3	1.9	3	23.6	2.2	2061	24.4	22506
HUT				68.8			64.7			61.7			

<< BELOW MINIMUM REPORTING STANDARDS
• CABLE NETWORK DID NOT TELECAST DURING THE ENTIRE DAYPART

B-3

COPYRIGHT 1992 NIELSEN MEDIA RESEARCH
PRINTED IN U.S.A.

FIGURE 13.7 A sample of the type of information Nielsen publishes about cable TV. (Courtesy of Nielsen Media Research.)

grams people are most likely to record with their VCRs, or the average number of channels on cable systems in each market.

One of the results of small ratings numbers for cable services has been the emphasis cable places on cumes rather than ratings or shares. Because it relates to the number of different people who watch a channel over time, a cume can show differentiation between two channels, both of which have a 2 rating. Advertisers are particularly interested in cumes because they often place their ads on cable channels a large number of times during one day. For example, a perfume manufacturer might have an ad run at 7:00, 7:35, 8:10, and 8:40 and want to know how many different people were exposed to that commercial. Services that are likely to have short-term viewing, such as CNN and The Weather Channel, particularly like to refer to their cumes.

Cable systems place less emphasis on ratings than do commercial systems because cable has two sources of income—advertisements and subscribers. Therefore, cable systems and networks often use subscriber statistics to evaluate their success—or lack thereof.

One of the important subscriber concepts for cable systems is "penetration." This refers to the number of homes in the cable system's territory that connect to the cable. If the number of households passed by the cable is 10,000, and the number of households that have asked to have the cable hooked to their TV sets is 6000, cable penetration is 60 percent. Cable systems hope for penetration of about 80 percent.

Cable also likes to keep its "churn" rate low. Churn refers to the number of people dropping the service even though they may later reconnect. For obvious financial reasons, pay cable networks like their subscribers to stay with them, not drop out for several months and then resubscribe. Cable systems, too, are averse to losing and regaining large numbers of customers. Areas with apartments that

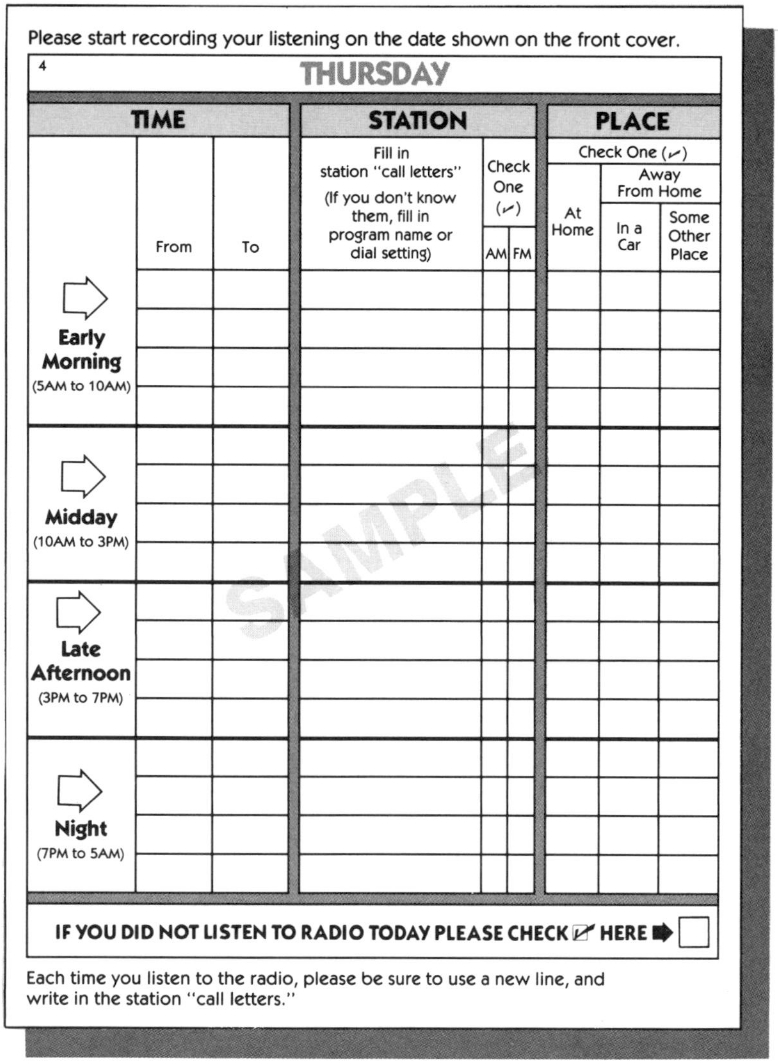

FIGURE 13.8
A sample page from an Arbitron diary. (Copyright 1991, The Arbitron Company. All rights reserved.)

turn over renters frequently are particularly subject to churn. Of course, worse than churn is the permanent disconnect—the subscriber who tries cable, doesn't like it, asks to have the service removed, and never comes back.

COMMERCIAL RADIO

As mentioned earlier, Nielsen dropped radio ratings in 1964, but Arbitron does undertake radio station audience measurement and is generally considered the primary radio ratings company. Other companies have entered (and left) the business. Collecting ratings data is expensive—a fact that has led to the demise of entrepreneuring companies. An industry-supported study, RADAR, measures radio network audiences.

Arbitron

Arbitron's radio ratings are determined entirely by diaries (not meters) and are based on people (not households). So much of the radio audience consists of one person listening to one radio outside of the home that meters placed on household radio sets would be very inadequate for measurement.

Samples are drawn on a household basis in a random manner that is similar to how households are selected for TV ratings. But, with radio, each person in the household over the age of 12 is asked to keep his or her own diary for one week and then mail it to Arbitron. No one is asked to keep a diary for longer than seven days. If the survey time period lasts longer than a week, different people are used for each week.

Radio ratings cover two different areas for local stations—metros and TSAs (Total Service Areas). Metros correspond to the government's Metropolitan Statistical Areas and are primarily individual cities. Usually a station's metro is the city to which it is licensed—an area in which it can be heard very clearly. TSAs are larger than metros and often include part of some other metro. The weaker stations of a metro area often cannot be heard clearly throughout the TSA. This difference in carrying power is what led Arbitron to break down ratings for the two areas. If ratings were reported only for the TSA, the lower-powered radio stations would be unjustly penalized.

Once the diaries are mailed back to Arbitron (the return rate is about 50 percent), they are edited, taking a variety of points into account. If someone has listed a dial setting or program name or personality rather than the call letters, an Arbitron employee finds and enters the correct call letters. Diaries that are postmarked before the survey period has ended are eliminated as are identical diaries from the same household or diaries that indicate that someone listened to the radio twenty-four hours a day for the whole week. As with TV, diaries from a particular demographic group may be weighted if that group as a whole returns a low percentage of diaries.[5]

The valid edited information is processed through a computer that calculates audience measurement statistics. Rating and share are important to radio, just as they are to TV, but the usual method for calculating these revolves around the average quarter hour (AQH). AQH ratings and shares are developed by adding the number of people who listen to a station for at least five minutes during a fifteen-minute period and then dividing that by the number of quarter hours in a particular daypart. For example, if 100 people are keeping diaries and three listen to KXXX for at least five minutes between 7:00 and 7:15 a.m., five listen between 7:15 and 7:30, seven between 7:30 and 7:45, ten between 7:45 and 8:00, three between 8:00 and 8:15, five between 8:15 and 8:30, four between 8:30 and 8:45, and three between 8:45 and 9:00, KXXX's AQH rating would be 5:

7:00—7:15— 3
7:15—7:30— 5
7:30—7:45— 7
7:45—8:00—10
8:00—8:15— 3
8:15—8:30— 5
8:30—8:45— 4
8:45—9:00— 3
Total 40

40 divided by 8 quarter hours = 5.

Average quarter hours are used because radio is not program structured as TV is. A great deal of button pushing occurs in radio listening, especially for all-news stations that people tend to dip into for brief periods of time. Measuring in small blocks helps find the audience.

The cume is also very important to radio. As with cable, advertisers are often interested in knowing how many different people their ads reach if the ads are placed at various times.

Arbitron surveys all of its radio markets at least once a year. The large markets are surveyed more intensively than the smaller markets; in fact, some very large markets are measured almost year round.

One of the interesting uses made of Arbitron data relates to the network-affiliate relationship. Radio networks look at ratings to decide what stations they wish to affiliate with. As Leon Cleveland of National Black Network points out: "We want to affiliate with stations that have high ratings. So we look at the ratings and then try to sell our service to stations that target our type of audience and have high ratings."

RADAR

Theoretically, radio network material can be extrapolated from Arbitron statistics. Local stations air network programming that is included in various quarter hours, so if a network knows when its material is being aired, it can find its ratings. However, the inconsistency with which network material is programmed makes such a process essentially impossible.

As a result, a group of about twenty networks have joined together to support Radio All-Dimension Audience Research (RADAR). These companies pay a company named Statistical Research, Inc., to track their programming by using telephone recall methodology.

For this methodology, trained interviewers phone randomly selected potential sample households and ask one person in each household to recall what radio programs he or she listened to yesterday and the day before yesterday. If the interviewee cannot remember specific programming, the interviewer asks for slogans, dial position, or anything else that might help define the material. The interviewer also assists the respondents with probes such as "Did you wake up to a clock radio?" or "Did you listen to the car radio while driving home from work?"

This type of phone research has a 63 percent response rate; in other words, 63 percent of people agree to be interviewed. The probable reason that this is about 20 percent higher than diary returns is that people can answer the questions instantaneously and do not have any burden placed upon them to remember to write down or mail anything. The drawback is that it relies on recall, which is often incomplete or inaccurate.

For the RADAR measurements, Statistical Research interviews about 6000 people nationwide and puts together reports for each fall and spring. Because such a large number of radio stations and radio networks exist, radio research usually yields low ratings and shares—numbers below 10.[6]

FIGURE 13.9 The layout for an Arbitron radio report showing average quarter hours. (Copyright 1992, The Arbitron Company. All rights reserved.)

ADI Target Audience

	MONDAY-FRIDAY 6AM-10AM			MONDAY-FRIDAY 10AM-3PM			MONDAY-FRIDAY 3PM-7PM			MONDAY-FRIDAY 7PM-MID			WEEKEND 10AM-7PM			MONDAY-SUNDAY 6AM-MID		
	AQH (00)	AQH RTG	CUME (00)	AQH (00)	AQH RTG	CUME (00)	AQH (00)	AQH RTG	CUME (00)	AQH (00)	AQH RTG	CUME (00)	AQH (00)	AQH RTG	CUME (00)	AQH (00)	AQH RTG	CUME (00)
+WAAA WRRR																		
P 12+	54	.3	376	51	.3	349	29	.2	352	25	.1	322	39	.2	272	37	.2	620
TEENS	8	.4	73	1		9	2	.1	47	4	.2	56	4	.2	42	4	.2	82
M 18-34	12	.4	94	7	.2	109	12	.4	115	13	.4	131	8	.2	59	10	.3	194
W 18-34	23	.7	114	32	.9	132	12	.4	118	7	.2	89	24	.7	141	18	.5	208
M 18-49	15	.3	129	9	.2	144	13	.2	161	14	.3	154	9	.2	65	11	.2	246
W 18-49	30	.5	161	38	.7	189	13	.2	137	7	.1	112	26	.5	165	21	.4	279
M 25-49	10	.2	105	7	.2	112	12	.3	137	10	.2	93	8	.2	41	8	.2	178
W 25-49	15	.4	76	18	.4	116	2		58			33	9	.2	68	8	.2	145
M 25-54	10	.2	105	7	.2	112	12	.3	137	10	.2	93	8	.2	41	8	.2	178
W 25-54	15	.3	82	18	.4	116	2		58			33	9	.2	68	8	.2	151
M 35-64	3	.1	35	2	.1	35	1		46	1		23	1		6	1		52
W 35-64	8	.2	60	9	.2	64	2	.1	26			23	2	.1	24	4	.1	84
WBBB																		
P 12+	73	.4	330	52	.3	282	46	.2	231	30	.2	173	64	.3	273	50	.3	452
TEENS	2	.1	7			7			7				1		8	1		16
M 18-34	15	.4	93	7	.2	77	7	.2	39	13	.4	77	23	.7	61	12	.4	93
W 18-34	3	.1	25	2	.1	14			8						12	1		31
M 18-49	39	.7	172	30	.5	145	26	.5	114	24	.4	122	32	.6	112	28	.5	183
W 18-49	14	.2	80	5	.1	40	8	.1	66	3	.1	25	11	.2	87	7	.1	132
M 25-49	25	.6	111	24	.6	84	21	.5	91	12	.3	61	9	.2	51	17	.4	122
W 25-49	14	.3	80	5	.1	40	8	.2	66	3	.1	25	11	.3	81	7	.2	126
M 25-54	28	.6	132	28	.6	95	22	.5	94	13	.3	68	12	.3	72	19	.4	147
W 25-54	17	.4	85	6	.1	45	9	.2	71	3	.1	25	14	.3	86	9	.2	131
M 35-64	29	.8	105	28	.8	91	20	.6	78	12	.3	52	12	.3	72	19	.5	127
W 35-64	19	.5	84	10	.3	61	13	.3	78	4	.1	37	19	.5	96	12	.3	146
WCCC																		
P 12+	25																.1	349
TEENS	9																.4	86
M 18-34	1																.1	72
W 18-34	5																.1	80
M 18-49	4																.1	121
W 18-49	12																.1	134
M 25-49	3																	85
W 25-49	11																.1	102
M 25-54	3																	88
W 25-54	11																.1	102
M 35-64	3																	52
W 35-64	7																.1	54
WDDD																		
P 12+	11																.1	342
TEENS																		17
M 18-34	4																.1	96
W 18-34	2																	75
M 18-49	5																.1	144
W 18-49	4																.1	118
M 25-49	4																.1	115
W 25-49	4																.1	104
M 25-54	4																.1	127
W 25-54	4																.1	104
M 35-64	1																.1	66
W 35-64	3																.1	62
TOTALS																		
P 12+	42690	27.6	134271	34055	22.0	104948	31303	20.2	118708	17075	11.0	93913	31634	20.4	117203	29445	19.0	149441
TEENS	2534	15.8	13413	1071	6.7	8007	3179	19.8	13155	2758	17.2	13135	2996	18.6	12709	2392	14.9	15512
M 18-34	6355	26.4	20192	6121	25.4	16178	5685	23.6	19159	3056	12.7	16101	4913	20.4	18108	4860	20.2	23316
W 18-34	6725	26.4	22597	6066	23.8	18457	5260	20.6	20947	3046	12.0	17956	5788	22.7	20866	4995	19.6	25020
M 18-49	11200	26.9	35494	9860	23.6	26887	9267	22.2	32737	4687	11.2	26361	8076	19.4	30508	8022	19.2	40153
W 18-49	12688	28.3	40025	10528	23.5	31948	8907	19.8	35239	4831	10.8	28273	9681	21.6	35119	8672	19.3	43970
M 25-49	9092	28.0	28072	7550	23.2	20292	7085	21.8	25400	3400	10.5	19855	6230	19.2	23386	6207	19.1	31268
W 25-49	10184	28.8	31561	8144	23.0	24922	6678	18.9	27059	3420	9.7	21067	7183	20.3	27064	6612	18.7	34608
M 25-54	10350	27.9	32179	8405	22.7	22992	7991	21.6	28817	3755	10.1	22200	7087	19.1	26800	6991	18.9	35693
W 25-54	11745	29.0	36161	9453	23.4	28432	7754	19.2	30976	3878	9.6	23649	8336	20.6	31057	7651	18.9	39482
M 35-64	8505	27.7	26800	6414	20.9	18539	6113	19.9	23119	2677	8.7	16841	5553	18.1	21406	5467	17.8	29370
W 35-64	10884	31.6	30885	8291	24.0	24468	6821	19.8	25426	3042	8.8	17290	7198	20.9	25818	6781	19.7	33369

ADI TARGET AUDIENCE

DEMOS P12+, TEENS, M&W 18-34, 18-49, 25-49, 25-54 & 35-64

GEOGRAPHY ADI ONLY

ESTIMATES AQH (00), AQH RATING AND CUME (00)

DAYPARTS PG 1: MON-FRI 6AM-10AM, 10AM-3PM, 3PM-7PM, 7PM,-MID, WKND 10AM-7PM AND MON-SUN 6AM-MID

PG 2: SAT 4 STANDARD, SUN 10AM-3PM AND 3PM-7PM

Footnote Symbols: * Audience estimates adjusted for actual broadcast schedule. + Station(s) changed call letters since the prior survey - see Page 5B.

ARBITRON RATINGS

RADIO MARKET, USA 236 WINTER 1987

PUBLIC BROADCASTING

Nielsen and Arbitron collect data regarding public broadcasting during the same surveys that involve commercial and cable television. AMOL has been a help in tracking public TV programming because PBS programs do not all run on all stations, and even those that do air on most stations are not always broadcast at the same time.

However, public broadcast stations do not place nearly the emphasis on ratings that commercial facilities do. As John Grant, vice president of scheduling and programming administration for PBS, says: "Rarely do we make decisions based on ratings. Viewership is important, and some programs should be seen by a large audience. However, for us, ratings are just a tool—a snapshot." Dan Everett, director of broadcasting at WGBH in Boston, echoes: "Audience measurement was designed for commercial broadcasting so is not of much use to us. We use it as one method of determining value, but we are not driven by it."

The actual size of the audience does not matter as much in the public broadcasting world as it does in the commercial arena. PBS and public television stations are delighted when they receive a rating as high as 9, which they did for the telecasting of "The Civil War," but no commercial rates are based on the size of the audience.

What public broadcasting is quite interested in is knowing whether or not its programs are serving the publics they were designed for. As PBS president Bruce Christensen has said, "While looking at an overall audience, we also want to know the makeup of that audience. Are children's shows being watched by kids and their families? Are programs we're producing for various age levels and categories of people being viewed by those people?"[7]

Nielsen publishes a special supplementary report on public TV on a market-by-market basis. PBS uses these figures, along with some from its own research department, to develop fact sheets and publications designed to show the quality of its audience in terms of such characteristics as level of education and involvement in community and political activities.

This material is used primarily to enhance the image of public broadcasting, but it is also used to attract underwriters and foundations wishing to serve the particular type of audience that public broadcasting attracts.

More important to PBS than ratings is an annual evaluation of programming undertaken by a subcommittee of the PBS board. It sends surveys to stations and gathers input from producers, minorities, underwriters, the press, and other people who have a stake in the national program service. Armed with the results of this survey, PBS programmers make decisions. They usually consider three major questions: What is the value of the program to the overall schedule? Is the program valuable to the overall goals of public broadcasting? Is the program meeting its own objectives?

Within public broadcasting, many other forms of evaluation surface. WGBH radio's Josie Patterson reports, "Often on shows where we want to get a sense of audience interaction, we offer something to the audience and then note how many requests we get." Director of Programming Roberta K. Smith of KOCE in Huntington Beach, California, says, "I depend more on phone calls and letters than ratings." NPR's Peter Pennycamp "concentrates mostly on audience numbers,

FIGURE 13.10
"The Civil War," which featured stories about people such as these Confederate volunteers of the First Virginia Militia, received much-acclaimed ratings as high as 9, but, in general, public broadcasting does not place high value on ratings. (Photo courtesy of Florentine Films.)

which are constantly going up." Says Pennycamp, "We really don't care that much about ratings."

HOW ACCURATE ARE THE RATINGS?

The main issue that a programmer has to deal with concerning ratings is determining how much weight to place upon them in relation to all the other factors involved in program decision-making. Should a commercial TV network programmer cancel a series if it ranks in the bottom half of the ratings? Should a cable TV network be removed from a cable TV system if it never achieves a rating higher than a 2? Should a radio disc jockey be fired if the ratings for his daypart fall from 5 to 4? Should "Sesame Street" be moved to a different time slot if the majority of the people watching it at 2:00 p.m. are men in their 30s?

One of the factors involved in determining how much weight to place on ratings revolves around whether or not the ratings are accurate—or perhaps, more appropriately, accurate enough.

Variables That Affect Rating Accuracy

Even the rating services do not claim that their numbers are totally accurate. Nielsen, in its booklet *What TV Ratings Really Mean,* writes: "Note that when we described the rating, we used the words 'statistical estimate.' That's because a rating is subject to a margin of statistical error."[8] But beyond statistical error, there are problems that can affect rating accuracy.

Sampling. Sample size is one of the aspects of ratings that is frequently questioned when it comes to determining rating accuracy. How can a company possibly report on what people are watching when only 4000 out of 93.1 million homes are sampled? No two people are alike, and even households with the same number of children and same income don't watch exactly the same television. Nielsen's answer to this:

> Try this interesting experiment (hypothetically—unless you happen to have 100,000 beads handy). Imagine 100,000 beads in a washtub: 30,000 red and 70,000 white. Mix thoroughly, then scoop out a sample of 4000. Even before counting, you'll know that not all beads in your sample are red. Nor would you expect your sample to divide exactly at 1200 red and 2800 white.
>
> As a matter of fact, the mathematical odds are about 20 to 1 that the count of red beads will be plus or minus 60 beads—or a range of 28.5 to 31.5 of the sample.
>
> So, in short, you have now produced a 'rating' of 30, plus or minus 1.5, with a 20 to 1 assurance of statistical reliability.
>
> These basic sampling laws wouldn't change even if you drew your sample of 4000 from 90 million beads instead of 100,000—assuming that the 90 million beads had the same ratio of red and white.
>
> This is a simple demonstration of why a small sample is just as adequate for a nation of 90 million households as for a city of 100,000.
>
> Yes, but people aren't beads. Of course not, but, then, neither do we attempt to measure people in all their complexities. In some ways, measuring a television audience is as simple in principle as counting beads. We're asking questions such as: "Is the set on?" and "If on, is it tuned to channel A, B, C, or D?" These questions are just as simple as asking if the bead is red or white. The answer in each case is a simple yes or no.[9]

Statisticians agree with Nielsen and other audience measurement companies that sampling does work. In fact, the sarcastic retort given to those who don't believe in the accuracy of sampling is: "The next time you go to the doctor to get blood drawn, tell the doctor to take it all because you don't believe that the small amount placed into the vial will give accurate results."

Statisticians have found that in general, the larger the sample, the smaller the sampling error. Unfortunately, doubling the sample size does not cut the sampling error in half; the sample size has to be quadrupled to do this. This type of increase is very costly. Another interesting statistical fact is that the size of the population being measured does not dictate the sample size. As Nielsen has pointed out, a sample size of 4000 is just as effective for a population of 90 million as it is for 100,000.

But even though sampling per se has been proved to work, problems encountered with sampling techniques can bias the sample. For example, when telephone numbers are used to draw a sample, households without telephones are excluded and households with more than one phone number have a better chance of being included. For many of the rating samples, people who don't speak either English or Spanish are eliminated. Probably worst of all are the people who eliminate themselves from the sample by refusing to cooperate. Although rating services make every attempt to replace these people with ones that are similar in demographic characteristics, the sample's integrity is somewhat downgraded.

Rating Techniques. Beyond sampling, there are many other problems that can creep in to the ratings process. People who know that their viewing or listening habits are being monitored may act differently than they ordinarily do. Then, in addition, each of the rating techniques has its own drawbacks.

Households with peoplemeters suffer from button-pushing fatigue, thereby artificially lowering ratings. Some groups of people are more likely to push buttons than others. When the peoplemeter was first introduced, sports viewing soared and children's program viewing decreased significantly. One theory was that men, who were watching the sports intently, were very reliable about the button pushing, perhaps some even out of fear that the TV would shut off if they didn't keep pushing that button. Children, on the other hand, were confused or apathetic about the button, therefore underrating children's programming. Another theory held that the women of the household usually kept the older diaries and weren't always aware of what their husbands were actually watching but were much more aware of what their children were watching. Therefore, under the audimeter/diary system the sports programming was underrated. At any rate, there were discrepancies between the ratings with peoplemeters and the ratings with the audimeter plus diaries.

But the audimeter has its own problems. It simply registers whether or not the TV set is on and not whether anyone is watching it. The TV set may be playing to an empty room but still the "viewing" is counted.

Diaries have innumerable problems. The return rate is low, furthering the problems of the number of uncooperative people in the sample. Even the diaries that are returned often have missing data—data that are filled in by a computer. To some ratings customers, this is an indefensible practice. Many people do not fill out the diaries as they watch TV or listen to the radio. They wait until the last minute and try to remember everything—perhaps aided by a copy of *TV Guide*. Some people are not honest about what they watch. Perhaps they don't want to admit to watching game shows or listening to radio sex/psychology shows.

With telephone interviews, people can be influenced by the tone or attitude of the interviewer or, again, they can be less than truthful about what they watched or listened to, out of embarrassment or in an attempt to project themselves in a respectable light. People are also hesitant to give information over the phone because they fear the person calling is really a salesperson.

One of the results of the problems with the different techniques is that competing audience measurement companies sometimes have different results. When numbers are large and spread out, these discrepancies are of little importance. For example, if Arbitron's rating for the top local newscast is 13.4 and Nielsen's is 12.9, and the nearest competitor is 9.1, the rating difference will not cause much of a stir. But recently a Rochester, New York, newscast received a 2.0 rating/4.0 share from Nielsen and a 1.0 rating/1.0 share from Arbitron. That's a difference of several hundred percent that could be very important in terms of advertising revenue.

Studies and Investigations

Because of the possibility for all these sampling and methodology errors, ratings have been subjected to numerous tests and investigations. In fact in 1963, the House of Representatives became so skeptical of ratings methodology that it held hearings. Most of the skepticism had arisen because of a cease and desist order from the Federal Trade Commission telling several audience measurement companies to stop misrepresenting the accuracy and reliability of their reports. The

FTC charged the rating companies with relying on hearsay information, making false claims about the nature of their sample populations, improperly combining and reporting data, failing to account for nonresponding sample members, and making arbitrary changes in the rating figures—pretty heavy accusations.

The main outgrowth of the House hearings was that broadcasters themselves established the Electronic Media Rating Council (EMRC) to accredit rating companies. This group periodically checks rating companies to make sure their sample design and implementation meets preset standards that broadcasters have agreed upon, to determine whether or not interviewers are properly trained, to oversee the procedures for handling diaries, and in other ways to make sure the ratings companies are compiling their reports as accurately as possible. All the major rating companies have EMRC accreditation.

Also as a result of the 1963 hearings, broadcasters undertook a number of basic research studies to determine the accuracy of ratings. They tested sample size and found, as mentioned, that the larger the sample the smaller the sampling error and that sample size did not bear on population size. They also found that people who cooperated with rating services watched more TV, had larger families, and were younger and better educated than those who would not cooperate. And they found that the telephone technique gets a 13 percent higher cooperation rate than diaries.[10]

Throughout the years a number of other studies have been undertaken to determine the reliability of ratings. As with the 1963 studies, the results are spotty. On the positive side, one study found that audimeters and telephone calls yield virtually the same ratings, and another found that Nielsen and Arbitron agree on program rankings 94 percent of the time. But a 1981 study undertaken for the cable industry indicated that diaries were a very unreliable means of determining cable TV viewing. And a study of Hispanics included in the Nielsen sample found that Nielsen's Hispanics watch less TV and have smaller families than Hispanics in general.

In 1989, broadcasters initiated a study to conduct a thorough review of the peoplemeter. The result was a list of recommendations to Nielsen that included changing the amount of time people participate from two years to one to eliminate button-pushing fatigue, metering all sets including those on boats and in vacation homes, and simplifying the procedures by which visitors log into the peoplemeter.[11]

Both electronic media practitioners and audience measurement companies want their ratings to be accurate, so both try to undertake testing to the extent they can afford it.

Technical and Sociological Considerations

In addition to methodological problems, ratings can be subject to technical problems—computers that go down, meters that don't function properly, cable TV systems that shift the channel numbers of their program services without notice, station antennas struck by lightning.

On top of this, rating methodology finds it very hard to keep up with technological and sociological changes in society. What should Nielsen and Arbitron do about the impact of the VCR on viewing habits? How should they take into account TV sets that can show four different channels at once? The portability of radio and out-of-home listening changed methodology for radio ratings, and now TV rating methodology faces the same challenge. How long should someone have to watch a program or listen to a station in order to be counted, especially in light of the fact that remote controls and push-button radios make channel switching an easy process? With radio being used more and more for background, should there be a difference between listening to a program and just being within earshot of it? Should there be a difference between viewing a program and just being in the room? If so, how should each be counted?

Even if ratings companies figure out ways to handle all these problems, new ones are bound to occur. How will an initiation of programming direct from a satellite to the home affect the ratings scene? If television becomes interactive so that people can order programs when they want to watch them, how will the audience measurement companies react?

Programming Aberrations

As if life weren't tough enough for the ratings companies, electronic media programmers often try for end runs around the ratings system. TV stations program their most sensational material during ratings periods: "Reverse circumcision. See how it's done. Tonight on channel 8." TV networks preempt their series and show star-loaded specials so that their affiliates will fare well in ratings and therefore can adjust their advertising rates upward. Cable networks show new programs as opposed to reruns. Radio stations run unique and financially rewarding contests that require listeners to "keep listening for further clues."

All of this, of course, negates the real purpose of determining which electronic media entities have the largest regular audience. It simply indicates which can program best for sweeps week.

What is at stake, of course, is advertising fees that can translate into millions of dollars. It is easy to understand why programmers who are under intense pressure from their managements to perform well during the sweeps period are tempted to do almost anything to maximize the numbers. The better PDs will stay within reason. The people who should really remain vigilant are the advertisers and their agencies. They are the ultimate losers if they pay rates that are based on a device that artificially stimulates the ratings and bears no resemblance to the programs in which the sponsor is actually investing.

How Programmers Should Use Ratings

Ratings are not foolproof, nor are they the only measurement of program success or failure. There are many other indications of a show's vitality and its potential for a productive run. Even Nielsen cautions, "Rarely . . . should a programming decision be made on just one ratings report; *repeated* measurements substantially reduce the range of statistical error that applies, as well as provide broadcasters with a vital sense of direction as to whether an audience is building or dropping off."[12]

Furthermore, the greater the fractionalization of the audience, the less precise the measurement. Nielsen admits it is dealing with numbers that 95 percent of the time are plus or minus one or two points from the real rating. The other 5 percent of the time the rating is even less accurate.

It is true that in commercial broadcasting ratings are the way the score is kept. But the savvy programmer will use them as an aid, not a crutch. Given the wide range of error that the rating services confess to, it would be foolish for a programmer to make a judgment on a show based on one or two rating points (to say nothing of a judgment made on one or two tenths). Also left out of the mix is the intensity with which viewers watch certain shows. An audience of 10,000,000 devoted fans is more valuable to a broadcaster and an advertiser than 12,000,000 viewers who are involved with two or three other activities while sitting in front of the set.

When looking at ratings, programmers should be more concerned about trends and tendencies than a single raw number. Even if a current rating is low, if it's part of a gradual increase over a period of time, it could have positive significance. On the other hand, a higher number that is part of a flat pattern may indicate a problem down the road.

The ratings-as-a-crutch approach to programming has a certain appeal. A number is easily understood; it permits facile comparisons; it has the look of science about it; and it offers a more compelling argument than such vague statements as, "I have faith in the show," "My gut tells me it will work," or the familiar laugh-getter, "Trust me." But good programmers are made of sterner stuff. They factor in the other considerations (trends, demos, mail pull, the "feel" of the show), and if there's enough there to rebut a number, they take a stand. As one network chieftain once said, "How can you sell your soul for a lousy tenth of a point?"

• • • • • • • • • • • • • • • •

But sometimes the numbers are bad, not by tenths of a point but by many points. Something has to be done. How programmers go about making the necessary changes is our next field of study.

NOTES

1. Nielsen publishes several booklets explaining its methodology. These include *Nielsen: The System for Success* (New York: Nielsen Media Research, 1988); and *Nielsen: An Overview of the Local Market Television Audience Measurement Service* (New York: Nielsen Media Research, 1990).
2. *Turning the Numbers into Sales Strategies* (New York: Arbitron Ratings Company, 1987): 7.
3. "Ratings for Syndication Start to Measure Up," *Adweek* (January 20, 1987): 22.
4. Arbitron's main book describing its television methodology is *Description of Methodology: Arbitron Ratings: Television* (New York: Arbitron Ratings Company, 1987).
5. Arbitron's main book describing its radio methodology is *Description of Methodology: Arbitron Ratings: Radio* (New York: Arbitron Ratings Company, 1987).
6. "Westwood, Sheridan Post Gains in RADAR 43," *Broadcasting* (August 19, 1991): 24.
7. "PBS's National-Local Balancing Act," *Broadcasting* (July 18, 1990): 37–38.
8. *What TV Ratings Really Mean* (New York: Nielsen Media Research, 1989): 8.
9. *What TV Ratings Really Mean*, 13–14.
10. Details of these studies can be found in *ARMS: What It Shows, How It Has Changed Radio Measurement* (Washington, DC: National Association of Broadcasters, 1966); *CRAM: Cumulative Radio Audience Method* (New York: National Broadcasting Company, 1966); and *Television Ratings Revisited: A Further Look at Television Audiences* (New York: Television Information Office, 1971).
11. "Study Urges TV Ratings Changes," *Variety* (December 14, 1989): 1.
12. *What TV Ratings Really Mean, 12.*

14 Changing and Canceling Programs

There are times a program will be painstakingly researched, carefully developed, launched with inspired promotion—and fail. In Chapters 7 and 8, we observed that experienced programmers know the general elements of successful show-making but are frequently defeated by taking wrong turns on some of the many decisions that must be made while attempting to attain those elements. When a show isn't clicking and begins to stumble, one of three changes may be necessary: interior adjustments in the program itself, a new time period, or cancellation.

COMMERCIAL TELEVISION

Before getting to specifics, we'd like to mention that shows in inventory are like backup quarterbacks. As every football fan knows, the most popular person in the stadium when things are going badly is the quarterback on the bench. He's always perceived to be the savior: fresh, exciting, gifted, and unflappable. Until he gets in the game. Then, more often than not, he proves why he's number two.

Generally the show in inventory was available at the time the schedule was set. It didn't get on because it was judged weaker than other available choices. In effect, many replacements are rejects that are being given a second chance. Programmers would do well to remember this fact before precipitously pulling the lineup apart.

Reasons for Change

There are many reasons for change, not all of which call for a cancellation. Programmers should remember that salvaged shows represent an enormous economic savings. Every effort should be made to keep them viable. But there are times when something has to be done, and the cause is generally one of the following.

Low Ratings. The vast majority of commercial television shows are canceled because they fail to achieve satisfactory ratings. The actual cause may lie elsewhere, but low audience levels is the reason for the cancellation. In many cases these shows didn't belong on the schedule in the first place. Their failure has nothing to do with time period, audience flow, promotion, or any of the hundreds of variables that go into programming.

The simple truth is that they were badly conceived and/or executed and had no right to be scheduled. They got there because of contractual commitments, an absence of alternatives, or bad judgment by a decision-maker. We need not dwell on these unfortunates; the faster they are dismissed, the better. But not all low-rated shows are essentially flawed. Their poor numbers may be the product of other forces.

Exhaustion. Unfortunately, all shows are mortal. Eventually, they just plain run out of gas. There are no more variations that can be worked on the theme; no more aunts, uncles, cousins, and babies who can be added to the cast without making the show look absurd. As gifted as the writers and producers of "The Cosby Show" were, there were no more arrows left in their creative quiver. Programmers must bow to the inevitability of fatigue and reluctantly say farewell to their old friends.

Loss of Focus. Programs can drift off course. Producers get wrapped up in new projects and forget to maintain the old ones; stars become "artistes"; writers get carried away and suddenly that fine little series becomes a bore, a distortion, or both. "Moonlighting," once a delightful, sophisticated comedy-adventure featuring witty, acerbic banter between Bruce Willis and Cybill Shepherd, lost its charm when internal friction and bizarre scripts diminished the magic between the leads.

In some instances a candid, air-clearing discussion among the principals can put the show back on track. Too often too much has transpired, and the program, sadly, is irretrievable.

Social Changes. "The Brady Bunch" was a harmless sitcom salute to traditional family values. Launched by ABC in 1969, it brought together a widower with three sons and a widow with three daughters. Their featherweight adventures were scheduled on Friday evening during the "family hour," and the show achieved reasonably strong ratings for three seasons.

However, by 1972 CBS was presenting such gritty hits as "All in the Family," "Maude," and "M*A*S*H," and such sophisticated and well-turned series as "The Mary Tyler Moore Show" and the "The Bob Newhart Show." As the audience responded to these adult themes, their appetite for the marshmallow episodes of "The Brady Bunch" diminished, and the show was off the air by August 1974. Changing tastes and the social climate had rendered it obsolete.

Ageing Demographics. The usual pattern with programs is that the longer they're on, the older their demographics become. Young viewers tend to be restless; they are the first to seek out new and different programs, and the first to leave them. Older viewers come along later but then become fiercely loyal and stay for years. At some crossover point the departure of the young audience makes the program unappealing to advertisers and a change is dictated. Sometimes the addition of new, youth-oriented elements in the program can slow down the departure rate. It is always worth a try if the alternative is cancellation.

Wrong Time Period. Some programs are placed in time periods where they don't have a chance. Take the example of "Man of the People," a situation comedy starring James Garner. In September 1991 NBC slotted it at 8:00 p.m. Sunday, opposite CBS's "Murder, She Wrote," a top ten series for a decade with enormous appeal to middle-aged to older adults; ABC's well-established "America's Funniest Home Videos"; and Fox's "In Living Color," a youth-oriented comedy series. In addition, "Man of the People" followed two brand-new half hours featuring virtually unknown performers while "Murder, She Wrote" inherited the audience (heavily adult) from "60 Minutes," perennially one of the five highest-rated shows in TV.

Garner, whose main appeal is to older adults, had nowhere to go to recruit viewers. "Murder" had locked up his primary audience, and the other two competitors siphoned off the kids and young adults. Garner's show expired quickly and ingloriously. The program ranked in the bottom ten of all rated shows and was yanked from the schedule before the November sweep period.

At first NBC planned to try it in a new period after the sweeps were concluded. But they reconsidered and finally just paid the producer for the unused episodes. The point is that the series had no hope from the day it was scheduled. The network wasted a potential asset by placing it in an unworkable time period.

Excessive Relocation. "Hill Street Blues" is a classic case of a good program being maneuvered into near extinction. In its first four months on the air, NBC shunted it around to three different time periods, much to the mounting fury of Grant A. Tinker, then the president of the producing company, MTM. Tinker recalls: "I watched them bounce the show around from pillar to post. [The NBC management] was crazed trying to find an audience. They were in a manic state even when they had something as good as they had with 'Hill Street.' So they put it on Thursday at 10:00, then Saturday at 10:00, then Tuesday at 9:00. Then they said, 'Make me two and we'll play it as two hours.' It was just no way to run a railroad."

Fortunately, the program finally settled in at Thursday, 10:00 p.m., and for five years was one of television's most popular series.

The Desire for Something New. When new managers arrive, changes generally occur. After all, they've been brought in to make things happen and standing pat won't get it done. New officers also bring their pet projects, attitudes, and personnel with them, and all of these usually find a place on the schedule. Sometimes the new ownership makes a dramatic across-the-board realignment of programming.

For example, when the Disney company took over channel 9 (KHJ) in Los Angeles in the late 1980s, it not only changed the call letters to KCAL, but it also changed the whole programming philosophy. The station abandoned all entertainment shows in the evening and presented nothing but news. This was an extreme form of a new broom sweeping clean. More often new management tries to retain the things that are working and discard the ones that are not.

Placing a Show in a New Time Period

If a program is compatible with its time period and the competition is just reasonably formidable, programmers are well advised to leave it alone. Remember, once the program is moved, the recruitment of viewers must start all over again. As we have seen in the "Hill Street Blues" example, unnecessary jockeying can be very counterproductive.

There are two valid reasons for relocation: 1) the program is underviewed and may prosper in a new period, and 2) the show has developed independent strength and is now ready to occupy a more important slot.

Programs, such as the earlier-cited "Man of the People," are frequently placed in inhospitable periods and have very little chance of success. They may be against solidly entrenched opponents and not have a fair opportunity to be sampled. Or the time period may not offer the best demographics for the show's primary appeal. Or both. In these instances, if the programmer still has faith in the show, a move should be tried.

Even a show as mighty as "M*A*S*H" couldn't succeed everywhere it was placed. It premiered on Sunday at 8:00 p.m., a dismal spot for a dark comedy about the madness of war, and was almost canceled after its first season. The following year CBS hammocked it Saturday night at 8:30 p.m. between the phenomenally strong "All in the Family" and the equally powerful "The Mary Tyler Moore Show." It zoomed to fourth in the season's ratings. Once it had established its following, the CBS management confidently moved it to Tuesday at 8:30 behind the popular "Good Times" where it continued to perform well. But one year later CBS inexplicably shuttled it to 8:30 p.m. on Friday to blunt the ratings of the then red-hot "Chico and the Man." The strategy backfired, and "M*A*S*H" was shifted to a 9:00 p.m. period on Tuesday for one year and then five smash years at the same hour on Monday.

The movement of a show to take on a more important role in an overall schedule can be a very effective strategy. If the timing and judgment are correct, the relocation can attract a "bonus" audience for the entire night. As we have just seen, "M*A*S*H" was nurtured in a hammock position, developed its own large and loyal audience, and then graduated to a key time period to anchor a night's schedule.

The key element in relocation is timing. It must be done early in the show's life when viewers are at their most fervent and will follow a favorite almost anywhere. The relationship between an audience and a time period is fragile. When the habit is disconnected, it is difficult to reconstruct. Only when a show is in its ascendancy is it prudent to risk a change. Two examples, a dozen years apart, make the same point.

As we noted in Chapter 11, ABC took "Laverne and Shirley" out of Tuesday, 8:30 p.m., where for three years it had been a successful extension of "Happy Days," and moved it to Thursday at 8:00 p.m. The show's rating plummeted. At mid-season it was restored to its original position, but it never recaptured its previous popularity.

In 1991 NBC attempted an even riskier change. For each of its first six seasons, "The Golden Girls," a comedy series about four 50+ women who live together, ranked among the top ten highest-rated programs. In order to strengthen its Saturday night lineup, the network moved the show from 9:00 p.m., where it had resided all its life, to 8:00 p.m. The goal was to have "The Golden Girls" start the night and feed its huge audience into the remainder of the schedule. It didn't quite happen that way. Instead, the program lost 21 percent of its audience, sank from tenth to thirty-third place in the season's final standings, and dissolved as a series.[1] In the following season it reemerged with an altered cast as "Golden Palace" and aired on a new night (Friday) on a new network (CBS).

On the other hand, "The Simpsons'" challenge of the "The Cosby Show" paid off because the timing was right. In its second season on the air, when Simpson mania was at its zenith, the animated show moved against the perennial ratings champ then beginning its seventh year. It not only held its own, it knocked Cosby out of the top ten. The following year the season-long rating for "The Simpsons" was less than two points behind Cosby and towards the end of the run was actually topping him on occasion.

Adjusting Program Elements

When a show opens well, but quickly begins to lose audience, it's a sign that viewers liked the idea but found something wanting. A change seems indicated. However, too often producers are content to make cosmetic modifications: redesign the set, create a new theme song, develop new graphics. These really aren't very meaningful; even though they may be improvements, they don't influence a viewer's decision to watch.

The fundamental problem, whatever it is, must be addressed. To discover it, programmers are advised to commission the most definitive research possible in the shortest span of time. A series of focus groups in a wide variety of markets throughout the country is a useful approach.

As soon as the problem is defined, programmers should move immediately to make the adjustment. Speed is essential. The longer the show appears with its damaging flaw, the greater the number of core viewers who will be alienated. The hardest job in television is to recapture viewers who have sampled a program and rejected it. They tried it, haven't liked it, and have gone on to other things.

If a crippled show has been on for more than a couple of months, it is probably best to take it off the air, repair it, and bring it back after sufficient time has elapsed for the audience to have had its recollections dimmed. It might also be wise to change the title to give the revised version the appearance of a fresh, new product. With strong creative promotion, the show has a chance to attract wide sampling.

One of a programmer's most perilous moves is to make a change in an ongoing hit. If it ain't broke, etc. The audience is very knowledgeable and will forgive necessary changes, e.g., the death of Coach in "Cheers," or the replacement of Valerie Harper in "Valerie" after a well-publicized contractual dispute. But the removal of a favorite performer for no other reason than an alleged freshening of the show is a catastrophe waiting to happen.

The most notorious such change occurred in 1989 on NBC's "Today" show. The venerable series, with co-hosts Bryant Gumble and Jane Pauley, was number one in the early-morning competition. A new network executive decided that the show's demographics could be improved if a younger, fresher face was introduced. He hired 30-year-old Deborah Norville to play a specially devised role that went beyond news inserts. The move was widely interpreted by the staff and public as an effort to replace the warm and motherly Pauley (age 39) with a "younger woman." The network denied the accusation, but after a brief, strained interval Pauley left and Norville became the co-host. Not only did NBC suffer a brutal PR lashing from the public and the press, but within weeks the show dropped 15 percent in its ratings and slipped back into second position. This plunge translated into a daily loss of 600,000 to 700,000 viewers and untold millions in revenue. It took almost two years and another change of co-host from Norville to Katie Couric before the "Today" show could once again challenge "Good Morning America" for ratings leadership.

Sometimes, however, a program needs a change of personnel. Performers can become arrogant or sloppy in their work and begin to alienate viewers. If counseling isn't effective, a change is indicated. On rare occasions backstage demands by one cast member may stir up such hostile feelings among other principals that decisive action is required.

The classic example is "Three's Company." The show premiered in 1977 and was an instant hit. In the summer of 1980 one of its three leads, Suzanne Somers, campaigned to have her salary raised from $30,000 to $150,000 per episode and to be given a piece of the profits.[2] This proposal suggested that she felt her contribution to the show's success was greater than her co-stars', John Ritter and Joyce DeWitt, a position not calculated to develop harmony on the set. The producers refused to renegotiate her contract (it still had one more year to go) and for one full season limited her appearances to sixty-second inserts at the end of each episode. These cameos took the form of telephone calls to other cast regulars and were taped separately when no other performers were present. When her contract terminated, she was dropped from the show. The program continued successfully until September 1984.

FIGURE 14.1 The departure of Jane Pauley from the "Today" show caused a precipitous decline in the program's ratings. Only after Katie Couric (left) was teamed with Bryant Gumble did the show start to become competitive again. (Photo courtesy of National Broadcasting Company.)

Cancellations

Ever since the early to mid-1980s, the rate of program cancellations has accelerated. Foremost among the reasons are the introduction of overnight ratings to virtually every major market and the expansion of the number of cable channels. These factors have intensified competition, thinned out the audience size for most shows, and supplied buyers with immediate rating information. In such a superheated environment buyers are disinclined to wait it out to see if a program can find an audience. In the area of network primetime programming the rate of cancellation has increased from 68 percent during the period from 1970 to 1984 to almost 80 percent from 1985 to the present.[3] The syndication story is much the same.

In view of such a mortality rate, it behooves programmers to handle the canceling process properly.

Giving a Cancellation Notice. The joy of the business is issuing pickups. Everyone is happy, hands are pumped, and optimism abounds. Buyers enjoy this function; it makes them heroes and reasserts their power. Unfortunately, there is a reverse side to the ceremony, the dreaded cancellation announcement. All shows come to an end, some after a few episodes, some after a decade. Whenever it happens, a member of the buying side must inform the people on the production side. Although always an uncomfortable, sometimes painful, process, there are certain procedures that, if properly observed, can ease the discomfort and form the basis for renewed relationships in the future.

The first thing representatives on both sides must remember is that the industry is small and time moves swiftly. Buyers need sellers and sellers need buyers. It is imperative that both parties separate in a manner that permits, even encourages, future associations. The cancellation conversation is no time for vindictiveness, vituperation, or the settlement of old scores. The brief pleasure such an outburst might afford can have disastrous consequences. Buyers who satisfy themselves by letting loose a diatribe against the producer may be cutting themselves off from future hits; sellers who unleash a blast are simply reducing by one the potential buyers for their next project.

The guidelines for cancellation are few but important:

1. The Head Person Should Do It

Too many program leaders try to duck the assignment. It is awkward and difficult and they'd rather not. That's a mistake. Production company executives admire the head buyer who has the courage to say the tough words, and it is somehow less painful to hear it from the summit.

It also makes it easier for the production executive to pass the word on to subordinates by being able to say: "Mike Johnson (president of the Entertainment Division) himself told me he thought we did a good job and he looks forward to working with us again as soon as possible."

When the cancellation notice is delegated to a lower level staff member, the production people are left with a second class feeling which, when added to their already depressed condition, can result in a deeply resentful attitude. Buyers should remember that their decision works a financial hardship on scores of people. They should soften the blow with expressions of concern and sensitivity. Not only is this decent, it is also good business. Canceled staff members will appreciate the gesture and more than likely bring the buyer their next program idea.

2. The Producers Must Be the First to Know

This is not always easy in a very visible business with a highly refined rumor-mill system. However, the quality buyer will always take every precaution against the premature leak of a cancellation decision. It is grossly unfair for a producer to read about his or her show's demise in a trade article or through some other unauthorized source. Nothing will guarantee a quicker and more long-lasting rupture between the buyer and seller than a clumsily handled cancellation that embarrasses the producer and the staff.

3. Tell the Right Person

This can vary from show to show. The prevailing rule is to notify the same person with whom the original deal was made. For the major production companies, the appropriate party is usually the head of television production at the studio. Chances are he or she is the one with whom the agreement was first reached, and it is fitting that the circle be completed with the same principals. Sometimes, because of personal relationships or other unusual dynamics, the right contact is the executive producer or producer.

With smaller production entities, the key person is often the agent representing the project. He or she might have put the package together, negotiated the deal, served as an intermediary in disputes, and generally been acknowledged as the pivotal business element in the enterprise. For these programs, it is appropriate to notify the agent first.

4. Make Additional Calls

After the chief person on the production side has been contacted and has had a reasonable opportunity to inform the talent and staff, it is wise and humane for the programmer to follow up with calls to cast members and key staff personnel. The personal touch will be appreciated and the conversations (brief is best) give the programmer an opportunity to express gratitude and establish a warm basis for future associations.

The cancellation dialogue is as much about the next show as it is about the defunct show. What's gone is gone. Both parties should conduct themselves in such a manner as to make another program not only possible but desirable.

But even with the best-intentioned people, a gracious let's-work-together-again separation isn't always possible.

WHAT IF TODAY'S CANCELED SHOW IS TOMORROW'S COMPETITOR?

From 1963 until 1979 "The Mike Douglas Show" enjoyed a successful run in syndicated television. The ninety-minute talk-variety series appeared in over 150 markets and vied with "The Merv Griffin Show" and "Dinah's Place" as the highest-rated light-conversation program in the marketplace. In 1972, when the contract between the affable host and his producer/distributor, Group W Productions, was up for renewal, a negotiation took place that had dramatic implications seven years later.

Douglas, who did not own any part of the show and was just a salaried performer, held out for $2,000,000 a year, a huge increase over his previous compensation. Group W balked at the new terms. Douglas immediately went out and found another distributor who was willing to produce a new series and pay the star the demanded amount. A clause in Douglas's existing contract permitted Group W to match any proposal made by another syndicator. Convinced that Douglas's new offer was real, and reluctant to lose the popular host, Group W signed him to a new seven-year deal on his terms. This experience was well-remembered and frequently recounted by Douglas and his people.

In 1979, about nine months before the contract was to terminate, Douglas's representatives presented to Group W their client's conditions for a renewal. They proposed an increase to $3,000,000 a year for the three additional years he would perform in the show, plus $500,000 a year for the ensuing five years during which time he would make himself available for appearances at Group W meetings and conventions. Since Group W held no conventions in which an out-of-work performer would be an attraction, the last five years amounted to a $2,500,000 bonus. The total package of $11,500,000 for three years' worth of programming almost doubled the host's current income.

Unfortunately for Douglas, his timing for such aggressive demands was bad. Fluffy, lighter-than-air entertainment was being replaced by shows with a harder edge. The ratings of Mike, Merv, and Dinah had eased off as the audience began to shift allegiance to Phil Donahue and other grittier hosts.

Furthermore, the CBS-owned and -operated stations in New York and St. Louis had already informed Group W that they would not sign for another season if Douglas remained the host. Because the show was offered to stations on a cash/barter basis (see Chapter 2), the cancellations had a double impact on Group W's revenue. Not only was the syndicator about to lose the stations' combined annual license fee of $2,000,000, but the loss of audience coverage of the two markets (about 10 percent of the United States) would also have a substantial negative effect on the barter income. As Group W saw it, the increased annual expense of $1,833,000 for Douglas's services plus the decreased revenue from the CBS defections ($2,000,000 in cash, $1,000,000 in barter) would cause a negative swing of almost $5,000,000 to the bottom line. In an era of declining interest in soft talk shows, the arithmetic was untenable.

Group W called in the performer's delegates, the same lawyer and personal representative with whom the 1972 deal had been made, and laid out the financial spreadsheets. The representatives sat through the presentation with growing impatience and when it was over dismissed the accounting report as a negotiating ploy. They then recounted with some relish the legendary saga of 1972. Unless the Group W executives wanted a lesson via deja vu, they would be well advised to come to terms now and spare everyone a lot of annoyance. The syndicator's assurances that this was no bluff went unheeded.

It was evident to Group W that no meaningful compromise was possible. Even if Douglas backed off his dollars somewhat, the loss of the CBS business was too great to make the show profitable. Another solution had to be found.

In the course of discussions with CBS officials, Group W learned that their O&O stations were very enthusiastic about John Davidson as a talk show host. When the distributor asked if the stations would stay with the show if Davidson replaced Douglas, they were told yes. Group W made an offer to the singer-actor-Vegas personality and found him most receptive.

Now the battle lines were drawn. Douglas was determined to have his demands accepted. But if Group W turned him down, he was prepared to sign with another distributor who would immediately contact all the stations carrying the Douglas show and ask them to come with Mike on a new series. If Group W didn't get to the stations first, the Davidson show would be left high and dry with no distribution. In this showdown Group W had one pivotal advantage; it controlled the timing. Douglas couldn't make his move until he received official word that he was being canceled.

As soon as the ink was dry on Davidson's contract, Group W summoned its entire sales staff to its West Coast office. On the following day at 11:00 a.m. New York time, Douglas's representatives were informed by the president of Group W Television that Mike's contract was not going to be renewed. One hour later, the Group W salespeople were instructed to contact the stations in their territories about the switch of hosts. The interim hour was to allow Douglas's people sufficient opportunity to inform their client of the cancellation but not enough time for a new distribution company, no matter how fast it worked, to reach station managers before the Group W staff. Within an hour Group W had talked to every station on the lineup and the vast majority agreed to the change. Only a handful requested more time to study the proposal.

Clearly this was a cancellation in which an amicable parting filled with warm sentiments about future relationships was impossible. The canceled show was about to become a deadly enemy. The situation called for Group W to observe the proper etiquette (decision delivered to the appropriate people by the highest-ranking official, notification to the canceled party prior to public release of the news, etc.), but the circumstances prevented any further discussion about doing future productions together.

To complete the story, Douglas found another distributor and launched a new series in the fall of 1980. The station lineup was weak with many marginal time periods, and after two money-losing seasons the show folded. Davidson enjoyed a reasonably prosperous first season but sagged badly in the second and went off the air at about the same time as did Douglas. The change in audience interests towards more controversial topics did them both in.

CABLE TELEVISION

The cable TV business is not much into cancellations or adjustments. Although there have been a few series that have had their runs shortened, generally the cable networks sign up for original or syndicated material and let it have its run. Of course, cable does not present as much original programming as the commercial networks do and therefore has less to cancel. Peter Hansen of A&E expresses the philosophy of most of the industry when he says: "We usually commit to thirteen shows and then decide whether to commit to thirteen more. Our general pattern is to do twenty-six shows. Producers usually have plenty of notice of cancellation."

As cable matures and builds up more advertiser support, it may find it necessary to succumb to some of the quick cancellation policies of the networks. But, at present, some of the cable services have enough trouble finding product in the first place that they are unlikely to remove it once they've gone to all the work of acquiring it.

Network Changes

What does get canceled and altered in cable is the networks themselves. Some of them don't survive economically while others can't make it on their own and must merge. Many of them have altered the overall concept a number of times in order to home in on something that works. HBO has recently

a

b

c

FIGURE 14.2
The Family Channel, which features wholesome entertainment such as "Big Brother Jake" (a), was originally a religious channel, Christian Broadcast Network. When Daytime and Cable Health Network merged, they changed the network name to Lifetime. The channel features primarily women's programming such as "It Figures," with Charlene Prickett (b). Comedy Central, formed from the merger of HA! and The Comedy Channel, went through several name changes (Comedy TV, CTV: The Comedy Network) before settling on Comedy Central. Shown here is Mo Gaffney, host of "Women Aloud" (c), described as a cross between "The Oprah Winfrey Show" and a women's stand-up comedy series. (Photos courtesy of The Family Channel (a); Lifetime (b); Comedy Central, Craig Blankenhorn, photographer (c).)

added more made-fors and reduced the number of theatrical films because it felt its audience was tiring of so many movies that had been played over and over. When E! Entertainment TV started in 1990, it programmed 24 hours a day of news about the world of entertainment. This didn't work as well as hoped, so the owners decided to broaden the scope and try a few series that were oriented toward show biz information.

When major concept changes occur, the networks often adopt different names—following the advice that it's better to start with a clean slate than have the remnants of the old failure left on the board. Christian Broadcast Network (CBN) was originally religious programming. When it decided to change to wholesome, family-oriented material, it became The Family Channel. Likewise, when Daytime and Cable Health Network merged, they changed the name to Lifetime. Viacom's HA! and Time Warner's The Comedy Channel merged to form Comedy Central.

Lineup Changes

Cable systems frequently make changes in their channel lineup. In some instances, the only thing that is altered is the channel number. MTV may be switched from channel 12 to channel 26. The reasons for this are multitudinous. Sometimes the system rebuilds to add channels and is then able to add more services. But rather than lumping all the new ones at the end, the system may want to give some rhyme or reason to channel numbers. For example, it may want to group all music services (MTV, VH-1, Country Music TV, The Nashville Network) together. If the only music service it had been carrying previously was MTV, grouping these services together may involve moving MTV. Sometimes channel moves are made in order to bring a whole region into conformance. Systems in a particular geographic area may decide that they should all place MTV on channel 26 so that MTV's programming can be promoted regionally without confusing the viewer.

Other times, cable systems remove certain services and replace them with other networks. Sometimes this is done from altruism—the desire to add a service that will serve the subscribers' needs more effectively. More often, however, the decision is economically based. One network's cost to the cable system will be less than another's. Or the system may feel it can make more from a pay-per-view service than a basic network. Rarely will a system replace one pay service with another, such as eliminating Showtime and adding Cinemax. Reeducating all of Showtime's subscribers that they should switch to Cinemax is costly—and generally not very effective. The main result is loss of subscriber income.

One of the chronic criticisms of cable systems is their inability (or lack of desire) to communicate channel changes to their subscribers. From the subscribers' point of view, the alterations look wanton. One day they turn on the TV and the channels are all flipped around—mass frustration. Systems are often wary of communicating exact details if the change involves eliminating some service. There are always people out there who will kick and fuss when something is taken away. What cable systems often do is send subscribers an innocent-looking channel card with their monthly bill. This lists the new lineup without indicating what has been eliminated.

COMMERCIAL RADIO

What is canceled most often on radio are entire formats and individual hosts or disc jockeys. As with commercial TV, the driving force behind both is ratings. If a station's Arbitrons are low period after period, doctoring on the scope of a major format change is called for.

Adjusting Program Elements

If a station does reasonably well overall in the ratings but one particular daypart is low, then it's time to look to the particular elements that make up that segment of time. Sometimes the problem is inherent in the clock—there may be too much news or too much chatter. Those problems are easy to correct. But more often the problem lies with the on-air personality—the disc jockey, talk show host, or anchor newscaster.

Focus groups and TV-Q-type tests (see

Chapter 6) can confirm this possibility. A disc jockey on a contemporary-oriented station may simply grow older and no longer relate to the teenage audience. When a personality must be fired, the executives should exercise the same care discussed regarding commercial TV show cancellations—the news should come from the top, and the person involved should be the first to know. Every on-air person has some supporters, and a messy firing can lead to negative PR for the radio station as a whole.

Another element that leads to program adjustment is burnout. The popularity of current songs does not last forever. Programmers must know when to remove a hit from airplay so that listeners do not turn to another station simply because they are tired of hearing the song. Watching the charts in the trade journals can give some indication of burnout as can the telephone research discussed in Chapter 6.

Programmers can also make rough predictions of when burnout will occur by considering the complexity of the music. Complicated music takes longer to burn out than simpler music. Songs with irregular rhythm and a large number of verses last longer than songs with regular rhythm and repetitive choruses. Also, the more often the song is played on the radio, the faster it is likely to burn out. Songs that are on their way to burnout should be played less often than songs that are at their peak.

Network Changes

Some radio networks produce programs that are subject to cancellation, but few are actually given the ax. Staples such as the shows produced by Casey Kassem, Dick Clark, and Paul Harvey have been on for years. The most common reason for short-term radio programs is that they have simply run their course. A special on Martin Luther King, produced by National Black Network, is intended to be one-shot only. So are concerts or programs such as ABC's "Fourth of July Celebration."

When a network does alter its programs, the change is usually not traumatic. As Ed Salamon, president of programming for Unistar, says: "All our programming is produced in-house. When something is canceled, people can move from one project to another within the company. Our main problems are with the advertisers. If an advertiser is on a show that isn't doing well and we need to cancel it, we have to decide what to do with the ads. Usually we can transfer them to another program. Also, if a station was dependent on the program we are canceling, we try to help them find something else."

Canceling and changing are not major issues for radio. When ratings (and hence dollars) are low, alterations are called for. But since some stations don't even use ratings as part of their sales pitch, a well-thought-out programming philosophy can keep advertisers and audience happy for long periods of time.

PUBLIC BROADCASTING

Public broadcasting is the most benign of all forms when it comes to cancellation. There are programs such as "Masterpiece Theatre," "The MacNeil/Lehrer Newshour," "Sesame Street," and "All Things Considered" that have been on for decades, virtually unchanged. In fact, the problems within public broadcasting have not revolved around cop-

FIGURE 14.3
Jim Lehrer (left) and Robert MacNeil have hosted an evening news show on PBS since October 1975. The program was originally called "The Robert MacNeil Report," and during its long tenure it has been upgraded from a half hour to an hour. (Photo courtesy of MacNeil/Lehrer Newshour.)

ing with change but rather around initiating any. Public broadcasting's audience members are loyal to their favorite programs and their time slots. Any changes cause an uproar.

Lack of Change

"We air 'Sesame Street' five times a day," reports Chris Fennimore, program director of WQED. "But each airing has a substantial audience. If we took it off one of those five times, the people watching it then would think we had taken it off entirely. We did take off 'Shining Time Station' so we had time to air Shari Lewis's puppet show. Lots of people complained." John Grant, vice president of scheduling and program administration for PBS, echoes, "We get crucified when we take something off."

Dan Jensen of APR notes, "We received a lot of complaints when we stopped the extensive Gulf War coverage—even though the war was over." Even worse for APR was the cancellation of Garrison Keillor's "Prairie Home Companion." Keillor left because he wanted to explore new vistas not related to radio. "The audience really complained," says Jensen. "We tried airing reruns of his programs, but that didn't work very well. When Garrison finally decided to return to public radio several years later, his following came back almost instantly."

FIGURE 14.4 "Sesame Street," which began in 1969, has had great longevity on public television. It draws top-name guest stars, such as Whoopi Goldberg. (Photo courtesy of Children's Television Workshop.)

Reasons for Change

With this as background, it should come as no surprise that the major reason public radio and TV programs leave the airwaves is that they have run their course. Peter Pennycamp, vice president of cultural programming for NPR, explains: "We look at many of our cultural programs as though they were exhibits with a set number of programs that will be aired over a set length of time. When they finish, they have achieved their purpose."

But on occasion programs are taken off the air for other reasons. Josie Patterson of WGBH Radio in Boston covers the waterfront when she says, "We take programs off the air for reasons that are usually idiosyncratic to the program: host retiring or not wanting to do the show anymore, lack of funding for the program, program receiving a very small audience relative to the kind of program we think would better serve our audience at that time, and desire to try something new."

Roberta K. Smith of KOCE in California talks about an unwanted reason for having to take shows off the air, one that has become a plague to public broadcasters. "We've had to take British programs off that we really wanted to keep running. But Arts and Entertainment bid more for them and received the exclusive right to air them. As a small station, we can't compete with the resources of A&E."

APR used to air a children's program called "Kids America." Only forty public radio stations aired it, so APR canceled it. "Of course, we got a lot of complaints from people in the forty cities where it was airing," reports Dan Jensen.

Rarely do public broadcasters change the airtime of particular shows. For the networks, this would be somewhat futile because the stations can put most programs on whenever they want. And then there is the audience to cope with. Move "Masterpiece Theatre" from

Sunday at 9:00 p.m. to Thursday at 8:00 p.m. in order to shore up the Thursday night schedule and you'll probably incite a civilian revolt!

Throughout the book we have concentrated on the creative decisions that confront the programmer every day on every project. However, on many occasions productions require another type of judgment, one that is just as important and frequently more difficult. It goes beyond the question "Is this an effective thing to do?" and asks instead, "Is this the right thing to do?" Our final chapter invites you to think about the ethical considerations that are involved in programming.

NOTES

1. "Final Standing: 1991–92 Season," *Variety* (April 15, 1992): 18.
2. Alex McNeil, *Total Television* (New York: Penguin Books, 1991): 763.
3. Susan Tyler Eastman, Sydney W. Head, and Lewis Klein, *Broadcast/Cable Programming* (Belmont, CA: Wadsworth Publishing Company, 1989): 155.

15 Programming Ethics

The way the story goes, the devil appeared to a network executive one day and said: "I'm going to offer you a deal. The next year I will arrange it so that you can have on your schedule Roseanne Arnold, Ted Danson, Bill Cosby, Peter Jennings, and the Simpsons. In exchange, you must turn over your soul to me for all eternity." The executive thought for a moment and said, "What's the catch?"

Many a joke has been leveled at the industry's cynical ethical behavior, partly because it's true, partly because it's a highly visible business, and partly because its employees have a gift for self-mockery. In fact the famed radio comedian Fred Allen once commented, "You can take all the sincerity in Hollywood, stuff it into a gnat's navel and still have room for six caraway seeds and the heart of an agent."

As an industry, the entertainment world is probably no worse than any other, which is far from a compliment. However, radio and television are the most far-reaching and persuasive media ever developed and, therefore, bear a greater ethical burden.

Unfortunately, the far-too-prevalent attitude is one of cynical realism. Get the deal first, sort out the sensitivities later. Don't worry about the people who got screwed; if they need you, they'll be back.

The hope is that the next generation of broadcasters will aim higher and do better. Towards that end we have included at the beginning of this chapter some guidelines for ethical behavior. We encourage readers to keep these in mind as they discuss and reflect upon the case histories that are presented at the end of the chapter.

THE MEANING OF ETHICS

Ethics is a code of moral principles. It is the value system relating to conduct by which the rightness or wrongness of actions is determined.[1] To be sure, there is a sizeable gray area in the evaluation of most ethical behavior. Many judgments are connected to a person's upbringing and training. People of good will can honestly differ on interpretations. U.S. Supreme Court Justice Potter Stewart acknowledged that he might not be able to define pornography, "but," he stated, "I sure know it when I see it." Similarly, broadcasters should be able to recognize ethical right and wrong. Although the issues may be murky, a disturbed conscience will help to light the way.

Ethics and Illegality

It is important to make a distinction between what is legal and what is ethical. The two are not always synonymous. It is against the law, and unethical, for a programmer to accept free tickets from an airline company to help defray the cost of a show and not reveal this arrangement on the air. It is unethical, but not illegal, for a news producer to show two-year-old footage of a forest fire while the anchorperson is describing a currently raging fire. If there is no indication that stock footage is being used, the viewer will conclude that the picture represents a live report—a deception but not an infraction. Programmers *must* observe the law; they are *urged* to behave ethically.

The Constituencies

For those in the electronic media, ethical conduct involves two different constituencies: the audience, and the people in the industry. The audience develops a feel for the character of a broadcaster by the messages that are put on the screen. Overly hyped and misleading promos, smut and excessive violence in newscasts, disregard for important but perhaps low-rated local events, wanton commercialism, cheap shabby productions, and "sweeps" specials drenched in sensationalism are clues to the viewer about the broadcaster's ethical standards.

Within the industry, producers, executives, and others build or harm their reputation by the honorableness of their behavior with suppliers, advertisers, and fellow employees. If the word gets out that you have to get in writing everything that "X" says, "X" is going to have a hard time getting people to deal with him. Nobody likes to do business with someone whose word is a bond only until a better offer comes along.

Ethics and Lying

In broadcasting, lying is probably the most common form of unethical behavior. The euphemist might prefer to call it "fudging the truth," but no matter how you perfume it, it is an intentional untruthful declaration to another person. German philosopher Immanuel Kant held that any lie, no matter how virtuous the intent, is unacceptable because it "vitiates the source of law."[2] Although few agree with his absolutism, the destruction of trust between humans is sad and serious.

Broadcasting is fast-paced. Much of the deal-making is done over the phone or at meetings in which the recording of agreements is either impossible or impractical. It is essential that both parties honor the commitments even if there are no other witnesses and no documents. Trust has to be the foundation of the contract. If one of the parties is lying, or later denies commitments, all future contact between the parties will be contaminated.

Some fine distinctions about ethics and lying need to be made about the buying and selling process. A certain tolerance is allowed for the hyperbole and enthusiasm of a seller. When a salesperson states, "This is the hottest show since 'M*A*S*H,' " the buyer knows not to accept that literally. It is merely an expression of faith in the product. Similarly, the buyer may say, "I've seen six shows this week that are better than that, and it's only Wednesday." The purpose is to inform the seller that whatever the asking price is going to be, it will be too high. The give and take of these sessions is understood by both parties, and considerable latitude is expected and extended.

However, the outright lie is impermissible. For a salesperson to state that "Candice Bergen, Sly Stallone, and Oprah Winfrey are all scheduled to guest star the first week," when it is known that all three have flatly refused to appear, is a breach.

The buyer-seller exchange is akin to the evasive moves of the ball carrier in a football game, the head fakes by the basketball star, the luring traps of the chess master. They are all designed to mislead—"lies" in a sense, but within an environment where such action is anticipated and accepted. The behavior this chapter addresses relates to words and actions that are presented as truth and intended to be accepted as such. Anyone who exploits that belief in order to misdirect and deceive is guilty of unethical conduct.

Ethical Guidelines

Most of the great philosophers—Hume, Mill, Kant, Augustine—have written extensively on the subject of ethics. Their works are available in libraries, and the reader is encouraged to consult them. It is not this book's intent to endorse any particular ethical system. It is simply the authors' hope that a heightened awareness of ethical considerations will help programmers to make honorable judgments and improve the moral environment of the industry.

A number of ethical codes have been drawn up by professional electronic media organizations in order to aid people in the industry. These codes often contain vague language and none of them are enforceable, but they serve as guidelines to ethical behavior. For example the Radio-Television News Directors Association (RTNDA) has drawn up the following standards, and its members have pledged to follow them.

1. Strive to present the source or nature of broadcast news material in a way that is balanced, accurate, and fair.

   A. They will evaluate information solely on its merits as news, rejecting sensationalism or misleading emphasis in any form.

B. They will guard against using audio or video material in a way that deceives the audience.

C. They will not mislead the public by presenting as spontaneous news any material which is staged or rehearsed.

D. They will identify people by race, creed, nationality or prior status only when it is relevant.

E. They will clearly label opinion and commentary.

F. They will promptly acknowledge and correct errors.

2. Strive to conduct themselves in a manner that protects them from conflicts of interest, real or perceived. They will decline gifts or favors which would influence or appear to influence their judgments.

3. Respect for dignity, privacy and well-being of people with whom they deal.

4. Recognize the need to protect confidential sources. They will promise confidentiality only with the intention of keeping that promise.

5. Respect everyone's right to a fair trial.

6. Broadcast the private transmissions of other broadcasters only with permission.

7. Actively encourage observance of this Code by all journalists, whether members of the Radio-Television News Directors Association or not.[3]

Other sources of ethical input revolve around established institutions such as universities or churches. For example, in October 1992 Cardinal Roger M. Mahony of Los Angeles issued a pastoral letter to all members of the entertainment business in which he declared that motion picture and television leaders have a moral obligation to communicate "human values" in their productions. He rejected as censorship any specific code of unpermissible behavior but offered instead a set of guidelines by which writers, directors, actors, producers, and others could judge the moral impact of their work. A list of nine areas of special concern were cited: relationships, sexuality, women, family, religion, work, possessions, authority, and violence. The pastoral devoted a section to each subject, preceded by reflections on what he said it means to be "truly human." For example, on sexuality, the cardinal commented, "If they [the characters] are saying something with their bodies they do not mean with their minds, hearts, and souls . . . is the picture honest about the inauthenticity, the terrible emptiness, the shallowness and self-deception of such a one-dimensional approach to human sexuality? Are the sexual scenes handled with the taste, delicacy and reverence the subject demands?" And on violence he asked: "1. Is the violence demanded by the story? Is it presented as a desirable way to resolve a conflict? Is it presented honestly? 2. Is non-violence, the way of dialogue, presented as a credible alternative?"[4] The cardinal's letter was well received by industry leaders. Jack Valenti, president of the Motion Picture Association of America, applauded the document for its avoidance of "censorship and rigidity" and for urging the industry to "enlighten as well as entertain."[5]

Probably the most effective guidelines are those that people of conscience develop for themselves. One veteran executive held himself to a simple rule of behavior. "I picture myself describing a situation to my wife and children at the dinner table. If I can tell them what I plan to do and I'm comfortable while I say it, I go ahead and do it. If I get itchy, I go back and rethink it." For those readers who do not have a spouse and children, use your mother.

CASE HISTORIES: ACTUAL INCIDENTS

Listed below are descriptions of a number of actual incidents that involve some very questionable behavior by electronic media practitioners. Readers are encouraged to think about how they would have acted in similar circumstances. As you ponder these case histories, assume that you have just left the boss's office where you have been told in very direct language that ratings must improve or heads will roll.

On-air Murder Confession

In June 1990 two deejays at station KROQ in Burbank, California, staged an event that triggered a nationwide police search for a nonexistent killer. On their morning show the two personalities, Gene Baxter and Kevin Ryder, developed a comedy element entitled "Confess Your Crime" in which listeners were encouraged to call in and reveal their transgressions on the air. To punch up the feature, they conspired with an Arizona radio performer, Doug Roberts, to fabricate a homicide. On June 13th, Roberts phoned the show and "confessed" that he had badly beaten his girlfriend. When one of the deejays asked, "Is there a chance that you killed her?" the caller responded quietly, "Yeah, I know I did."

Over the next few days the station was inundated with calls and faxes from people trying to help solve the crime. The network TV series "Unsolved Mysteries" twice broadcast a tape of the conversation and was deluged with responses. Burbank police coordinated their efforts with police agencies around the country to investigate the hundreds of leads that had been generated. The most persistent callers were a mother and father in North Carolina who were desperate to unravel the mystery surrounding the death of their daughter. In April 1991 the deejays admitted that it was all a hoax, undertaken for the sake of increasing the ratings.

Under the rules then in effect, the FCC had limited options for punishing hoaxers. It could either admonish a station or revoke its license. Revocation could be enforced only if there was proof of management participation or perjury. In the KROQ case, executives of Infinity Broadcasting, owners of the station, claimed that the deejays did it on their own and nobody else knew about it. Their testimony held up under the FCC's scrutiny, and the station's only penalty was a four-page letter of admonishment for "deliberate distortion of programming." The commission also stated that the letter would be entered in the station's file and could be taken into account when its license comes up for renewal.

The deejays were suspended for a week without pay and were forced to reimburse the police investigative costs, approximately $12,000. Within a year, the FCC crafted and adopted a rule that allowed it to fine stations up to $250,000 if they broadcast "false information concerning either a crime or a catastrophe if it is foreseeable that broadcast of such information will cause substantial public harm, and broadcast of the information does in fact directly cause such harm."[6]

Radio Station Jeopardizes Bank Executive

Late one evening, three masked men broke into the home of a bank manager, tied his mother to a chair, and fitted him with a belt that was said to be loaded with dynamite that could be detonated by remote control. The next morning the bank manager was driven to work and instructed to withdraw a large sum of money from his branch office and deliver it to a certain location. If he did not, he was told, both he and his mother would be killed.

Upon entering the building, the manager directed the security guards to contact the police. Within minutes the officers arrived and put the building under surveillance. At noon a local radio station learned of the unfolding drama and broke the media's voluntary agreement to withhold public announcement. As soon as the report was broadcast, the bomb squad rushed into the bank.

"We had every reason to believe the belt contained real dynamite," said the police director. "But we had no choice once the report came over the radio." Fortunately, the explosives were dummies, and the mother was freed without injury. The police director lashed out at the media for reporting the crime while it was still in progress, saying that the station had endangered the lives of numerous people.[7]

The Killer of Laura Palmer Revealed

One of the more intriguing story lines of 1990 was the question of who killed Laura Palmer in the primetime series "Twin Peaks." The drama, reminiscent of "Who shot J. R.?" in the "Dallas" serial a decade earlier, had generated considerable interest among American audiences.

The following year the series was sold in international syndication to RTL Plus, a German network. After just three episodes had aired, a competing network, SAT 1, revealed to its viewers the identity of the murderer along with other plot elements and further suggestions that RTL might in some way alter the series from its original story line.

A Hamburg court ruled that the action of SAT 1 was "immoral" and forbade the unrepentant network from making any other announcements about the series. Said the head of legal affairs for RTL Plus, "No upstanding businessman tries to profit from such methods at the cost of the competition."[8]

A Sad Story Made Ghoulish

A Los Angeles man was accused of killing a neighbor's dog, then dismembering and barbecuing it. One TV station in the market reported the story showing pictures of the suspect's arraignment. A second station felt compelled to add to the incident's gruesomeness by televising the open barbecue grill on which the deed was said to have been committed. A third station, not satisfied that the event was sufficiently graphic, presented footage of the decapitated dog's head mounted on a post.

No formal or informal action was taken against any of the stations.[9]

Checkbook Journalism?

During 1991, when the William Kennedy Smith rape trial was big news, the syndicated show "A Current Affair" paid $40,000 to one of the key witnesses to appear on the show and discuss the case. This payment was questioned by media watchers, but a spokesperson for "A Current Affair" said, "Sometimes we compensate for stories of interest to the public, and in so doing, we are no different from any other network or syndicated show of this type."

Av Westin, executive producer of a similar program, "Inside Edition," agreed that fees are often paid to people, sometimes for exclusive rights to their story and sometimes as reimbursement for travel expenses or lost work days. "We would have no problem offering [her] money," he said. David Bartlett, president of the Radio-Television News Directors Association, stated that compensation has become a sticky issue. "Is there a substantive difference between paying $150 for a limo drive or $40,000 for an interview? Yes, about $39,850. But does that constitute an ethical difference? I'm not sure I'm prepared to make that call."[10]

Suicide on TV

In January 1987 Pennsylvania State Treasurer R. Budd Dwyer, who was being investigated for fraud, called a news conference. After making a brief statement, he pulled out a gun and shot himself—while TV cameras taped away. Because the footage was available, stations and networks had to decide whether or not to air it.

CNN, ABC, NBC, and CBS all chose not to show the actual suicide, but they did broadcast Dwyer talking at the press conference before he shot himself. A number of Pennsylvania TV stations showed the entire event—suicide and all.[11]

Should a Private Choice Be a Public Issue?

Bree Walker, an anchorwoman on Los Angeles station KCBS-TV, has a rare genetic condition termed "ectrodactyly," in which the bones of the fingers and toes are partially fused. Although she already had a 3-year-old daughter with the same condition, she and her husband, fellow anchor Jim Lampley, decided to have another child.

Radio talk-show host Jane Norris invited her listeners to phone in and offer their opinions about the Lampleys' choice. In the course of her program, Ms. Norris made it quite clear that she felt the decision was improper and unfair to the child. When advised to mind her own business, Norris replied that the behavior of a prominent newsperson was "everybody's business." Ms. Walker objected to the suggestion that her decision was irresponsible. "No one has the right to predict or judge the quality of another person's life."[12]

The Right to Privacy?

Arthur Ashe was one of the finest American tennis players ever. For a decade he generally ranked within the top ten, and in 1975 he won the Men's Singles Championship at Wimbledon. In 1984 Ashe underwent surgery for a heart condition that involved some blood transfusions. The blood was contaminated and he contracted the HIV virus. Although his condition required constant monitoring, all those connected with his medical treatment respected his request to keep the condition from becoming public knowledge. Ashe and his wife were raising a young daughter, and they felt her childhood would be far happier if her father's condition remained unpublicized.

In 1992 a reporter for the newspaper *USA Today* learned of Ashe's infection and asked him for a confirmation. Fearing the imminent publication of the information, Ashe called a news conference and emotionally announced that he was HIV positive.

The newspaper vigorously defended its actions on the basis of the public's right to know. Others disagreed. Ashe was no longer an active athlete and therefore not a legitimate news figure; he was not an officer of a publicly traded company whose health might affect stockholders' investments; and he did not hold public office where his illness might prevent him from representing his constituency effectively. Why, many asked, did the public have to know about his personal anguish? Where, they further inquired, should print and electronic media reporters draw the line when their pursuit of a story conflicts with an individual's right to privacy?[13]

CASE HISTORIES: FICTIONAL EXAMPLES

To provide the reader with examples of ethical problems in other areas, we have listed below some fictional case histories. Once again, to heighten the realism of the incidents, the reader is asked to assume that he or she has just been told to improve company performance "or else."

An Agent's Assurances

Fred Roberts, a top agent, has always assured his client, Tom Martin, that as long as he's the highest-rated star on the XYZ network, he'll be the highest paid. Suddenly Roberts finds himself in position to negotiate an even greater figure for another of his clients, Gloria Anson. This deal will not only fatten Roberts' income, it will boost his stature in the agency. Troubled by the dilemma, Roberts discusses the problem with a lawyer friend of his. "No problem," says the lawyer. "You write the contract and call for a lower figure than Martin. But you prepare a side letter that commits to a huge bonus that will lift her past Martin. If the contract is ever subpoenaed, you're in clover; it'll show that you were true to your word with Martin." Should Roberts follow the advice?

Assuming Roberts does as the lawyer suggests, consider the following variation on the story. As part of his regular duties at the agency, Bill Bruton, one of the junior agents with the company, has become privy to the deals with the two stars. Shortly after the contract with Ms. Anson has been signed, Bruton accepts an offer to work for a rival agency. Should he contact Tom Martin and tell him that his agent is deceiving him and that Ms. Anton now makes more than he does? Should he further suggest that Martin leave Roberts' agency and join his?

A Trashy Story

Radio station WAAA has heard a rumor that competing station WZZZ is planning a really big promotion for sweeps week. The general manager of WAAA suggests (in a joking manner) to two of the staff underlings that they could benefit their station immensely if they rooted through the trash of WZZZ and uncovered some paperwork detailing what the big promotion was going to be. That way WAAA could counter the promotion with an even bigger one.

The two WAAA staff members take the general manager up on his "suggestion," and at midnight rummage through the dumpster outside WZZZ. They find a very sketchy memo stating that more details about the promotion will be forthcoming shortly. They vow to return the next night.

The next morning a WZZZ employee notices that the station's trash is in a state of disarray. Sensing foul play, the WZZZ staffers decide to retaliate. They write a totally bogus memo that alludes to details about the promotion and mentions that the main details will be on computer discs distributed confidentially to top executives. They place this memo in the trash, along with a virus-laden computer disc.

The two WAAA staffers return at midnight and take the bait. Not only do they fail to find out about the promotion, but they render WAAA's computer system inoperable for a week.

Settlement Is Better Than a Suit

A producer agrees to pay $50,000 to a performer to hold himself out of the marketplace for six months. During that period, an effort will be made to get a series idea off the ground that will star the actor. The allotted time goes by, no sale is made, and the performer demands his money. The producer tells him he's not going to receive it, and if he's not happy about that, "sue me." The producer is sure the performer will settle for half the amount rather than pay a lawyer $400 an hour and wait years for the money.

Selling What You Don't Own

In a little-known literary journal, a programmer finds a summary of a novel that might be salable as a made-for-TV movie. The executive has two choices: 1) locate the publisher, express interest in the work, and negotiate the right (pay money) to represent the material; or 2) save the money and quietly meet with a few buyers to see if there is any interest. If there's none, move on to something else. If there is a potential buyer, go to the publisher and try to make a deal.

A CHECKLIST OF ETHICAL DILEMMAS

Here are a few other ethical situations to consider.

1. Should you, as a news producer badly needing a scoop to boost your ratings, let the air out of the tires of a competing station's news van so that you can get to the scene of the story sooner?
2. Should you, as a radio station programmer, pay people to call a competing station and request dying (or dead) songs in hopes that the listeners will switch to your station for a better music selection?[14]
3. Should you release the name of a rape victim (as NBC did in the William Kennedy Smith trial)?[15]
4. Should you show live coverage of someone being administered a controversial AIDS treatment (as CNN did)?[16]
5. Should you reenact news footage of a spy handing over a briefcase because the actual footage you had of the event was technically poor (as ABC once did)?[17]
6. Should you show graphic depictions of crimes that can easily be duplicated (copycat crimes) by people in the viewing audience (as just about every network and station does)?[18]
7. Should you pay college students to moonlight at research companies and report on opponents' projects?
8. Should you, as a programmer, allow to be built for free that swimming pool you've always wanted in your backyard, because it will serve as a set for one scene in a miniseries you have just commissioned?
9. If you get a tip that a presidential candidate is going to spend the weekend at a motel with a starlet, should you send a cameraperson to hide in a nearby tree to capture "appropriate photographic material"?
10. If you have agreed verbally to hire a Susan Soso as the lead in a miniseries, should you withdraw your verbal agreement because Sally Superstar becomes available hours before you are scheduled to sign the written contract for Ms. Soso?
11. Should you give airtime to terrorists who threaten to kill hostages unless they can broadcast their message? (You know in this case that giving the time may encourage other off-balance individuals to ask for airtime, but not giving it may result in the death of innocent people.)[19]

QUOTES ABOUT ETHICS

Many people have expressed their opinions about ethics in the electronic media business. See which of the following ideas you agree with.

According to Michael Josephson, president of the nonprofit Joseph and Edna Institute on Ethics, "[The movie and TV business] is definitely among the lower practitioners of ethics. The moral code of anything-goes-as-long-as-it-sells is totally wrong . . . People just don't trust each other in the industry, and trust is the most fundamental of ethical issues . . . If people would stop squinting and open up their eyes, they'd see that what appears to be shades of gray is really black."[20]

An anonymous Hollywood agent feels: "Most of the moral players are ethical people of strong moral fiber. But they're not beyond stretching the rules if they're in a tight spot so they can compete."[21]

Jeff Wyatt, program director of KPRW-FM, says: "In the heat of battle anything short of a felony is fair game to make a radio property success . . . This is the radio business, which is perceived to be made up of a bunch of adults who don't want to grow up. Therefore, while these things [pranks] aren't appreciated, they're sort of expected."[22]

Producer Steve Tisch feels: "There's a poorly defined set of rules, an ongoing conflict ten hours a day. We're all part of a murky system."[23]

According to screenwriter Naomi Foner, it is possible to retain one's integrity. "The trick is to compromise without compromising what you set out to do. Maybe I'm stuck in the '60s, but I still am trying to find some magic, a way of reaching people through films. I call it the 'trickle up' theory—making people think about a movie afterwards."[24]

Producer Tom Musca says: "You should have total honesty when it comes to bridge-building over great gorges where lives are at stake. But . . . this business isn't nuts and bolts and shouldn't be."[25]

"CONCLUSIONS" ABOUT ETHICS

Ethical considerations in the electronic media too frequently receive short shrift. The urgency to complete the deal and make the profit tends to overwhelm the moral considerations. When this attitude pervades the entire industry, the reliability of all within it suffers. Every handshake leaves a doubt, every commitment becomes suspect. Only if broadcasters make the rightness and the wrongness of decisions a priority, only if they create a climate of honor can the industry become a workplace of trust. Admittedly the ground rules are vague and many of the calls are borderline. Sometimes the only guideline is that pesky inner voice that doesn't want to go away. Listen to it. How else are we going to make it better?

RED CHANNELS: A TEST THAT TELEVISION FLUNKED

With the disintegration of the Soviet Union now well behind us, it is difficult for Americans in the 1990s to realize how frightened people were in the 1950s of the perceived imminent spread of Communism. This national fear provided the television industry with an early test of its integrity, and it failed badly.

At the conclusion of World War II, a hysteria gripped the United States. Nazism had been defeated, but a new and potentially more destructive opponent had surfaced—Communism. When the Soviets exploded an atomic device in 1948, thereby ending America's exclusive hold on nuclear armament, the country was terrified over the prospect of an Armageddon-like showdown with the "Red Menace."

Fueling these fears was Senator Joseph McCarthy, who claimed to have knowledge of Communists, "pinkos," red sympathizers, and dupes in the United States Cabinet, the Army, the State Department, and all levels of national and state government.

Citizens desperately attempted to establish their own patriotic purity and expose the traitorous intent of others.

This paranoid environment had a devastating effect on the entertainment industry. Since radio, television, and motion pictures were capable of reaching many millions with their messages, the super- patriots resolved to prevent any possibility of sinister influence by denying employment to anyone even remotely tainted with Communist sympathies. Thus was born the pernicious policy of "blacklisting."

In 1947 some former FBI employees formed a newsletter titled "Counterattack," a publication designed to do battle with Communist infiltration in the United States. Shortly thereafter they published the first edition of "Red Channels," a booklet containing the names of all those in the entertainment media who were considered politically unacceptable.

A person could be listed for a variety of reasons: membership in an organization deemed to be subversive, a relationship with someone with Communist links, attendance at a leftist function, signing a telegram congratulating the Moscow Art Theater on its 50th anniversary, etc. Much was based on hearsay and unsupported evidence, but once a name made the list, it was virtually impossible to have it removed.

The booklet was distributed to networks, stations, advertisers, and their agencies. The implications were clear: any broadcaster or sponsor who hired someone on the list was knowingly aiding the cause of Communism in the United States.

Other organizations supplemented "Red Channels" in the search for offenders. The American Legion's Commission on Americanism developed "Firing Line," a publication that identified dupes and red-fronters, and a supermarket operator in Syracuse, Laurence A. Johnson, founded the Syracuse Crusade. Whenever a manufacturer advertised on a program that employed a "listed" person, Johnson threatened to hang a sign on the shelves where the offender's products were displayed stating that the company supported subversives.

These "patriotic" efforts were dismayingly effective. The networks and the agencies instructed producers that no performer, writer, or director was to be hired without undergoing a clearance procedure. If the person's name appeared in "Red Channels" or any of the other lists, he or she was to be rejected. However, the existence of a blacklist was never to be acknowledged. Producers were told to use such phrases as "not right for the part," or "we've found someone a little more qualified." This led some very talented people to despair, financial ruin, and even suicide.

In some instances, the malevolence of the practice reached ludicrous levels. On one occasion, Mark Goodson, television's foremost producer of game shows, was stunned to learn that a panelist on one of his shows had been declared unacceptable by the advertising agency and was to be dropped immediately. She was an English actress named Anna Lee, and she had no known political affiliations and a seemingly innocent history. Goodson dug into the matter and discovered that she had been confused with another Anna Lee who wrote occasional articles for the *Daily Worker,* a leftist newspaper. When he confronted the agency with this information, they said: "Yes, we know that, but she still has to go. We're already receiving letters." Goodson exploded and stormed out of the meeting. He was later told that as soon as the door closed someone asked, "Is he a pinko?" Ms. Lee stayed on. So did the show. And so did blacklisting, at least until the late 1950s.

Ironically, one of the main contributing forces to the end of the practice was television. The famed TV journalist Edward R. Murrow presented a program that exposed the evils of Senator Joseph McCarthy and his methods. This program is generally cited as the beginning of the senator's declining influence. In 1954 network TV covered Congressional hearings in which McCarthy accused U.S. Army personnel of having red sympathies. He and his aides bullied witnesses to such an extent that the public became outraged and the Senate itself voted to censure McCarthy.

In this day and age it seems unlikely that censorship and character assassination of this magnitude will ever again plague the television industry. But lesser versions are a constant possibility. They can only be avoided by vigilance and courage.

NOTES

1. Some excellent books on communication ethics are: Louis A. Day, *Ethics in Media Communications: Cases and Controversies* (Belmont, CA: Wadsworth Publishing Company, 1991); Carl Hausman, *Crisis of Conscience: Perspectives on Journalism* (New York: Harper Collins Publishers, 1992); James A. Jaksa and Michael S. Pritchard, *Communication Ethics: Methods of Analysis* (Belmont, CA: Wadsworth Publishing Company, 1988); Philip Patterson and Lee Wilkins, *Media Ethics: Issues and Cases*

(Dubuque, IA: Wm. C. Brown Publishers, 1991); Philip Meyer, *Ethical Journalism: A Guide for Students, Practitioners, and Consumers* (New York: Longman, 1987); Deni Elliott, ed., *Responsible Journalism* (Beverly Hills, CA: Sage, 1986); Clifford G. Christians, Kim B. Rotzoll, and Mark Fackler, *Media Ethics: Cases and Moral Reasoning* (New York: Longman, 1983); Conrad Phillip Kottak, *Prime-Time Society* (Belmont, CA: Wadsworth Publishing Company, 1990); and John L. Hulteng, *The Messenger's Motives: Ethical Problems of the News Media* (Englewood Cliffs, NJ: Prentice Hall, 1986).

2. Sissela Bok, *Lying: Moral Choice in Public and Private Life* (New York: Vintage Books, 1979): 40.
3. *Code of Broadcast News Ethics* (Washington, DC: Radio-Television News Directors Association, n.d.).
4. "Mahony Urges 'Yeoman Values' in Films, TV," *Los Angeles Times* (October 1, 1992): A-1.
5. Ibid.
6. "Will KROQ Get Away With Murder?" *Los Angeles Times* (April 22, 1991): F-1; "FCC Panel Clears Infinity, KROQ Management in Hoax," *Variety* (November 26, 1991): 3; and "FCC Adopts $25,000 Fine for Hoaxes," *Broadcasting* (May 18, 1992): 5.
7. "Banker Forced to Wear Belt of 'Dynamite' in Robbery Bid," *Toronto Globe and Mail* (December 19, 1991): A-1.
8. "SAT 1 Chastised for Killing Laura Palmer Suspense," *Variety* (October 15, 1991): 6.
9. "Going for the Ghastly, This Time at KCBS," *Los Angeles Times* (September 27, 1991): F-1.
10. "Checkbook Journalism Bounces Back," *Broadcasting* (December 9, 1991): 5.
11. David E. Tucker, "Teaching Broadcast Ethics," *Feedback* (Spring 1989): 18.
12. "Whose Baby Is It, Anyway?" *Newsweek* (October 28, 1991): 73.
13. "Ashe Case Raises Fame vs. Privacy Debate," *Los Angeles Times* (April 10, 1992): A-26.
14. "'Dirty Tricks' Define Edge of Radio Ethics," *Broadcasting* (January 14, 1991): 104.
15. "Naming Names," *Newsweek* (April 29, 1991): 26–32.
16. "Medical Progress—Live! On CNN!" *Time* (June 25, 1990): 50.
17. Paul Messaris, "Ethics in Visual Communication," *Feedback* (Fall 1990): 3.
18. "TV on Trial," *Broadcasting* (November 5, 1984): 35.
19. Julius K. Hunter and Lynne S. Gross, *Broadcast News: The Inside Out* (St. Louis: The C. V. Mosby Company, 1980): 260.
20. "How Blind Is Hollywood to Ethics?" *Los Angeles Times Calendar* (April 22, 1990): 9.
21. Ibid.
22. "Radio Executives Examine 'Unconventional' Tactics," *Broadcasting* (January 14, 1991): 102.
23. "Grappling With Ethical Problems in Hollywood," *Los Angeles Times* (October 22, 1990): F-12.
24. Ibid.
25. "How Blind Is Hollywood to Ethics?" 10.

Glossary

Actuality An event that actually occurs or an interview with someone who talks about what actually happened.

Ad hoc network An interconnection of stations for a specific program or event.

Addressability The ability of a cable system to feed its programming on a home-by-home basis.

ADI See "area of dominant influence."

Advertising agency A company or firm that places its clients' commercial messages on radio or television or in print media.

Affiliated stations Radio and TV stations that are aligned with a network.

AMOL See "Automated Measurement of Lineups."

AQH See "average quarter hour."

Area of dominant influence A term used by Arbitron to denote the individual parts of the country that it surveys.

Ascertainment A process that required stations, as part of their license renewal process, to interview community leaders to learn what they believed were the main problems in the community.

Audience measurement A term used to describe methods that endeavor to determine the number and kind of viewers watching a program or listening to a station.

Audimeter Nielsen's metering device that was attached to TV sets to measure minute-by-minute viewing of sample households.

Auditorium testing A research technique in which a large number of people are brought into one place to record their reactions to TV or radio programming material.

Automated Measurement of Lineups Nielsen's method of picking up special codes from programs so that it can determine what programs are on what channels in each market.

Average quarter hour An audience measurement calculation based on the average number of people listening to a particular station for at least five minutes during a fifteen-minute period.

Barter The sales practice in which a station forfeits a certain amount of commercial time to a supplier in exchange for the right to broadcast a program.

Basic networks Those programming services that cable systems offer subscribers for a fairly low price.

Bible A detailed report that describes a program's essential elements, such as the premise, storyline, main characters, locations, and future story growth.

Blue Book A 1946 FCC report that pronounced the criteria on which the Commission would base its license renewal decisions.

Blunting A competitive strategy that attempts to minimize the effectiveness of the early episodes of an opponent's programs.

Bridging A programming strategy in which the start time of a competitor's show is arched by an ongoing program.

Broadcast standards and practices A department at a network that rules on the acceptability of program content.

Broadcasting The dissemination of radio and TV programs to the public at large.

Burn When a song that is being played on a radio station starts to lose its appeal to the station's listeners.

Cable system Locally based organizations that supply programming to households through wires.

Cable-based research A testing process in which the program to be examined is transmitted on a cable channel to a preselected universe of homes.

Call out A telephone survey conducted by a radio station, most commonly used to find out which musical selections should be taken off a radio station and which should be added, but also sometimes used for other elements such as determining the popularity of a new disc jockey.

Cash plus barter A sales arrangement whereby a station pays a supplier cash and forfeits an agreed-upon amount of commercial time for the right to broadcast a program.

Cash sale The amount paid by a station for the right to broadcast a program for a fixed period.

Channel A television band that provides an exclusive outlet for the transmission of a station or cable service.

Churn The rate of turnover of cable subscribers as a result of disconnects and new customers.

Classics Music from ten to twenty years ago.

Clearance An agreement by a station to transmit a program.

Clear and present danger A situation that would threaten the security of the country if certain information were made public.

Clock A radio station's program schedule in which each hour is divided into carefully timed units.

Co-production An agreement between two or more companies to combine in the underwriting, development, and production of a program.

Compression A technology that allows more audio and/or video information to be carried through the airwaves or on a cable TV system.

Counterprogramming The strategy of scheduling a program whose appeal is to an audience not being reached by rival shows.

Coverage area The geographic reach of a station's signal.

Crossprogramming The tactic of having performers in one series make guest appearances on another program.

Cume Cumulative audience, i.e., the total nonduplicated audience for one program or commercial message or a series of programs or messages.

DAB See "digital audio broadcast."

Dayparting The practice of dividing the broadcast day into segments according to the size and composition of the available audience.

DBS See "direct broadcast satellite."

Demographics The classification of audiences based on social and economic conditions such as sex, age, income, education, ethnic origin, and occupation.

Deregulation Any action by Congress or the FCC to reduce or remove restrictions governing any portion of the industry over which they have regulatory authority.

Designated market area A term used by Nielsen to define a geographic area in which the local stations receive the largest audience share.

Development The process by which a program is nurtured from initial concept to pilot, or sample show; in public broadcasting, the process of trying to obtain funding for a program or series.

Diary A questionnaire on which respondents are asked to record their listening or viewing activities for a specified period of time.

Digital audio broadcast A proposed method of radio transmission in which digital sound would be transmitted directly from satellite to receivers.

Direct broadcast satellite A satellite service whose signal is delivered directly to a viewer's home via the use of the viewer's own earth station, or dish.

Distant signal importation The process by which a station from another market is brought in by a cable system and carried locally by it.

DMA See "designated market area."

Double pump The device of scheduling a new program twice in the same week in order to widen its audience exposure.

Encore A euphemism for the repeat broadcast or cablecast of a program.

Equal time A provision in Section 315 of the Communications Act that states that if a broadcast licensee permits use of its facilities by a political candidate, it shall offer an equal opportunity to all other legally qualified candidates for the same office.

Ethnic weighing The policy of some audience measurement organizations of counting diaries returned by members of certain ethnic groups more heavily than diaries returned by members of other ethnic groups because the percent of diaries returned by one group is much lower than the return rate by the other group.

Exit polls The research technique of interviewing voters immediately after they leave the polling booths.

Fairness Doctrine A series of FCC rulings and court cases that required a broadcaster to afford a reasonable opportunity for a discussion of conflicting views on issues of public importance.

Family hour A voluntary effort by broadcasters to present programs in the 7 to 9 p.m. period that would be suitable for all members of the family.

Fin-syn see "financial interest-domestic syndication."

Financial interest-domestic syndication An FCC ruling that prohibited the three networks from owning or controlling the rebroadcast rights of the shows they presented on their airwaves.

First-run Programs that make their initial appearance in domestic syndication.

Focus group A collection of sample listeners or viewers, usually numbering no more than ten to twelve, who offer their opinions about programs or performers under the guidance of a moderator.

Format The dominant type of programming an electronic media operator (particularly a radio station) selects, such as all-news, country music, or children's programming.

Franchise An agreement between state and/or local government and a cable television service that awards the right to install a coaxial cable or fiber optics in a community in order to deliver cable TV programming.

Group owner An organization that owns two or more radio and/or television stations.

Hammocking A scheduling strategy that places a new program between two established shows.

Households using television The percentage of homes in a survey area with one or more sets in use during a specific time period.

HUT See "households using television."

In-house Programs produced by the same company that broadcasts or distributes them.

Indecency An offense against recognized standards of propriety and good taste in a particular community.

Independent stations Those stations that have no network affiliations. (Stations allied with Fox are still called independents because that was their status before the new network originated.)

Indie Another name for an independent station; also, a small production company that supplies programs to networks.

Infomercial A commercial message in the form of a full program, usually 30 minutes but sometimes longer.

Invasion of privacy An illegal or insensitive intrusion into the life of another person.

Kinescope recording A record of a television show made by filming it off a television monitor.

Libel A statement about a person, broadcast or printed, that is unfavorable and false.

License Formal permission from the FCC for a broadcaster to operate a station.

License fee The amount of money a network or distributor pays to a production company to make a program; or, the amount of money a station or cable operator pays to a distributor for the right to broadcast a syndicated program.

Local origination Programs produced by a cable system for distribution within that cable system.

Lottery The involvement of chance, prize, and consideration (money) for a game or contest barred from broadcasting by section 1304 of the United States Criminal Code.

Made-for Movies produced expressly for television.

Magazine concept A sales policy in which advertisers buy commercials within a show purely on the basis of its audience size and with no involvement in its content.

Major A term applied to the handful of prominent companies that produce television shows, e.g., Paramount, Time Warner, Universal.

Metro An area, usually an individual city, that is used to determine radio station ratings.

Minipilot A film or tape that presents a few key scenes of a sample show rather than a completed episode.

Minitheater research The testing of a program before an audience in a small auditorium of thirty-five to forty seats.

Monte Carlo Audience Ascription A practice some audience measurement companies have of filling in missing diary data by making educated assumptions about what would have been filled in had the diary keeper completed the diary correctly.

Monthly balance Programming a pay cable channel in such a way that sometime during the month it pleases all subscribers so that they will, therefore, subscribe for another month.

MSO See "multiple system owner."

Multiple system owner A company that owns more than one cable system.

Must carry An FCC ruling that states that cable operators are compelled to present the programming of the TV stations that are broadcasting in the system's local area.

Narrowcasting Directing programs towards specific demographic audiences or limited interest groups.

Network A group of television or radio stations or cable systems linked by telephone lines, coaxial cable, satellite, or microwave in order to broadcast programs simultaneously.

Nielsen Television Index A service of the A. C. Nielsen Company that provides audience estimates for all sponsored U. S. network television programs.

NTI See "Nielsen Television Index."

O&O See "owned and operated station."

Obscenity Actions or language that depict sexual acts in an offensive manner, appeal to prurient interests of the average person, and lack serious artistic, literary, political, or scientific value.

Off-net Programs that have first appeared on a network and are then available for rebroadcast on other outlets.

Oldies Music from thirty or forty years ago.

Operation Prime Time A consortium of independent stations that came together in 1976 to finance, produce, and own ambitious movies and miniseries to compete with the networks' primetime programs.

OPT See "Operation Prime Time."

Overnights Nielsen household ratings and shares provided to clients the morning following the day or evening of telecast.

Owned and operated station A station that is owned and managed by a network or by one of the large station groups such as Group W or Tribune Company.

Package A collection of disparate parts combined into a single unit, e.g., a movie package (a number of motion pictures offered for one overall sales price).

Package commission A fee exacted by an agent or agency based on the cost of an entire production as opposed to commissions on just the people in the production whom the agent or agency represents.

Pay cable Any of a number of program services for which cable subscribers pay a monthly charge in addition to the basic cable subscription fee.

Pay-per-view A user fee system in which subscribers pay on a program basis rather than on a monthly subscription basis.

Payola The practice by radio personalities of taking cash payments in exchange for playing specified records.

Penetration The proportion of television or cable households to total households in an area.

Persons using television The percentage of all persons or of total persons within a given demographic category in the survey area who are viewing television during a specific time period.

Peoplemeter A mechanical device developed by the A. C. Nielsen Company that contains a button for each individual in a household to engage whenever he or she is watching television. The device records all the information provided by the standard audimeter plus the demographics of the viewer.

Perceptual call-outs Phone calls made to radio listeners to ask them questions about elements of radio other than music, e.g., the likability of a talk show host or the recognizability of a station logo.

Petition to deny A detailed complaint to the FCC that spells out the reasons why the complainant believes a station operator's license should not be renewed.

Pillars Long-lasting, popular public TV programs that occupy the same time spot for many years.

Pilot A sample episode of a projected series.

Pitch A presentation by program creators to prospective buyers, the goal of which is to obtain a commitment for further development.

Play list The musical selections played by a radio station.

Plugola The payment of cash or a bestowing of some other favor on producers, performers, disc jockeys, or others connected with a program in exchange for a visual or verbal promotion of a product or service.

Preemption The substitution of a special production for a program normally scheduled in the time period.

Premiere The first airing of a new series, movie, or other program.

Primetime access rule A 1971 FCC ruling that prohibits affiliated stations in the top fifty markets from scheduling more than three hours of network programming between the hours of 7:00 and 11:00 p.m.

Prior censorship When an official agency such as a court or the FCC punishes a station or tells it not to air something before it actually airs it; such a practice is outlawed by the Communications Act of 1934.

Production companies Those firms that specialize in the making of television programs.

Promise versus performance A procedure whereby a station promised what it would do during a license period and then was judged on the fulfillment of its promises when its license came up for renewal.

PTAR See "primetime access rule."

Public access Programming planned and produced by members of the public for local cable TV channels.

PUT See "persons using television."

Quota A limitation on the amount of TV programming that is produced in a foreign country that can be shown on a national TV system.

Rating The estimate of the size of a television or radio audience relative to the total group sampled, expressed as a percentage; for example, the rating of a national television show is the percentage of 93,100,000 households that are tuned to the program.

Ratings See "audience measurement."

Recurrents Music that is no longer current but is not yet considered old.

Reruns The airing of a program beyond its initial play.

Rigging Controlling the outcome of a game show by supplying answers or other guidance to a preferred contestant.

Rights fees The amount paid to the owner of an event or property for the privilege of broadcasting it.

Roll out The pace at which a program is distributed to the public.

Rotating schedule A cable TV term for showing the same program at many different times.

Sampling The selection of one or more elements (individuals or households) from a universe to represent the universe.

Satellite A space vehicle that receives radio and television signals and then transmits them back to Earth.

Satellite news gathering The collection and distribution of current events via orbiting space vehicles and trucks equipped with uplinks.

Scarcity theory The belief that broadcasters should be rather heavily regulated because there are not enough station frequencies available for everyone to have one, and that the people who own them should, therefore, act in the public interest.

Section 315 See "equal time."

Share The percent of the households or people using radio or television that are tuned to a specific program or station in a specified area at a specified time.

Shield laws State regulations that allow reporters to keep their sources of information confidential.

SNG See "satellite news gathering."

SPC See "Station Program Cooperative."

Spin-off A new series developed out of an existing one, usually by lifting out secondary characters and making them leads.

Spot A commercial message placed within or between programs in such a way that the company is not directly associated with the program.

Stacking A scheduling strategy in which a number of programs of similar appeal and demographics are placed consecutively for all, or the bulk, of a daypart.

Station compensation The money paid by a network to its affiliated stations for carrying its programs.

Station groups Broadcast companies that own two or more stations.

Station breaks The time between programs when the local station identifies itself and airs commercials and/or promotional announcements.

Station Program Cooperative PBS's former program selection procedure in which all public TV stations participated in selecting and cooperatively funding many of the programs on the national schedule.

Strip Term used for a program series that appears in the same time period five or more days a week.

Stripping The strategy of scheduling a series in the same time period throughout the week.

Stunting The use of imaginative scheduling or programming devices to produce immediate ratings increases.

Superstation A station whose signal is available via satellite to cable systems around the country.

Sustaining show A program without commercial sponsorship.

Sweeps A period, usually of four weeks' duration, during which all local TV markets are simultaneously measured and reported by a rating service.

Syndicated exclusivity An FCC ruling stating that if a cable system imports a program on a distant channel that is already being broadcast in a market, the cable system must black out the show and substitute other programming.

Syndication A program delivery system through which shows are provided directly to stations or cable networks by the production company or its distributor.

Syndex See "syndicated exclusivity."

Target audience The principal viewers or listeners to whom a program is directed.

Telephone research The obtaining of information via question-and-answer sessions over the telephone.

Tentpoling The placement of a well-established show in a pivotal time period to provide strength to the programs around it.

Theming The grouping of programs of similar content or tone into a schedule.

Tiering The act of charging customers separately for different cable services; a syndication buying practice among TV stations in which prices and other sales arrangements are established for a wide range of time periods.

Toll station The early radio technique of charging a fee to anyone who wished to use a station's facilities to present a message.

Topless radio The presentation of salacious material by radio personalities or by listeners who called in to talk shows.

Total service area Arbitron's division for radio station ratings of large geographic areas that may not receive all stations clearly.

Treatment A synopsis of a program's storyline that may also include other pertinent data such as descriptions of principal characters and key locations.

TSA See "total service area."

TV-Q A research technique that attempts to determine the intrinsic popularity and recognizability of performers and programs as opposed to their ratings.

Umbrella deal A contractual arrangement between a creative person (producer, director, writer) and a production company in which the company supplies money and services to the creator in exchange for the exclusive right to his or her output.

Underwriter A company that pays for the production or distribution of public broadcasting programming.

Units of good A public broadcasting term used to designate the inherent, uplifting worth of a program.

Wheel See "clock."

Window The period of time between when a movie is shown in the theater and when it is made available to various television media forms.

Zippers Short musical interludes in public radio programming that allow stations to cut away and program something else.

Suggested Readings

Auletta, Ken. *Three Blind Mice: How the TV Networks Lost Their Way.* New York: Random House, 1991.

Ball Rokeach, Sandra, and Muriel G. Cantor, eds. *Media, Audience, and Social Structure.* Newbury Park, CA: Sage Publications, 1986.

Barnouw, Erik. *The Golden Web.* New York: Oxford University Press, 1968.

Barnouw, Erik. *The Image Empire.* New York: Oxford University Press, 1970.

Barnouw, Erik. *A Tower in Babel.* New York: Oxford University Press, 1966.

Bedell, Sally. *Up the Tube: Primetime TV and the Silverman Years.* New York: The Viking Press, 1981.

Beville, Hugh Malcolm. *Audience Ratings: Radio, Television, and Cable.* Hillsdale, NJ: L. Erlbaum Associates, 1988.

Blum, Richard A., and Lindheim, Richard D. *Primetime Network Television Programming.* Boston: Focal Press, 1987.

Broughton, Irv. *Producers on Producing: The Making of Film and Television.* Jefferson, NC: McFarland, 1986.

Brown, Lester L. *Les Brown's Encyclopedia of Television.* Detroit: Visible Ink Press, 1992.

Buzzard, Karen. *Electronic Media Ratings.* Boston: Focal Press, 1992.

Cantor, Muriel G., and Pingree, Suzanne. *Prime-Time Television: Content and Control.* Beverly Hills, CA: Sage, 1980.

Carey, John; Gherardi, Thomas; Kapppes, Harold; and Moss, Mitchell. *Modularization and Packaging of Public Television Programs.* Washington, D. C.: Corporation for Public Broadcasting, 1983.

Carroll, Ray, and Davis, D. M. *Electronic Media Programming: Strategies and Decision-Making.* New York: McGraw-Hill, 1991.

Cassate, Mary, and Skill, Thomas. *Live on Daytime Television: Tuning-In American Serial Drama.* Norwood, NJ: Ablex, 1983.

Christians, Clifford, Rotzoll, Kim, and Fackler, Mark. *Media Ethics: Cases and Moral Reasoning,* 2nd ed. White Plains, NY: Longman, 1987.

Clift, Charles, and Greer, Archie, eds. *Broadcasting Programming: The Current Perspective.* Washington, D. C.: University Press of America, 1981.

Day, Louis A. *Ethics in Media Communications: Cases and Controversies.* Belmont, CA: Wadsworth Publishing Company, 1991.

DeLong, Thomas A. *The Mighty Music Box.* Los Angeles: Amber Crest Books, 1980.

DeLuca, Mariann. *Survey of Radio Stations.* New York: Torbet Radio, 1983.

Dominick, Joseph R., and Fletcher, James E. *Broadcasting Research Methods.* Boston: Allyn and Bacon, 1985.

Eastman, Susan Tyler, Head, Sydney W., and Klein, Lewis. *Broadcast/Cable Programming: Strategies and Practices,* 3rd ed. Belmont, CA: Wadsworth Publishing Company, 1989.

Eliot, Marc. *Television: One Season in American Television.* New York: St. Martin's Press, 1983.

Elliott, Deni, ed. *Responsible Journalism.* Beverly Hills, CA: Sage, 1986.

Fletcher, James E., ed. *Handbook of Radio and TV Broadcasting: Research Procedures in Audience, Program, and Revenues.* New York: Van Nostrand Reinhold, 1981.

Fletcher, James E., and Wimmer, Roger D. *Focus Group Interviews in Radio Research.* Washington, D. C.: National Association of Broadcasters, 1981.

Forbes, Dorthy, and Layng, Sanderson. *The New Communicators: A Guide to Community Programming.* Washington, D. C.: Communications Press, 1980.

Frank, Ronald E., and Greenberg, Marshall G. *Audiences for Public Television.* Beverly Hills, CA: Sage, 1982.

Gitlin, Todd. *Inside Prime Time.* New York: Pantheon, 1983.

Gross, Lynne Schafer. *The New Television Technologies.* Dubuque, IA: Wm. C. Brown Publishers, 1990.

Grossman, Gary H. *Saturday Morning TV.* New York: Dell, 1982.

Hall, Claude, and Hall, Barbara. *This Business of Radio Programming.* New York: Billboard Publications, 1977.

Halliwell, Leslie, with Purser, Philip. *Halliwell's Television Companion.* London: Granada Publishing Ltd., 1982.

Halper, Donna L. *Full-Service Radio.* Boston: Focal Press, 1991.

Hausman, Carl. *Crisis of Conscience: Perspective of Journalism Ethics.* New York: Harper Collins Publishers, 1992.

Heeter, Carrie, and Greenberg, Bradley S. *Cableviewing.* Norwood, NJ: Ablex, 1988.

Howard, Herbert H., and Kievman, Michael S. *Radio and TV Programming.* New York: Macmillan, 1983.

Hulteng, John L. *The Messenger's Motives: Ethical Problems of the News Media.* Englewood Cliffs, NJ: Prentice Hall, 1986.

Jaksa, James A., and Pritchard, Michael S. *Communication Ethics: Methods of Analysis.* Belmont, CA: Wadsworth Publishing Company, 1988.

Keirstead, Philip A. *All-News Radio.* Blue Ridge Summit, PA: TAB Books, 1980.

Keith, Michael C. *Radio Programming: Consultancy and Formatics.* Boston: Focal Press, 1987.

King, Larry, with Occhiogrosso, Peter. *Tell It to the King.* New York: Putman, 1988.

Kottak, Conrad Phillip. *Prime-Time Society.* Belmont, CA: Wadsworth Publishing Company, 1990.

Levin, Murray B. *Talk Radio and the American Dream.* Lexington, MA: Lexington Books, 1987.

Levinson, Richard, and Link, William. *Stay Tuned: An Inside Look at the Making of Prime-Time Television.* New York: St. Martin's Press, 1981.

Lull, James. *Popular Music and Communication.* Beverly Hills, CA: Sage, 1987.

MacFarland, David T. *Contemporary Radio Programming Strategies.* Hillsdale, NJ: Lawrence Earlbaum Associates, 1990.

MacFarland, David T. *The Development of the Top 40 Format.* New York: Arno Press, 1979.

McNeil, Alex. *Total Television.* New York: Penguin Books, 1991.

Matelski, Marilyn. *Broadcast Programming and Promotions Worktext.* Boston: Focal Press, 1989.

Matelski, Marilyn J. *Daytime Television Programming.* Boston: Focal Press, 1991.

Matelski, Marilyn J. *TV News Ethics.* Boston: Focal Press, 1991.

Metz, Robert. *CBS: Reflections in a Bloodshot Eye.* Chicago: Playboy Press, 1975.

Meyer, Philip. *Ethical Journalism: A Guide for Students, Practitioners, and Consumers.* New York: Longman, 1987.

Meyrowitz, Joshua. *No Sense of Place: The Impact of Electronic Media on Social Behavior.* New York: Oxford University Press, 1985.

Newcomb, Horace, and Alley, Robert S. *The Producer's Medium: Conversations with Creators of American TV.* New York: Oxford University Press, 1983.

Oringel, Robert S. *The Access Management's Handbook: A Guide for Managing Community Television.* Boston: Focal Press, 1987.

Orlik, Peter. *Critiquing Radio and Television Content.* Boston: Allyn and Bacon, 1988.

Paiva, Bob. *The Program Director's Handbook.* Blue Ridge Summit, PA: TAB Books, 1983.

Passman, Arnold. *The Deejays.* New York: Macmillan, 1971.

Patterson, Philip, and Wilkins, Lee. *Media Ethics: Issues and Cases.* Dubuque, IA: Wm. C. Brown Publishers, 1991.

Routt, Ed, McGrath, James B., and Weiss, Frederic A. *The Radio Format Conundrum.* New York: Hastings House, 1978.

Schneider, Cy. *Children's Television: The Art, the Business, and How It Works.* Lincolnwood, IL: NYC Business Books, 1987.

Schultze, Quentin. *Television: Manna from Hollywood.* Grand Rapids, MI: Zondervan Publishing, 1986.

Shane, Ed. *Programming Dynamics.* Overland Park, KS: Glovecom, 1984.

Sklar, Robert. *Prime Time America.* New York: Oxford University Press, 1980.

Sklar, Rick. *Rocking America: How the All-Hit*

Radio Stations Took Over. New York: St. Martin's Press, 1984.

Smith, V. Jackson. *Programming for Radio and Television.* Washington, D. C.: University Press of America, 1980.

Thayer, Lee, ed. *Ethics, Morality, and the Media.* New York: Hastings House, 1980.

Williams, Huntington. *Beyond Control: ABC and the Fate of the Networks.* New York: Atheneum, 1989.

Wimmer, Roger D., and Dominick, Joseph R. *Mass Media Research: An Introduction.* Belmont, CA: Wadsworth Publishing Company, 1987.

Index